Neuroinformatics

METHODS IN MOLECULAR BIOLOGY™

John M. Walker, SERIES EDITOR

436. Avian Influenza Virus, edited by *Erica Spackman, 2008*
435. Chromosomal Mutagenesis, edited by *Greg Davis and Kevin J. Kayser, 2008*
434. Gene Therapy Protocols: Volume 2: *Design and Characterization of Gene Transfer Vectors* edited by *Joseph M. LeDoux, 2008*
433. Gene Therapy Protocols: Volume 1: *Production and In Vivo Applications of Gene Transfer Vectors,* edited by *Joseph M. LeDoux, 2008*
432. Organelle Proteomics, edited by *Delphine Pflieger and Jean Rossier, 2008*
431. Bacterial Pathogenesis: *Methods and Protocols,* edited by *Frank DeLeo and Michael Otto, 2008*
430. Hematopoietic Stem Cell Protocols, edited by *Kevin D. Bunting, 2008*
429. Molecular Beacons: *Signalling Nucleic Acid Probes, Methods and Protocols,* edited by *Andreas Marx and Oliver Seitz, 2008*
428. Clinical Proteomics: *Methods and Protocols,* edited by *Antonio Vlahou, 2008*
427. Plant Embryogenesis, edited by *Maria Fernanda Suarez and Peter Bozhkov, 2008*
426. Structural Proteomics: *High-Throughput Methods,* edited by *Bostjan Kobe, Mitchell Guss, and Huber Thomas, 2008*
425. 2D PAGE: Volume 2: *Applications and Protocols,* edited by *Anton Posch, 2008*
424. 2D PAGE: Volume 1:, *Sample Preparation and Pre-Fractionation,* edited by *Anton Posch, 2008*
423. Electroporation Protocols, edited by *Shulin Li, 2008*
422. Phylogenomics, edited by *William J. Murphy, 2008*
421. Affinity Chromatography: *Methods and Protocols, Second Edition,* edited by *Michael Zachariou, 2008*
420. Drosophila: *Methods and Protocols,* edited by *Christian Dahmann, 2008*
419. Post-Transcriptional Gene Regulation, edited by *Jeffrey Wilusz, 2008*
418. Avidin-Biotin Interactions: *Methods and Applications,* edited by *Robert J. McMahon, 2008*
417. Tissue Engineering, *Second Edition,* edited by *Hannsjörg Hauser and Martin Fussenegger, 2007*
416. Gene Essentiality: *Protocols and Bioinformatics,* edited by *Svetlana Gerdes and Andrei L. Osterman, 2008*
415. Innate Immunity, edited by *Jonathan Ewbank and Eric Vivier, 2007*
414. Apoptosis in Cancer: *Methods and Protocols,* edited by *Gil Mor and Ayesha Alvero, 2008*
413. Protein Structure Prediction, *Second Edition,* edited by *Mohammed Zaki and Chris Bystroff, 2008*
412. Neutrophil Methods and Protocols, edited by *Mark T. Quinn, Frank R. DeLeo, and Gary M. Bokoch, 2007*
411. Reporter Genes for Mammalian Systems, edited by *Don Anson, 2007*
410. Environmental Genomics, edited by *Cristofre C. Martin, 2007*
409. Immunoinformatics: *Predicting Immunogenicity In Silico,* edited by *Darren R. Flower, 2007*
408. Gene Function Analysis, edited by *Michael Ochs, 2007*
407. Stem Cell Assays, edited by *Vemuri C. Mohan, 2007*
406. Plant Bioinformatics: *Methods and Protocols,* edited by *David Edwards, 2007*
405. Telomerase Inhibition: *Strategies and Protocols,* edited by *Lucy Andrews and Trygve O. Tollefsbol, 2007*
404. Topics in Biostatistics, edited by *Walter T. Ambrosius, 2007*
403. Patch-Clamp Methods and Protocols, edited by *Peter Molnar and James J. Hickman 2007*
402. PCR Primer Design, edited by *Anton Yuryev, 2007*
401. Neuroinformatics, edited by *Chiquito J. Crasto, 2007*
400. Methods in Membrane Lipids, edited by *Alex Dopico, 2007*
399. Neuroprotection Methods and Protocols, edited by *Tiziana Borsello, 2007*
398. Lipid Rafts, edited by *Thomas J. McIntosh, 2007*
397. Hedgehog Signaling Protocols, edited by *Jamila I. Horabin, 2007*
396. Comparative Genomics, Volume 2, edited by *Nicholas H. Bergman, 2007*
395. Comparative Genomics, Volume 1, edited by *Nicholas H. Bergman, 2007*
394. Salmonella: *Methods and Protocols,* edited by *Heide Schatten and Abraham Eisenstark, 2007*
393. Plant Secondary Metabolites, edited by *Harinder P. S. Makkar, P. Siddhuraju, and Klaus Becker, 2007*
392. Molecular Motors: *Methods and Protocols,* edited by *Ann O. Sperry, 2007*
391. MRSA Protocols, edited by *Yinduo Ji, 2007*
390. Protein Targeting Protocols *Second Edition,* edited by *Mark van der Giezen, 2007*
389. Pichia Protocols, *Second Edition,* edited by *James M. Cregg, 2007*
388. Baculovirus and Insect Cell Expression Protocols, *Second Edition,* edited by *David W. Murhammer, 2007*
387. Serial Analysis of Gene Expression (SAGE): *Digital Gene Expression Profiling, edited by Kare Lehmann Nielsen, 2007*
386. Peptide Characterization and Application Protocols, *edited by Gregg B. Fields, 2007*
385. Microchip-Based Assay Systems: *Methods and Applications, edited by Pierre N. Floriano, 2007*
384. Capillary Electrophoresis: *Methods and Protocols, edited by Philippe Schmitt-Kopplin, 2007*
383. Cancer Genomics and Proteomics: *Methods and Protocols, edited by Paul B. Fisher, 2007*
382. Microarrays, Second Edition: *Volume 2, Applications and Data Analysis, edited by Jang B. Rampal, 2007*
381. Microarrays, Second Edition: *Volume 1, Synthesis Methods, edited by Jang B. Rampal, 2007*
380. Immunological Tolerance: *Methods and Protocols, edited by Paul J. Fairchild, 2007*

METHODS IN MOLECULAR BIOLOGY™

Neuroinformatics

Edited by

Chiquito Joaqium Crasto

Yale Center for Medical Informatics and Department of Neurobiology, Yale University School of Medicine, New Haven, CT

Current Address: Department of Genetics, University of Alabama at Birmingham, Birmingham, AL

Foreword by

Stephen H. Koslow

Biomedical Synergy
West Palm Beach, FL

HUMANA PRESS TOTOWA, NEW JERSEY

999 Riverview Drive, Suite 208
Totowa, New Jersey 07512

www.humanapress.com

This publication is printed on acid-free paper. ♾
ANSI Z39.48-1984 (American Standards Institute) Permanence of Paper for Printed Library Materials

Cover illustration: The figure describes the role of information technology represented by binary digits in the filed of Neuroscience represented by the silhouette of a human brain.

Production Editor: Christina M. Thomas
Cover design by: Chiquito Joaqium Crasto

For additional copies, pricing for bulk purchases, and/or information about other Humana titles, contact Humana at the above address or at any of the following numbers: Tel.: 973-256-1699; Fax: 973-256-8341; E-mail: humana@humanapr.com; or visit our Website: www.humanapress.com

Printed in the United States of America. 10 9 8 7 6 5 4 3 2 1

Library of Congress Control Number: 2007928342

For Cherie

Contents

Preface .... *ix*

Foreword .... *xi*

Contributors .... *xix*

Acknowledgments .... *xvii*

Color Plates .... *xxiii*

PART I: NEUROSCIENCE KNOWLEDGE MANAGEMENT

1. Managing Knowledge in Neuroscience
   Chiquito J. Crasto and Gordon M. Shepherd ***3***
2. Interoperability Across Neuroscience Databases
   Luis Marenco, Prakash Nadkarni, Maryann Martone, and Amarnath Gupta ***23***
3. Database Architectures for Neuroscience Applications
   Prakash Nadkarni and Luis Marenco ***37***
4. XML for Data Representation and Model Specification in Neuroscience
   Sharon M. Crook and Fred W. Howell ***53***
5. Creating Neuroscience Ontologies
   Douglas M. Bowden, Mark Dubach, and Jack Park ***67***

PART II: COMPUTATIONAL NEURONAL MODELING AND SIMULATION

6. Model Structure Analysis in NEURON
   Michael L. Hines, Tom M. Morse, and N. T. Carnevale ***91***
7. Constructing Realistic Neural Simulations with GENESIS
   James M. Bower and David Beeman ***103***
8. Simulator for Neural Networks and Action Potentials
   Douglas A. Baxter and John H. Byrne ***127***

9. Data Mining Through Simulation
William W. Lytton and Mark Stewart **155**

10. Computational Exploration of Neuron and Neural Network Models in Neurobiology
Astrid A. Prinz **167**

PART III: IMAGING

11. Brain Atlases and Neuroanatomic Imaging
Allan MacKenzie-Graham, Jyl Boline, and Arthur W. Toga **183**

12. Brain Mapping with High-Resolution fMRI Technology
Nian Liu **195**

13. Brain Spatial Normalization
William Bug, Carl Gustafson, Allon Shahar, Smadar Gefen, Yingli Fan, Louise Bertrand, and Jonathan Nissanov **211**

14. Workflow-Based Approaches to Neuroimaging Analysis
Kate Fissell **235**

15. Databasing Receptor Distributions in the Brain
Rolf Kötter, Jürgen Maier, Wojciech Margas, Karl Zilles, Axel Schleicher, and Ahmet Bozkurt **267**

PART IV: NEUROINFORMATICS IN GENETICS AND NEURODENEGERATIVE DISORDERS

16. An Informatics Approach to Systems Neurogenetics
Glenn D. Rosen, Elissa J. Chesler, Kenneth F. Manly, and Robert W. Williams **287**

17. Computational Models of Dementia and Neurological Problems
Włodzisław Duch **305**

18. Integrating Genetic, Functional Genomic, and Bioinformatics Data in a Systems Biology Approach to Complex Diseases: Application to Schizophrenia
F. A. Middleton, C. Rosenow, A. Vailaya, A. Kuchinsky, M. T. Pato, and C. N. Pato **337**

19. Alzforum
June Kinoshita and Timothy Clark **365**

Index *383*

Preface

> If the human brain were so simple that we could
> understand it, we would be so simple that we couldn't
>
> Emerson M. Pugh

The inception of the Human Brain Project (HBP) in the early 1990s gave birth to a new field of study—Neuroinformatics. Neuroinformatics sought to bring together computational scientists and neuroscientists to further understand the complex processes that govern the nervous system. Although the field of Neuroinformatics has provided myriad results, most will agree that we have merely begun to scratch the surface of what remains to be discovered.

I decided to give the areas of research that received support from the HBP funding agencies preferred representation in this volume. My work in the development of "NeuroText" to mine relevant neuroscience information from the biomedical literature came under the auspices of the HBP, as did the development of the SenseLab project at the Yale University School of Medicine, where I worked.

The four sections of this volume focus on: (1) concepts used in the development of databases and the dissemination of neuroscience knowledge; (2) presenting new developments and facets of imaging research as applied to the neuroscience; (3) developing computational methodologies to simulate neuronal processes through modeling, simulation, and the use of neural networks; and (4) applying neuroinformatics to neurogenentics and the understanding of neurological disorders.

As a domain of research and discovery, neuroinformatics is limitless. Just as the evolution of neuroinformatics will result in new avenues of discovery, this volume, one hopes, will find an extended readership. Experimental neuroscience workers should find this volume important as an introduction to how information science can be applied to their fields of study. The sections in every chapter in this volume are formatted (with bullet points and/or enumeration) for a point-by-point explication of the topics discussed. The writing style is simple enough to be appreciated by undergraduate and graduate students.

Several techniques and methodologies presented in this work, while developed for neuroscience, are easily extensible to other biomedical domains of research.

Over the last few years, neuroinformatics has grown to be a vital part of neuroscience. It mirrors the importance that information technology has had on almost every aspects of modern life. This volume, I hope, will be informative and yet make the reader want to know more about this complex system we call the nervous system—as much as it does neuroinformatics researchers.

Chiquito Joaquim Crasto

Foreword

Neuroinformatics heralds a new era for neuroscience research. As evidenced by the chapters ahead, the growing field of Neuroinformatics has met the first set of challenges in assuring its longevity and success: the creation of a web-based infrastructure for neuroscience data sharing, integration, and analysis. Once populated, these databases will provide the entire research community with impossibly rich sets of data for reanalysis, integration, and computational analysis. For the first time, neuroscientists will be able to rapidly integrate, analyze, and optimally synthesize vast neuroscience data from diverse heterogeneous modalities under investigation, greatly enhancing our collective and individual understanding of brain function and, ultimately, the human nervous system.

Clearly, the next significant litmus test for Neuroinformatics will be how quickly, and how completely, the neuroscience community populates these databases. As Neuroinformatics continues to build momentum as a field of research and a critical path to unraveling one of science's greatest mysteries, monographs such as *Neuroinformatics* will play a critical role in providing this community with the tools and impetus to share their research and papers with their colleagues around the globe by offering insights, information, and, importantly, compelling examples of success on which we can all build.

Background

Neuroinformatics combines neuroscience and informatics research to develop and apply innovative tools and approaches essential for facilitating a major advancement in understanding the structure and function of the brain. Neuroinformatics research is uniquely placed at the intersections of biomedical, behavioral sciences, biology, physical, mathematical sciences, computer science, and engineering. The synergy from combining these disciplines will enable the acceleration of scientific and technological progress, resulting in major medical, social, and economic benefits ***(1)***.

The end goal of Neuroinformatics is to understand the human nervous system, one of the greatest challenges of the twenty-first century. The accumulation of data on nervous system development and function across the life span of many species has been impressively rapid. There has been an explosive growth in both the number of neuroscientists and the rate of data collection; the latter has occurred primarily through the development of various high-throughput technologies, the completion of the mapping of the human genome, and the new capabilities of discovery approaches using computational analysis. However, current insights into the integration and functional significance of the data are not optimal. Neuroinformatics should enhance our understanding of the detailed and fine grain heterogeneous data available, as well as facilitate the efforts of discovery neuroscience through the sharing of data and the use of computational models ***(2,3)***.

The field of Neuroinformatics was created in 1993 through the efforts of a number of US funding agencies: National Institutes of Health, National Science Foundation, National Aeronautics and Space Administration, and the Department of Energy. The genesis of Neuroinformatics was a study by the National Academy of Science (NAS) commissioned by these same Federal agencies to evaluate: (1) the need and desirability of the field of Neuroscience research to share data and (2) the capability of the information technology (IT) field to handle the diversity, complexity, and quantity of data currently available and to become available in the future. This question was timely given the start of the Human Genome project, whose goal is to share broadly all genome data. The NAS 2-year study found that a neuroinformatics program is critical to understand brain development and function and to understand, treat, and prevent nervous system disorders ***(4)***.

For Neuroinformatics to succeed, the field of neuroscience, in cooperation with other research fields, would need to develop an interoperable infrastructure and analytical capabilities for shared data. This would include interoperable web-based neuroscience data and knowledge bases, analytical and modeling tools, and computational models.

Upon launch, the initial US program was called the Human Brain Project. As it expanded globally, the moniker morphed into Neuroinformatics.

Neuroinformatics is timely, appearing nearly a decade and a half after the Human Brain Project launched. As the neuroscience community is beginning to harvest the results of this inaugural phase, Neuroinformatics progenitors and advocates are already looking to re-evaluate progress, challenges, and new directions for the field given rapid developments in IT and neuroscience. This publication should help refocus and enhance some of the research in this area, as it informs new and current practitioners of Neuroinformatics.

Neuroinformatics is conceptually divided into four sections:

Neuroscience Knowledge Management has outstanding chapters dealing with the critical issues germane to computer science as applied to neuroscience, namely managing knowledge, databases interoperability, database architectures, data representation, and model specification, and neuroscience ontologies. A significant challenge in knowledge management is extracting text from the scientific literature. NeuroText is a new automated program with these capabilities that Crasto et al. describe with examples of its utility, extracting information relevant to the SenseLab databases at Yale. Interoperability is one of the greater challenges in the field of Neuroscience. Marenco et al. take on this difficult problem, providing a substantive overview of ongoing efforts, and have constructive suggestions on approaches to overcoming this pivotal issue. Nadkarni and Marenco review the important principles of database architectures and introduce and explore the value of the entity-attribute-value model. Crook and Howell do an excellent job showing the benefits of XML data representation and how it is used in Neuroinformatics. Bowden et al integrate some of the major issues in knowledge management, using real-life examples influencing interoperability, controlled vocabularies, and data heterogeneity, and show how to manage this problem using a neuroanatomical nomenclature: NeuroNames.

Computational Neuronal Modeling and Simulations presents in-depth expert summaries on specific computational models and simulations as well as approaches to data mining. Bower and Beeman, the creators of Genesis, provide expert descriptions of this simulator along with information on its routine use and abilities to create new simulations. Lytton and Stewart discuss the use of their new data-mining tool as an adjunct to the simulator Neuron, bringing both novice and expert to new levels of understanding in computational neuroscience. The chapter by Prinz provides an excellent demonstration and discussion on unraveling the components of neural networks into their component parts.

Imaging focuses on informatics representation and approaches to the structural complexity of the brain using a variety of both traditional and non-invasive imaging methods. The challenge here is accurate representation of the data in either 2D or 3D space. Ultimately, the data are best represented in four dimensions, a true, additional challenge. Two of the premier groups working in this area present two different approaches, each valuable in its own right, for the creation of data-rich probabilistic atlases. Toga and his team present the Minimum Deformation Atlases as used to capture the variations of brain structure because of genetic manipulation. Nissanov and his colleagues approach the issue by extracting and placing data onto a normalized atlas and providing a query system to extract the integrated data sets. Both chapters provide the reader with a detailed explanation of these two different approaches, along with their advantages and limitations—certainly a must read for work in this area.

The two chapters dealing with fMRI, while technically robust, are also invaluable for working with and creating databases of functional brain data. Liu presents an excellent detailed approach to creating an fMRI database, which in this case is for mapping the olfactory brain areas. Fissell, on the contrary, presents in her chapter a complete and efficient analysis of structural or functional MRI data that are applicable to most research. Kötter presents a highly detailed review of his receptor database, a critical tool for reviewing and understanding the distribution of brain receptor systems.

Neuroinformatics in Genetics and Neurodegenerative Diseases will be of great interest to individuals interested in genomic approaches to understanding the nervous system and its disorders. The reader will benefit greatly from the integrated material. These chapters demonstrate the value of using components of Neuroinformatics as a way to understand the complex disorders of Dementia, Schizophrenia, and Alzheimer's disease and demonstrate how to use informatics in a systems approach to unravel the intricate interactions between networks of genes. Rosen et al. on trait analysis, Duch on dementia and other neurological disorders, Middleton et al. on Schizophrenia, and Kinoshita and Clark on Alzheimer's disease all present excellent cases of Neuroinformatics, successfully leading to the discovery of neuroscience. The four examples provide the reader with a gestalt of how this all comes together to provide a better understanding of brain function and malfunction.

Future Trends and Challenges

Has the field of Neuroinformatics developed sufficiently to sustain itself and to be a placeholder for other areas of Biology and Biomedical Research? Is it a rational approach? This monograph and others do an excellent job of reviewing and demonstrating the advantages and success of the developing field of Neuroinformatics ***(5,6)***.

There is plentiful evidence elsewhere as well that the field is firmly established and flourishing. First, there are new field-specific journals providing a platform for cutting-edge research, such as *Neuroinformatics* and the *Journal of Integrative Neuroscience*. That these publications have quickly developed, following is a true testament to broad interest in the field, particularly given the great abundance of journals already covering neuroscience.

There are indications from the National Institutes of Health as well that the field is considered critical and sustainable. The original NIH program to stimulate and support initial research on Neuroinformatics, which received significant funding over the past 15 years, has concluded; scientists seeking support for new Neuroinformatics research programs can now do so through traditional NIH research mechanisms, through specific disease-oriented research institutes. And NIH continues to invest in support infrastructure to develop the field. In October 2006, NIH announced a new Road Map program, a 5-year, $3.8 million grant to the NIH Blueprint for Neuroscience Research to build an Internet-based clearinghouse to promote the adoption, distribution, and evolution of neuro-imaging tools, vocabularies, and databases.

Neuroinformatics is receiving support and attention from other US institutions as well. The Society for Neuroscience (SfN) has recently established a Neuroinformatics committee composed of international neuroscientists. Their goal is to monitor both the informatics needs of the community and the centralized gateway for neuroscience databases and to advise the SfN leadership on opportunities to enhance data sharing. Complete information on the functions of this committee is available on their Web site *(7)*. In Seattle, Washington, the Allen Institute for Brain Science is freely sharing through the Web an enormous quantity of mouse brain anatomical genomic expression data, at the cellular structural level *(8,9)*.

Globally, there is also long-term commitment to Neuroinformatics. Many countries involved in neuroscience research have initiated Neuroinformatics funding programs and have jointly created the International Neuroinformatics Coordinating Facility (INCF) *(10)*. The INCF's responsibility is to coordinate the Neuroinformatics activities of its member countries, thereby creating an integrated global capability for neuroscientists.

The milestones and accomplishments cited above, along with the publication of this new monograph, confirm that Neuroinformatics is a thriving field of research with a bright future ahead. Individuals interested in keeping up with the latest issues in Neuroinformatics and/or learning about and joining this field of research will greatly benefit from the outstanding and timely material and concepts drawn together by these Neuroinformatics experts and presented in *Neuroinformatics*.

Stephen H. Koslow

References

1. Beltrame, F. and Koslow, S.H. (1999) Neuroinformatics as a megascience issue. *IEEE Transactions on Information Technology in Biomedicine*, **3(3)**, 239–240.
2. Koslow, S.H. (2002) Sharing primary data: a threat or asset to discovery? *Nature Reviews Neuroscience* **3(4)**, 311–313.

3. Koslow, S.H. (2000) Should the neuroscience community make a paradigm shift to sharing primary data? *Nature Neuroscience* **3(9)**, 863–865.
4. Pechura, C.M. and Martin, J.B. (1991) Mapping the *Brain and Its Functions. Integrating Enabling Technologies into Neuroscience Research*. National Academy Press, Washington, D.C.
5. Koslow, S.H. and Subramanian, S. (Eds) (2005) *Databasing The Brain: From Data to Knowledge*. Wiley Press, New York.
6. FinnArup Nielsen. (2001) *Bibliography on Neuroinformatics 2001*. http://www2.imm.dtu.dk/~fn/bib/Nielsen2001BibNeuroinformatics.pdf#search=%22FinnArup%20Nielsen%20define%20Neuroinformatics%22
7. http://www.sfn.org/index.cfm?pagename=committee_Neuroinformatics
8. Bhattacharjee, Y. (2006) News focus. *Science* **313(5795)**, 1879.
9. www.brain-map.org
10. http://www.neuroinf.org/incf/index.shtml

Acknowledgments

To series editor, Professor John M. Walker of the University of Hertfordshire, UK. Professor Walker advised me on different aspects of this volume's development. For the opportunity to edit this book and his availability when I asked for help, I am very grateful. Considering the breadth and comprehensiveness of this series (Methods in Molecular Biology), Professor Walker deserves praised for what can only be described as a stupendous achievement. Thanks also to Mrs. Walker, who was involved during all correspondence with Professor Walker. Mrs. Walker, I suspect, helped as a shadow entity.

To Dr. Stephen H. Koslow, acknowledged by most in the field as the "Father of Neuroinformatics." The Human Brain Project was inspired by Dr. Koslow, as director of the National Institute of Mental Health at NIH. Throughout its evolution, Dr. Koslow worked tirelessly, shepherding the Human Brain Project, to ensure that its end-goals were within sight and funding was available to those workers who deserved it. I feel privileged that he agreed to write a Foreword for this volume.

To Professor Gordon Shepherd at the Yale University School of Medicine (also of SenseLab) whose guiding hand was always there, both personally and professionally.

To my colleagues at the Yale Center for Medical Informatics, who contributed to research and development in the SenseLab project and in my own personal, intellectual and professional development: Professors Perry Miller, Prakash Nadkarni, Luis Marenco, Kei Cheung, and to Drs. Nian Liu and Thomas Morse.

To the editorial staff at Humana Press (now a part of Springer Science), in no order of importance: Kavitha Kuttikan (Integra India), Christina Thomas, Patrick Marton, Donna Niethe, Saundra Bunton, Lisa Bargeman, Patricia Cleary and Mary Jo Casey, whose dedication, attention to detail, hard work and patience (with me) have made this volume possible.

To my parents, Joaquim and Janet Crasto, my sister Professor Selma Alliex and my brother, Mario Crasto, whose affection and support have continued to sustain me.

To my wife Cherie, whom I turn to for everything.

Thank you.

Contributors

Douglas A. Baxter • *Department of Neurobiology and Anatomy, The University of Texas Medical School at Houston, Houston, TX*
David Beeman • *Department of Electrical and Computer Engineering, University of Colorado, Boulder, CO*
Louise Bertrand • *Laboratory for Bioimaging and Anatomical Informatics, Department of Neurobiology and Anatomy, Drexel University College of Medicine, Philadelphia, PA*
Jyl Boline • *Laboratory of Neuro Imaging, Department of Neurology, University of California, Los Angeles, CA*
Douglas M. Bowden • *Department of Psychiatry and Behavioral Sciences, University of Washington, Seattle, WA*
James M. Bower • *Research Imaging Center, University of Texas Health Science Center and Cajal Neuroscience Center, University of Texas, San Antonio, TX*
Ahmet Bozkurt • *Department of Plastic, Reconstructive and Hand Surgery, Burn Center, University Hospital, RWTH, Aachen, Germany*
William Bug • *Laboratory for Bioimaging and Anatomical Informatics, Department of Neurobiology and Anatomy, Drexel University College of Medicine, Philadelphia, PA*
John H. Byrne • *Department of Neurobiology and Anatomy, The University of Texas Medical School at Houston, Houston, TX*
N. T. Carnevale • *Department of Psychology, Yale University New Haven, CT*
Elissa J. Chesler • *Oak Ridge National Laboratory, Oak Ridge, TN*
Timothy Clark • *Harvard University Initiative in Innovative Computing & Department of Neurology, Harvard Medical School, Cambridge, MA*
Chiquito J. Crasto • *Yale Center for Medical Informatics and Department of Neurobiology, Yale University School of Medicine, New Haven, CT; Current Address: Department of Genetics, University of Alabama at Birmingham, Birmingham, AL*
Sharon M. Crook • *Department of Mathematics and Statistics and School of Life Sciences, Arizona State University, Tempe, AZ*

MARK DUBACH • *Department of Psychiatry and Behavioral Sciences, University of Washington, Seattle, WA*
WŁODZISŁAW DUCH • *Nicolaus Copernicus University, Department of Informatics, Torun, Poland*
YINGLI FAN • *Laboratory for Bioimaging and Anatomical Informatics, Department of Neurobiology and Anatomy, Drexel University College of Medicine, Philadelphia, PA*
KATE FISSELL • *Learning Research and Development Center, University of Pittsburgh, Pittsburgh, PA*
SMADAR GEFEN • *Laboratory for Bioimaging and Anatomical Informatics, Department of Neurobiology and Anatomy, Drexel University College of Medicine, Philadelphia, PA*
AMARNATH GUPTA • *San Diego Supercomputer Center, University of California, San Diego, CA*
CARL GUSTAFSON • *Laboratory for Bioimaging and Anatomical Informatics Department of Neurobiology and Anatomy, Drexel University College of Medicine, Philadelphia, PA*
MICHAEL L. HINES • *Departments of Neurobiology and Computer Science, Yale University, New Haven, CT*
FRED W. HOWELL • *Textensor Limited, Edinburgh, Scotland UK*
JUNE KINOSHITA • *Alzheimer Research Forum, Waltham, MA*
ROLF KÖTTER • *Section Neurophysiology & Neuroinformatics, Department of Cognitive Neuroscience, Radboud University Nijmegen Medical Centre Nijmegen, The Netherlands; and Vogt Brain Research Institute and Institute of Anatomy II, University of Düsseldorf, Düsseldorf, Germany*
ALLAN KUCHINSKY • *Genotyping and Systems Biology, IBS Informatics, Agilent Technologies, Santa Clara, CA*
NIAN LIU • *Center for Medical Informatics and Department of Anesthesiology, Yale University School of Medicine, New Haven, CT*
WILLIAM W. LYTTON • *Departments of Physiology, Pharmacology and Neurology, SUNY Downstate Medical Center and Department of Biomedical Engineering, Downstate/Polytechnic University, Brooklyn, NY*
ALLAN MACKENZIE-GRAHAM • *Laboratory of Neuro Imaging, Department of Neurology, University of California, Los Angeles, California*
JÜRGEN MAIER • *Vogt Brain Research Institute, University of Düsseldorf, Düsseldorf, Germany*

KENNETH F. MANLY • *Department of Anatomy and Neurobiology, Department of Pediatrics, University of Tennessee Health Science Center, Memphis, TN*

LUIS MARENCO • *Yale Center for Medical Informatics, Yale University School of Medicine, New Haven, CT*

MARYANN MARTONE • *National Center for Microscopy and Imaging Research and Department of Neurosciences, University of California, San Diego, CA*

FRANK A. MIDDLETON • *Center for Neuropsychiatric Genetics, Department of Psychiatry and Department of Neuroscience & Physiology, Upstate Medical University, Syracuse, New York, NY*

TOM M. MORSE • *Department of Neurobiology, Yale University School of Medicine, New Haven, CT*

PRAKASH NADKARNI • *Yale Center for Medical Informatics, Yale University School of Medicine, New Haven, CT*

JONATHAN NISSANOV • *Laboratory for Bioimaging and Anatomical Informatics, Department of Neurobiology and Anatomy, Drexel University College of Medicine, Philadelphia, PA*

JACK PARK • *NexistGroup, Menlo Park, CA*

CARLOS N. PATO • *Center for Neuropsychiatric Genetics and Department of Psychiatry, Upstate Medical University, Syracuse, NY; Department of Psychiatry, University of Southern California, Los Angeles, CA; and Department of Veteran's Affairs Medical Center, Washington, DC*

MICHELE T. PATO • *Center for Neuropsychiatric Genetics and Department of Psychiatry, Upstate Medical University, Syracuse, NY; Department of Psychiatry, University of Southern California, Los Angeles, CA; and Department of Veteran's Affairs Medical Center, Washington, DC*

ASTRID A. PRINZ • *Department of Biology, Emory University, Atlanta, GA*

GLENN D. ROSEN • *Department of Neurology, Beth Israel Deaconess Medical Center and Harvard Medical School, Boston, MA*

CARESTON ROSENOW • *Genotyping and Systems Biology, IBS Informatics, Agilent Technologies, Santa Clara, CA*

AXEL SCHEICHER • *Vogt Brain Research Institute, University of Düsseldorf, Düsseldorf, Germany*

GORDON M. SHEPHERD • *Department of Neurobiology, Yale University School of Medicine, New Haven, CT*

MARK STEWART • *Department of Physiology and Pharmacology, SUNY Downstate Medical Center, Brooklyn, NY*

ARTHUR W. TOGA • *Laboratory of Neuro Imaging, Department of Neurology, University of California, Los Angeles, CA*
ADITYA VAILAYA • *Genotyping and Systems Biology, IBS Informatics, Agilent Technologies, Santa Clara, CA*
ROBERT W. WILLIAMS • *Department of Anatomy and Neurobiology, Department of Pediatrics, University of Tennessee Health Science Center, Memphis, TN*
KARL ZILLES • *Vogt Brain Research Institute, University of Düsseldorf, Düsseldorf, Germany; Institute of Medicine, Research Center Jülich, Jülich, Germany*

Color Plate

The following color illustrations are printed in the insert.

Chapter 18

Fig. 1: Region of the velocardiofacial syndrome (VCFS) deletion on chromosome 22 detected by comparative genome hybridization (CGH). The figure shows a screenshot of CGH Analytics software. The three windows show all chromosomes (bottom window), chromosome 22 (middle window), and the hemizygous microdeleted region with its associated genes. Red and green points are probes on the array with measured intensities. The deletion is clearly identified on the q-arm of chromosome 22, in the 17–20 Mb region. A gene list can be saved from all the genes in the deleted region and imported into GeneSpring GT or GeneSpring GX.

Fig. 3: Workflow for integrating gene expression profiling and genotype analysis. **(A)** Combined gene expression and genotype gene list-based pathway enrichment analysis identifies four canonical pathways. **(B)** Agilent Literature Search tool extracts associations from literature for disease-related genes and proteins. **(C)** Literature-based extended association network for the four canonical pathways identified by combined genotype and gene expression datasets. **(D)** "Interesting" sub-network identified from the extended network in **(C)** using combined single-nucleotide polymorphism (SNP) LOD score and gene expression *P*-values.

Fig. 4: GeneSpring Network builder. Schematic representation of the workflow for the plug-in. A gene list is selected and the selected databases are searched for gene–gene interactions. A network is built using the Cytoscape viewer.

Fig. 5: The Huntington and ERK5 gene lists from the gene expression, genotyping, and comparative genome hybridization (CGH) experiments are used to build an interaction network using the GeneSpring Network builder. New and known interactions can be identified. *GRB2* is one of the connector genes between the genes of the ERK5 pathway and the Huntington pathway.

I

Neuroscience Knowledge Management

1

Managing Knowledge in Neuroscience

Chiquito J. Crasto and Gordon M. Shepherd

Summary

Processing text from scientific literature has become a necessity due to the burgeoning amounts of information that are fast becoming available, stemming from advances in electronic information technology. We created a program, NeuroText (http://senselab.med.yale.edu/textmine/neurotext.pl), designed specifically to extract information relevant to neuroscience-specific databases, NeuronDB and CellPropDB (http://senselab.med.yale.edu/senselab/), housed at the Yale University School of Medicine. NeuroText extracts relevant information from the Neuroscience literature in a two-step process: each step parses text at different levels of granularity. NeuroText uses an expert-mediated knowledgebase and combines the techniques of indexing, contextual parsing, semantic and lexical parsing, and supervised and non-supervised learning to extract information. The constrains, metadata elements, and rules for information extraction are stored in the knowledgebase. NeuroText was created as a pilot project to process 3 years of publications in *Journal of Neuroscience* and was subsequently tested for 40,000 PubMed abstracts. We also present here a template to create domain non-specific knowledgebase that when linked to a text-processing tool like NeuroText can be used to extract knowledge in other fields of research.

Key Words: Information extraction; natural language processing; text mining; neuroscience; neuroinformatics; indexing; supervised and unsupervised learning; algorithmic development.

1. Introduction

As information technology makes larger amount of information available electronically, the need for rapid and efficient knowledge dissemination increases commensurately. This is no more evident than in the neurosciences,

From: *Methods in Molecular Biology, vol. 401: Neuroinformatics*
Edited by: Chiquito Joaquim Crasto © Humana Press Inc., Totowa, NJ

a domain that spans several sub-disciplines and extends from biosciences to the clinical domain. CellPropDB and NeuronDB were created as part of the SenseLab suite of databases housed at the Yale University School of Medicine (http://senselab.med.yale.edu/senselab) ***(1,2)*** to make available information related to neuronal membrane properties (*see* Chapter 2 for interoperability techniques between NeuronDB and other neuroscience databases, and Chapter 3 for a description of SenseLab's database architecture). These curated databases contain annotated, first-principle descriptions of information related to three essential elements of rapid neuronal signaling for a fixed set of neurons: neurotransmitters, neurotransmitter receptors, and intrinsic ion channels (*see* **Fig. 1**). The information in CellPropDB (neuronal properties of a cell as a whole) and NeuronDB (properties that have been experimentally localized to specific neuronal compartments) includes bibliographic citations to

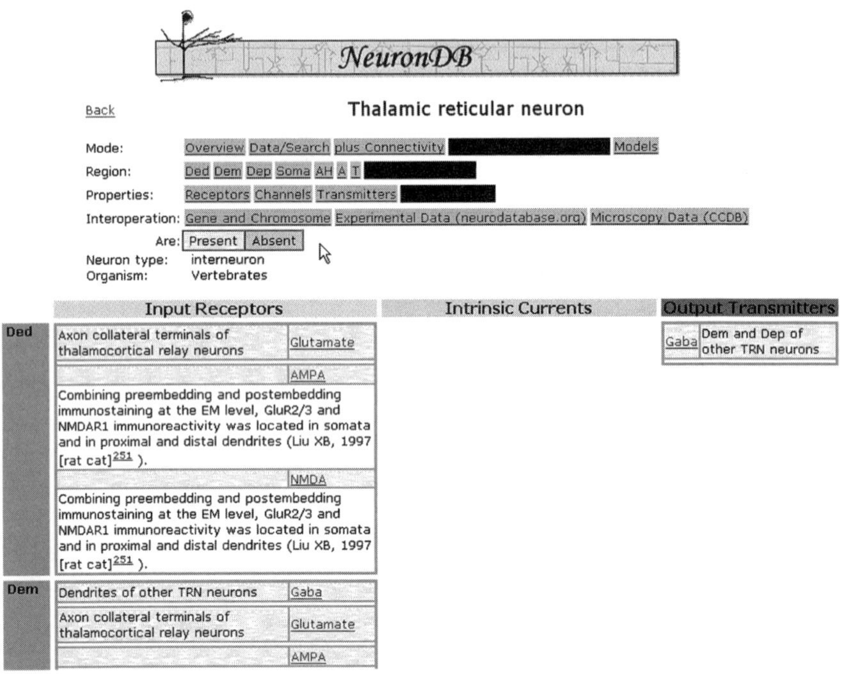

Fig. 1. The figure shows a page in NeuroText which contains first principle descriptions of receptor, neurotransmitter, and current properties of the reticular neuron in the thalamic region of the brain. The figure shows annotations to citations for properties expressed in the dendritic compartments of the thalamic reticular neuron.

articles in the neuroscience literature. Populating these databases had initially been done manually *(3)*; the rapidly expanding literature, however, requires the development of automated procedures that offer significant help to the expert. One such automated procedure would involve creating an information extraction tool that processes the text of neuroscience literature and retrieve knowledge that would be relevant to CellPropDB and NeuronDB.

1.1. Steps to be Considered During Information Extraction

Our approach in extracting information from unstructured text to populate databases involved the following key steps:

- Creating a knowledgebase that contained specific information (metadata from the database to be populated) that would help identify relevant keywords in the correct hierarchy (if information in database to be populated is hierarchical). This knowledgebase would contain synonyms that could be mapped back to the original keywords. Additionally, a variable scoring for keywords would be necessary if certain keywords are deemed more contributory to the relevance of an article.
- Examining the text to determine whether keywords occurred in a context relevant to the database domain. Scanning and identifying key lexical and semantic relationships between keywords to correctly identify relationships that would match the hierarchy in which the information extracted was to be stored.
- Using a lexical scanner to ascertain the affirming/negating context of the text. Correctly identifying journal articles that specifically refute the presence of a property would be both important and necessary.
- Accommodating the evolution of information in the database domain. The knowledgebase would be dynamically updated when new keywords, synonyms, and relationships were identified, as well as when existing knowledgebase information became irrelevant.
- Ensuring that such a program would be extensible to other domains without significant algorithmic modifications.
- Validating the information extracted.

We developed NeuroText (*see* **Fig. 2**), an expert-mediated, knowledgebase-supported information extraction tool, to identify articles relevant to NeuronDB and CellPropDB *(4)*.

2. Background

Natural language understanding involves the development of computational systems to process and assimilate unstructured, non-annotated written and spoken language in a fashion that mimics human understanding. Natural

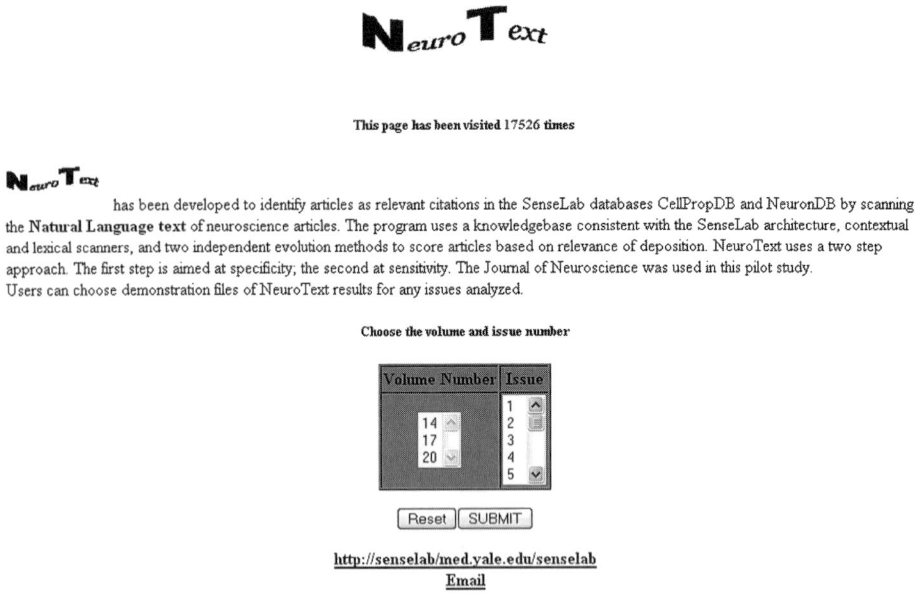

Fig. 2. Front page of the Web-accessible results of NeuroText following the processing of approximately 3000 abstracts from *Journal of Neuroscience*. Results can be accessed by choosing a volume and issue number from *Journal of Neuroscience* from the scrolling lists. This page can be accessed at http://senselab.med.yale.eud/textmine/neutotext.pl.

language processing (NLP), an information science endeavor for well over 20 years, includes projects in many areas, e.g., word-indexing, syntactic parsing of sentences, separating relevant keywords from random noise in a sentence, restructuring retrieved information to and from databases ***(5)***, interfacing programs with audio media, and translating documents between languages.

A wide range of NLP tools have been created in projects focusing on information science ***(6–11)***. NLP techniques have also been used in the clinical and biological fields. Methods have been reported where gene names can be extracted from the literature by using the techniques of BLAST searching ***(12)***. Several groups have created NLP systems for molecular biology ***(13–16)***. Systems such as BRENDA, which presents information related to enzymes ***(17)***, and NTDB, a database for thermodynamic properties of nucleic acids ***(18)***, similarly rely on published scientific literature as their sources. NLP methods have also been used in the clinical domain ***(19–24)***.

3. Methods

NeuroText does not address the full natural language understanding of problems for general, unstructured text as studied by researchers in the areas of artificial intelligence and computational linguistics *(25)* (Chapter 17 describes the use of NLP techniques in studying neurodegenerative disorders). NeuroText is an information extraction tool. It can process text from neuroscience articles with a focused end-goal in mind: populating a database with specific information about neuronal membrane properties. The use of a pre-designed knowledgebase is what distinguishes NeuroText from typically employed NLP methods.

3.1. NeuroText and Information Extraction

The nature of the information that had to be deposited into CellPropDB and NeuronDB necessitated the creation of NeuroText.

Specific features of NeuroText's approach included the following:

- Keyword counting for potential relevance: NeuroText was premised on the straightforward assumption that researchers presenting information will mention concepts and keywords related to that information more frequently than other non-related or distantly related concepts. Elementary counting statistics differentiated keywords and concepts more relevant to an article than a tangential mention.
- Using contextual constraints to refine potential relevance: The context of the occurrence of relevant keywords, however, precluded reliance on the mere counting of database keywords. Context was defined by the constraints of the specific database domain and of neuroscience. CellPropDB and NeuronDB present information (structured in the region-neuron-property hierarchy) about a subset of normal, healthy neurons from in vivo or in vitro studies. Work related to diseased cells or neurological disorders and work pertaining specifically to in vitro cell cultures, for example, was considered contextually irrelevant. BrainPharm, a new database in development in SenseLab (http://senselab.med.yale.edu/senselab/BrainPharm/) extends the membrane property paradigm (that constitutes the data in CellPropDB and NeuronDB) to neurological disorders. To identify articles in BrainPharm, NeuroText's knowledgebase could be modified to identify keywords, concepts, and contexts specific to epilepsy, Alzheimer's disease and Parkinson's disease.
- Recognizing relevance of keywords to primary subject in the publication: When conflicting but relevant keywords occurred in the same article, NeuroText sought to separate keywords that were contributory versus keywords that occurred in the article in a random and contextually irrelevant manner by variably scoring keywords and ensuring the "brain region-neuron-membrane property" hierarchy.

- Identifying important relationships between keywords: Identifying relationships (in the text) between regions, neurons, and the properties they express was challenging, especially if more than one property had been identified with more than one neuron (or neuronal compartment). NeuroText incorporated semantic relationships (neuroscience and lexical) to define fine-grained relationships between keywords. These semantic relationships were manually identified using a training set of articles that were already manually deposited in NeuronDB. Additional relationships were added through an unsupervised learning step.
- Easy automated updating of the knowledgebase: NeuroText analyzed articles whose context and content are unrestricted, encompassing all of neuroscience (*Journal of Neuroscience*). Concepts hitherto unidentified in the knowledgebase needed to be incorporated to guide the system's operation. A supervised learning step during processing would aid in this automated updating of the knowledgebase.
- The domain expert needs to make the final decision: A survey of the efficacy of knowledge engineering studies reveals that successful programs identify target articles approximately 70% of the time—determining the parameters *precision* and *recall*, which relate to specificity and sensitivity of the retrieval strategy, respectively ***(26–28)***. Because the eventual aim of our study is to populate databases with authoritative information, the expert/curator is charged with the responsibility of depositing articles with 100% accuracy.
- Presenting the results of NeuroText's analysis to the expert for validation and automatic deposition: The final, necessary step was to present the results of text mining in an interface to the expert. The interface highlighted keywords, concepts of interest, and affirming and negating sentences if these are potentially decisive in determining whether an article is citable. The interface also provided the expert with the tools to dynamically override erroneous results of the program. The interface allowed the expert to store validation results for continuous monitoring of the efficacy of the algorithms.

3.2. NeuroText Operation

This section gives an overview of NeuroText's operation and design. **Figure 3** provides a process flow diagram for NeuroText. A single PERL server-side script runs the entire NeuroText program—including the pre- and post-processing steps. The principal steps in NeuroText's operation are as follows:

3.2.1. Processing for Sensitivity

The sensitivity search uses a commercial indexing program DTSearch© ***(29)***. Database keywords and synonyms from the NeuroText knowledgebase are dynamically incorporated into the DTSearch control files. The search is designed to identify, if possible, whether each abstract processed contained a

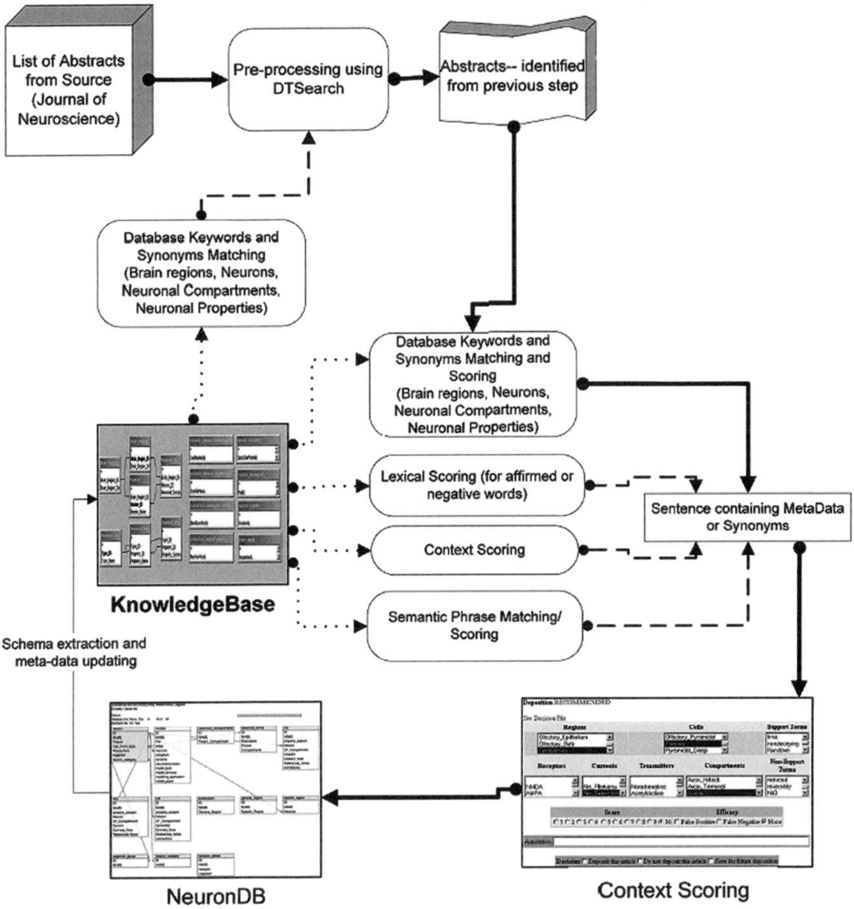

Fig. 3. Flow diagram shows the entire NeuroText process as described in **Subheading 3**. This figure also appears as Figure 2 in the **ref.** *4*. Adapted from **ref.** *4*.

general description of the hierarchy: at least one region, one neuron, and one property. The aim of this step is to identify as many articles as possible that contained keywords or concepts associated with database keywords (metadata) in the correct hierarchy. During this step, NeuroText does not seek to relate region, neuron, and property. Every abstract-search that meets the sensitivity search criterion is combined into a report that became the starting point for further analysis. This step specifically limits false-negatives.

3.2.2. Post-Processing for Specificity

Post-processing is designed to help ensure the specificity of the neuroscience abstract for deposition into CellPropDB or NeuronDB: the neuron mentioned in the article does belong to a specific region in the brain, and the property is (or is not) associated with that neuron and/or its compartments. Post-processing helps ensure that the contextual and lexical constraints of information being suggested for deposition into NeuronDB and CellPropDB are adhered to. The information and rules for processing are stored in the NeuroText knowledgebase. The text is processed thus:

- Re-identification of database keywords and synonyms: This includes identifying the occurrence of CellPropDB and NeuronDB keywords. These include keywords for brain regions, neuron names, intrinsic currents, neurotransmitters, and neurotransmitter receptors. Synonyms for these keywords if present are mapped back to the original keywords stored in the database. A simple count to measure the occurrence of these keywords is then initiated.
- Neuroscience context: Every word in the entire sentence where a keyword occurs is scanned against the context table to identify contextual patterns that might enhance or reduce the score each keyword receives. If no discernible context is identified, the initial score for each keyword is retained. During the context search, NeuroText also searches for additional keywords from NeuronDB, namely, concepts related to neuron compartments (axonal, somatic, and dendritic) that are not part of the sensitivity search. For example, certain fiber pathways and interneurons, which provided key connectivity information as to their originating and terminating neurons, can also be identified. Consider the keyword "mossy fibers." These terms apply to axons that arise from dentate granule cells and terminate on CA3 cells in the hippocampus ***(30)***, and axons that arise from inferior olive cells and terminate on Purkinje cells in the cerebellum; NeuroText counts the occurrence of the keyword "mossy fibers" in the text of an article as a score increment for dentate granule axons and the presynaptic axon input to CA3 pyramidal neurons.
- Affirming and negating context: Each sentence of the abstract's text that contains keywords is also scanned for affirming or negating contexts. Sentences in the text where either concept occurred are flagged. Only sentences that contain keywords for properties (receptors, currents, and transmitters) and neuronal compartments are scored if they contain affirmed or negated associations. If a sentence contains both an affirmed and negated word, then it is marked up differently so as to bring it to the attention of the domain expert. A tabulated list of affirmed and negative concepts was created using Roget's Thesaurus list by expanding word relatedness beginning with "certain" for affirmed and "uncertain" for negated concepts.
- Hierarchy matching: It is relatively easy to associate cells with regions in the brain. For example, the olfactory mitral cell is the principal neuron in the olfactory

bulb *(31)*. In making these associations, NeuroText discriminates ambiguities that stem from the occurrence of keywords of brain regions and unrelated neurons. In order to handle complicated situations that arise due to neurons associated with two different regions (granule cells arise in both the "cerebellum" and "dentate gyrus"), we provided links to the full-text article so that the curator can ascertain the proper hierarchy.

- Semantic phrases for fine-grained parsing and unsupervised learning: During post-processing, when a "neuronal property" keyword occurs in a sentence by itself or with another keyword, the sentence is scanned against the "semantic phrase" component of NeuroText's knowledgebase. If a relevant phrase is identified, the keyword score is enhanced. It is also flagged as related to the neuron (or compartment) or region. The sentence is also scanned for an affirming or negating tone. If a negating word or concept is identified with the property, the score for that property was not enhanced. Each property (or neuronal compartment) keyword was thus scored differently from a region or a neuron.

  If potential "relation" phrases are not identified, the sentence containing the keyword matches is stripped of database keywords and extraneous noise words and then included in the semantic relationship table in the knowledgebase. This is the unsupervised learning step. Every new article can then avail itself of this new semantic relationship, possibly enhancing or negating a score. This enables a fine-tuned identification of neuron-property matches.
- Scoring for relevance: After the abstract text has been scanned, NeuroText determines the scores for each neuroscience keyword identified. The maximum score for keywords of each type is calculated. Brain regions are associated with neurons. Regions and neurons that did not match are discarded. If more than one region–cell pair or property or neuronal compartment exists in the text, keywords whose count is less than a fourth of the count for the maximum of that class of keywords are discarded as likely "random" occurrences.
- Interface design: Every abstract processed by NeuroText is presented to the expert for validation. **Figure 4 A and B** shows examples of articles that are deemed (by NeuroText and agreed on by the domain expert) fit "for deposition" and "not for deposition," respectively. This interface, which is generated as the post-processing step proceeds, is a dynamically generated PERL (CGI) script that performs information presentation and data deposition. An expert can access this interface remotely on the Internet, make decisions, validate the results of text processing, and deposit relevant data. The file presents the abstract with relevant words and sentences highlighted. Database keywords are enlarged; enhancing concepts are boldfaced; non-support concepts have "strikethroughs" in them. Negating words are italicized; affirming words have a different font. Sentences with negating words are white text on black background; sentences that point to lexically affirming tones have a gray background. Sentences with conflicting negating and affirming tones or no discernible tones are not highlighted.

819.pdf-Link to Full-Text of Article 819.xml XML-Link

Volume 17, Number 2, Issue of January 15, 1997 pp. 819-833

Cellular, Subcellular, and Subsynaptic Distribution of **AMPA-Type Glutamate** Receptor Subunits in the **Neostriatum** of the Rat

Received Sept. 4, 1996; revised Oct. 29, 1996; accepted Nov. 4, 1996.

Véronique Bernard , Peter Somogyi , and J. Paul Bolam

Medical Research Council, Anatomical Neuropharmacology Unit, University Department of Pharmacology, Oxford University, Oxford OX1 3TH, United Kingdom

Glutamate released in the basal ganglia is involved in the expression of clinical symptoms of neurodegenerative diseases like Parkinson's or Huntington's. **Neostriatal** neurons are the targets of glutamatergic inputs derived from the cortex

the **thalamus** acting via **AMPA-type** as well as other **glutamate** receptors. To **determine** the location of subunits of the **AMPA** subclass of **glutamate** receptors (GluR) in the rat **neostriatum,** we applied multiple immunocytochemical techniques

using **anti-peptide** antibodies *against* the GluR1, GluR2/3, and GluR4 subunits at both the light and electron microscopic levels. All medium spiny efferent neurons, some of which were identified as striatonigral neurons, displayed

immunoreactivity for GluR1 and GluR2/3 subunits. Double immunofluorescence revealed that at least 70-90% of parvalbumin-immunopositive **GABAergic** interneurons were immunoreactive for each of GluR1, GluR2/3, or GluR4 subunits and that at

least 40% of choline acetyltransferase-immunopositive **cholinergic** interneurons were immunopositive for GluR1 or GluR4 subunits. The majority of nitric oxide synthase-immunopositive neurons had no detectable immunoreactivity for any of the

AMPA receptor subunits. Electron microscopic analysis confirmed the presence of immunoreactivity for GluR1 and GluR2/3 in the perikarya of spiny neurons and interneurons and GluR4 in perikarya of interneurons only. GluR1 and GluR2/3

subunits were detected in dendrites and spines. A **significant** population of extrasynaptic receptors was revealed by pre-embedding ~~immunogold~~ labeling along the plasma membranes of perikarya, dendrites, and spines. Receptors were

concentrated in the postsynaptic membrane specialization of asymmetrical synapses, as revealed by the postembedding ~~immunogold~~ method. Quantitative analysis demonstrated that immunoreactivity for the GluR1 and GluR2/3 subunits is higher

the periphery than at the middle of the postsynaptic membrane specialization.

Our results demonstrate that **AMPA** receptor subunits are distributed widely and heterogeneously among **striatal** neurons and are concentrated on the postsynaptic membrane of asymmetrical synaptic specializations, although extrasynaptic

are also present.

Neostriatum; Medium_Spiny; ; AMPA; Glutamate;

Deposition RECOMMENDED

1226.pdf-Link to Full-Text of Article 1226.xml XML-Link

Volume 17, Number 4, Issue of February 15, 1997 pp. 1226-1242

N -arachidonoyl PE in adult brain tissue and the enzyme pathways that underlie its biosynthesis are, however, still undetermined. We report here that rat brain tissue contains both anandamide (11 ± 7 pmol/gm wet tissue) and N -arachidonoyl

PE (22 ± 16 pmol/gm), as assessed by gas chromatography/mass spectrometry. We describe a N -acyltransferase activity in brain that **catalyzes** the biosynthesis of N - arachidonoyl PE by transferring an arachidonate group from the sn -1

of phospholipids to the ~~amino~~ group of PE. We also show that sn -1 arachidonoyl phospholipids are present in brain, where they constitute ~0.5% of total phospholipids. N - acyltransferase activity is **Ca** 2+ **dependent** and is enriched in brain

and testis. Within the brain, N -acyltransferase activity is highest in brainstem; intermediate in cortex, **striatum, hippocampus,** medulla, and **cerebellum;** and lowest in **thalamus,** hypothalamus, and **olfactory** bulb. Pharmacological **inhibition**

of N -acyltransferase activity in primary ~~cultures of~~ **cortical** neurons prevents **Ca** 2+ -stimulated N -arachidonoyl PE biosynthesis. Our results demonstrate, therefore, that rat brain tissue contains the complement of enzymatic activity and

lipid substrates necessary for the biosynthesis of the anandamide precursor N -arachidonoyl PE. They also suggest that biosynthesis of N -arachidonoyl PE and formation of anandamide are tightly coupled processes, which may concomitantly be

stimulated by elevations in intracellular **Ca** 2+ occurring during neural activity.

Deposition NOT RECOMMENDED

Fig. 4. **(A)** A result of "Deposition Recommended" for a sample abstract. NeuroText found occurrences of keyword "medium spiny neuron" in the neostriatum, which expressed alpha-amino-3-hydroxy-5-methyl-4-isoxazolepropionic acid (AMPA) and glutamate receptors. The metadata elements have been "marked-up" in large and boldface fonts. The words with strikethroughs are contextually negating. The sentence with a black background was found by NeuroText to be lexically negating. **(B)** A result of "Deposition Not Recommended" for a sample abstract. NeuroText found several instances of "brain region" keywords but did not find any discernible relationships with neurons. The "Ca^{2+} ion channel" was not associated with a specific neuron. The metadata elements from the databases are in boldface and large. The words with strikethroughs are contextually negating.

- NeuroText's assessment decision: After analyzing the abstract, NeuroText presents one of three decisions in dynamically generated Web page to the expert. A "Deposition Recommended" decision is made if a region, a neuron in that region, and a property identified with the neuron or its compartment are clearly identified. "Deposition Not Recommended" is the NeuroText decision if the score for keywords in an abstract during the sensitivity search is nullified, if the sentences are identified by the context that would negate the scores, or if identified regions are not associated with identified neurons. "Deposition Under Advisement" decision is made if a region–cell pair is properly identified in NeuroText but no specific property (from the database) is identified, or if the property scores are negated but the region–neuron pairs are not. NeuroText assumes a possible relation between a neuron (and its compartments) and its properties and directs the expert to take a closer look before making a decision.

  A link to a dynamically generated "decision tree" is included after the results of processing each abstract. This decision tree contains a step-wise presentation of how and why each keyword is scored—when a score is enhanced and negated. This Web-based file also shows how the scores for keywords were tabulated and how the decision was made to inform the domain expert whether an article was deemed fit for deposition or not.
- Form for validating and depositing NeuroText results: The "marked-up" abstract in the results Web page is followed by a deposition form (*see* **Fig. 5**) containing

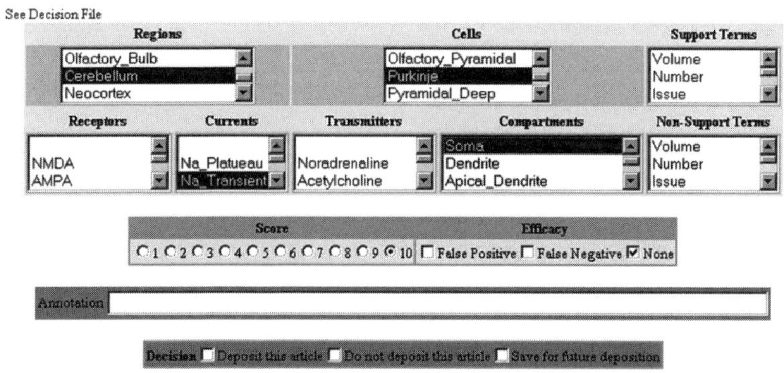

Fig. 5. The form that accompanies NeuroText's result after each abstract has been processed. The form shows that NeuroText has identified a sodium transient current in the Purkinje cell of the cerebellum. The form is provided so that the domain expert can override any erroneous findings during validation. The scrolling lists of support terms and non-support terms are provided so that the domain expert can enhance the contextual constraints in the knowledgebase to increase the efficiency of processing subsequently scanned articles. The form also provides the expert with an opportunity to validate the results.

tabulated, scrolling lists in which keywords for regions, neurons, receptors, ionic currents, neurotransmitters, and neuronal compartments are identified. The expert can also ascertain if the decision as a whole or in part contains false-positives or false-negatives and record these while scoring the search efficacy. The expert can override NeuroText based on their assessment of an abstract by making changes in this information before it is deposited into the SenseLab database (by clicking on the correct information in the scrolling lists).

- Supervised learning: Supervised learning is included in NeuroText to allow the system to update and modify the knowledgebase in a continually evolving fashion. To allow such supervised learning, two additional scrolling lists, besides those associated with keywords, are presented in the interface to the expert. These word-concept lists are identical. The lists are created from an array of words from the abstract's text after stripping away every word that already exists in the knowledgebase. If a concept not present in the knowledgebase is deemed necessary to enhance the score or diminish the score of keywords, the expert can click on that word or concept in the appropriate list. When the information is submitted for deposition, the negating concepts are added to the file containing non-support concepts in the knowledgebase. Any phrases or sentences containing these words are also removed from the phrase knowledgebase and placed in an archival table. Similarly, new support concepts identified by the experts are placed in the file containing support terms. The program also scans the archival tables to retrieve any phrases associated with this affirmed concept that might have been previously placed in the archive.

4. NeuroText Results and Discussion

Journal of Neuroscience abstracts from the years 1994, 1997, and 2000 (numbering, approximately 3000) were first downloaded (http://www.jneurosci.org). Each abstract was also independently scanned by two domain experts. The results of NeuroText post-processing are available at the SenseLab Web site at http://senselab.med.yale.edu/textmine/NeuroText.pl. (*see* **Fig. 1**). Of the 177 article abstracts identified for post-processing, (1) 29 articles were deferred by NeuroText for final decision to the experts, (2) 126 articles were correctly identified, (3) 13 articles were incorrectly identified (false-positives), and (4) 9 articles were identified by the experts that NeuroText judged "Deposition Not Recommended" (false-negatives). Thus, NeuroText identified 126 articles correctly (in agreement with the experts) and 22 incorrectly, for an accuracy of 85%. Similarly, the proportion ***(32,33)*** for identifying true-positives was 72% and for true-negatives 90%. Alternatively, using the odds-ratio test ***(34)***, the odds ratio of a correct identification of an article (as citable or not citable) by NeuroText is approximately 26:1.

Subsequent analysis of the results revealed that almost all articles that NeuroText deferred (deemed as "Under Advisement") required that the experts consult the full text (available as a link in NeuroText results) before deciding whether to cite the article. Most of NeuroText false-positives were deemed as possible "weak" citations—that the abstracts did not contain novel information. Most of the false-negatives were due to inadequacies of the knowledgebase. With a subsequently enhanced knowledgebase, the number of falsely identified articles decreased significantly. The experts did not call into question the algorithmic details or the search and scoring strategies for any of the articles analyzed. Every volume of *Journal of Neuroscience* contains approximately 1000 articles. NeuroText's time for processing 1000 abstracts in a volume is less than 2 h.

Using an enhanced version of NeuroText following the above described pilot study, 40,000 abstracts from the PubMed housed at the National Library of Medicine were extracted and processed. Searches in PubMed were conducted using the search phrase "Brain Region AND neuron." The pre- and post-processing steps of NeuroText took approximately 48 h to process these abstracts. Nine hundred fifty abstracts were identified as potential citations and were presented to domain experts for final validation. This demonstrated NeuroText's enhanced parsing abilities. It also ensures that relevant, up-to-date and comprehensive information can be made available to users of NeuronDB and CellPropDB.

As mentioned earlier, it is impossible to create an all-encompassing knowledgebase. We anticipate that NeuroText's knowledgebase will continually evolve and that the domain expert will make the final decisions about deposition into the SenseLab databases. The advantage of presenting the interface to the experts with the tools to allow the dynamic modification of NeuroText decisions while making modifications to enhance the knowledgebase ensures that the information deposited is accurate. By changing the knowledgebase, most of NeuroText failures can be remedied such that subsequent articles scanned would benefit from these changes. This would result in better agreements between expert and the computer program.

In the analysis of the results, the expert often termed an article as too general or too specific to be deposited. This qualitative determination can be corrected neither by NeuroText results nor by a careful perusal of the decision tree. The telling detail in these NeuroText failures is that they are not consistently false-positives or false-negatives. The failures encompass both in equal measure—indicative of information that might be inferred from the abstracts in the absence of keywords or a concept that could enhance or negate the scores for the keywords, if present.

5. Template for Extending NeuroText Methodologies to Other Domains

NeuroText as an information extraction tool was designed to be extensible to domains other than neuroscience. None of the algorithmic features in NeuroText were neuroscience specific. Instead, NeuroText relied on a knowledgebase that contained tables of metadata (keywords and synonyms that mapped back to the keywords) and domain-dependent contextual and linguistics lexical constraints. The rules for extracting information were determined by the domain experts and were embedded into the knowledgebase.

The validation–deposition interface in the NeuroText result is linked to the SenseLab databases, which use a specific architecture *(35)*. Naturally, such links would not be useful to populate databases with different articles on different platforms. To address this issue, every NeuroText result also contains a dynamically generated XML file in whose nested fields are embedded the mined information (*see* Chapter 4 for detailed description of XML use in neuroscience). The XML files are created with a view to interoperability. Researchers who wish to use the NeuroText tool would have to simply create an XML parser (XML parsers are available for different platforms and programming languages) to extract relevant data and link (or post) it to a database or storage medium of their choice. Resource Description Framework (RDF) (http://www.w3.org/RDF/) is now becoming the information exchange medium of choice. NeuroText results could alternatively be output in the RDF format.

Presented here is a template for the creation of a Web-based tool that will allow a domain expert to dynamically create a knowledgebase pertaining to an area of study and embed in this knowledgebase preset (and easily modifiable) rules for information extraction:

- A database will be created first. This database will serve as the knowledgebase. The tables, relationship connectivities, and other features in this database are instituted irrespective of whether an expert wishes to use that feature. These tables will be dynamically populated. The additional (and depending on the curator's choice—unnecessary) information and rules are to allow the domain expert the flexibility of modifying the knowledgebase at a later time.
- The expert will choose a training set of articles relevant for his or her purposes. A fast indexing algorithm to create unique word-concept lists. These unique words will be stored in the Objects table of the database and presented to the expert in an interface. The desired keywords, which the expert will "click" on and which will be stored in a special sub-table of the Objects table, will form the basis for a simple keyword search. Additionally, most biomedical keywords are associated with

Medical Subject Headings (MeSH) terms in PubMed and terms in Unified Medical Language Service (UMLS). These selected concepts will be identified with MeSH and UMLS identifiers.

- The knowledgebase (KB) creation tool is Web-based. A domain expert can create and design knowledgebases remotely, if necessary. The first page of this interface will query the user as to what features are desired. At any point during the knowledgebase creation process, the expert will be able to modify the rules for retrieval. The dynamic and adaptive nature of this process will allow the modification of future KB creation steps to reflect those changes. The expert will be queried as to need for the following features:
  - Whether synonyms will be needed for keywords selected (each synonym would then be matched to its primary keyword)
  - Whether keywords were to be stored in specific hierarchies
  - Whether the keyword list should also be scanned for context words and phrases and the sources of these context words and phrases
  - Whether lexical negators or affirmers would be used, and the source of these words
  - What scheme (if any) should be applied to score for relevance of retrieval or final decision making
- As soon as these primary decisions are made, the expert is prompted with representations (lists) of selected words in as many columns as there are hierarchical relations. The expert will click on keywords that will be used to serve as a primary filter for relevant literature.
- The expert would then click on keywords that would meet certain hierarchies. For example, NeuroText recognizes keywords in a two-step hierarchy "brain region" and "neuron." The neuron "Purkinje" exists only in the brain region, "cerebellum." The expert would click one word in the first column and its hierarchical descendant in another column. This information would be stored as a relationship. If, in the front page, the expert chooses not to represent information in hierarchies, then the knowledgebase creation tool will skip this step.
- Based on whether the expert chooses to enter synonyms, a prompt containing a word list and several blanks next to it would dynamically appear on the next page. Here, the expert would be able to enter a synonym or chose a synonym if it were already in the word list. Each synonym clicked or manually typed in will be mapped on to the original keyword or concept.
- Contextual keywords and phrases will be extracted from "windows" around keywords (words within a particular distance on either side of the keyword of interest). These will be presented in two lists: support concepts and negating concepts. If, for example, the expert does not wish articles with "computational modeling" results, the expert can allot words associated with computational modeling into the non-supportive lists. Depending on what numerical score is assigned to this

word, the text-mining program will either determine that an article bears zero relevance or reduce the occurrence of a keyword and hence its random occurrence. Contextual associations can also disambiguate polysemy—identical words with different meanings.

- Word lists of lexical affirmers and negators (created using the connectivities of Roget's thesaurus) will be presented to the expert to identify words for lexical parsing.
- As the keywords for different levels of parsing are chosen, the tool will run in the background, rescanning the training sets of articles identifying sentences where these words or phrases occur. In the event that more than one keyword is identified in the same sentence, a phrase relating the two keywords will be extracted and stored in a list as a putative semantic phrase. This list will be made available to the expert, who can then choose appropriate semantic phrases. The expert will "click on" phrases based on relevance. The chosen phrases will be stored. As test sets of articles are scanned, concepts associated with or joined by semantic phrases matches will receive enhanced scores.
- Any information removed by the expert will be archived for future reinstatement into the knowledgebase. Information selected in the form of concepts, keywords, phrases, and parsers will be stored in the database. Information that is not identified will also be archived, as will word lists from articles in the test sets as they are being scanned for retrieval. Archiving will allow the expert to "call up" information if the expert deems that the knowledgebase needs to be modified by adding more information. Thus, the tool will allow the evolution of the knowledgebase to match progress and new developments in domain of (AD) research. An expert can also enhance the knowledgebase by allowing a new training set of articles, especially if he or she feels that an area of research has not been adequately represented. On the contrary, the expert can also delete information that has already been stored in the knowledgebase. The knowledgebase will be thus updated; and consequently, the text-mining efficacy will improve.
- The tool is designed to be used by experts as knowledge management solutions to retrieve information by simple keyword searches (from the first word list) or by using all the complexities allowed. The tool will also be modifiable if the curators decide that the level of complexity needs to be enhanced so that information can be entered into the knowledgebase efficaciously.

6. Conclusion

The knowledge retrieval process described here provides an advantage over typical NLP methods because the rules of information extraction and the knowledge-nuances within a domain, as required by the domain expert, can be embedded into the knowledgebase. Such a methodology is also useful when the results of information retrieved have to be deposited into a database with high accuracy. Tools such as NeuroText are charged (by domain experts) with identifying information in the appropriate domain context, ensuring also that lexical negation does not provide information that is the opposite of what is desired. NeuroText also provides for the recognition of database hierarchical constructs and ensures that random occurrences of otherwise relevant keywords are not considered. The supervised and unsupervised learning steps coupled with the variable scoring of metadata elements ensure that the knowledge retrieval process can be made more flexible and extensible.

Acknowledgments

This work was supported in part by NIH grant P01 DC04732 and by NIH grants P20 LM07253, T15 LM07056, and G08 LM05583 from the National Library of Medicine and NIH grant P01 DC04732-03. The authors thank Dr. Michele Migliore, Dr. Prakash Nadkarni, and Dr. Perry Miller for their useful comments in the development of this manuscript and NeuroText.

References

1. Marenco, L., Nadkarni, P. M., Skoufos, E., Shepherd, G. M., and Miller, P. L. (1999) Neuronal database integration: the SenseLab EAV data model. *Proc Am Med Inform Assoc Symp* 102–06.
2. Shepherd, G. M., Mirsky, J. S., Healy, M. D., Singer, M. S., Skoufos, E., Hines, M. S., Nadkarni, P. M., and Miller, P. L. (1998) The Human Brain Project: neuroinformatics tools for integrating, searching and modeling multidisciplinary neuroscience data. *Trends Neurosci* **21,** 460–68.
3. Migliore, M., Morse, T. M., Davison, A. P., Marenco, L., Shepherd, G. M., and Hines, M. L. (2003) ModelDB: making models publicly accessible to support computational neuroscience. *Neuroinformatics* **1,** 135–40.
4. Crasto, C. J., Marenco, L. N., Migliore, M., Mao, B., Nadkarni, P. M., Miller, P., and Shepherd, G. M. (2003) Text mining neuroscience journal articles to populate neuroscience databases. *Neuroinformatics* **1,** 215–37.
5. Crasto, C. J., Marenco, L., Miller, P. L., and Shepherd, G. M. (2002) Olfactory receptor database: a meta-data driven automated population from sources of gene and protein sequences. *Nucleic Acids Res* **30,** 354–60.
6. Justeson, J. S., and Katz, S. (1995) Technical terminology: some linguistic properties and an algorithm for identification in text. *Nat Lang Eng* **1,** 9–27.
7. Prager, J. M. (1999) Linguini: Language Identification for Multilingual Documents. *Proc 32nd Hawaii Ins Sys* **2,** 1–11.
8. Verity.com (2000), Vol. 2002, Verity.com.
9. CindorSearch.com (2002), Vol. 2002, Cindorsearch.com.
10. Textwise.com (2002), Vol. 2002, Textwise.com.

11. Lagus, K. (2000) Text mining with the WEBSOM. *Acta Polytech Scand Math Comput* **110,** 1–54.
12. Krauthammer, M., Rzhetsky, A., Morozov, P., and Friedman, C. (2000) Using BLAST for identifying gene and protein names in journal articles. *Gene* **259,** 245–52.
13. Friedman, C., Jra, P., Yu, H., Krauthammer, M., and Rzhetsky, A. (2001) GENIES: a natural-language processing system for extraction of molecular pathways from journal articles. *Bioinformatics* **17,** S74–84.
14. Karp, P. D., Riley, M., Paley, S. M., Pellegrini-Toole, A., and Krumenacker, M. (1999) EcoCyc: encyclopedia of Escherichia coli genes and metabolism. *Nucleic Acids Res* **27,** 55–58.
15. Iliopouos, I., Enright, A. J., and Ouzounis, C. (2001) TextQuest: document clustering of MEDLINE Abstracts for Concept Discovery in Molecular Biology. *Pac Symp Biocomput*, 374–83.
16. Iliopoulos, I., Enright, A.J., and Ouzounis, C.A. (2001) TextQuest: document clustering of Medline abstracts for concept discovery in molecular biology. *Proc Int Conf Intell Syst Mol Biol*, 60–67.
17. Schomburg, I., Chang, A., and Schomburg, D. (2002) BRENDA, enzyme data and metabolic information. *Nucleic Acids Res* **30,** 47–49.
18. Chiu, W. L. A. K., Sze, C. N., Ip, L. N., Chan, S. K., and Au-Yeung, S. C. F. (2001) NTDB: thermodynamic database for nucleic acids. *Nucleic Acids Res* **29,** 230–33.
19. Friedman, C., Alderson, P. O., Austin, J. H., Cimino, J. J., and Johnson, S. B. (1994) A general natural language text processor for clinical radiology. *J Am Med Inform Assoc* **1,** 161–74.
20. Hersh, W. R., Crabtree, M. K., Hickman, D. H., Sacherek, L., Freidman, C., Tidmarsh, P., Mosbaek, C., and Kraemer, D. (2002) Factors associated with success in searching MEDLINE and applying evidence to answer clinical questions. *J Am Med Inform Assoc* **9,** 283–93.
21. Aronson, A. (2001) Effective mapping of biomedical text to the UMLS Metathesaurus: the MetaMap program. *Proc AMIA Symp*, 17–21.
22. Kim, W., Aronson, A. R., and Wilbur, W. J. (2001) Automatic MeSH term assignment and quality assessment. *Proc Am Med Inform Assoc Symp*, 310–23.
23. Weeber, M., Mork, J., and Aronson, A. R. (2001) Developing a test collection for biomedical word sense disambiguation. *Proc Am Med Inform Assoc Symp*, 746–50.
24. Mutalik, P. G., Deshpande, A., and Nadkarni, P. (1999) Use of general-purpose negation detection to augment concept indexing of medical documents. *J Am Med Inform Assoc* **8,** 598–609.
25. Baeza-Yates, R., and Ribeiro-Neto, B. (1999) *Modern Information Retrieval*, Addison-Wesley, New York.

26. Korfhage, R. R. (1997) *Information Storage and Retrieval*, John Wiley and Sons, New York.
27. Raghavan, V. V., Jung, G. S., and Bolling, P. (1989) A critical investigation of recall and precision as measures of retrieval system performance. *ACM Trans Inform Sys* **7,** 205–29.
28. Tague-Sutcliffe, J. (1992) Measuring the informativeness of a retrieval process. *Proc 15th Ann Intern ACM SIGIR Conf Res Dev Inform Retrieval*, 23–36.
29. DTSearch (1999), Arlington, VA.
30. Claiborne, B. J., Amaral, D. G., and Cowan, W. M. (1986) A light and electron microscopy study analysis of the mossy fibers of the rat dentate gyrus. *J Comp Neurol* **246**(4), 435–58.
31. Mori, K., Nowycky, M. C., and Shepherd, G. M. (1981) Electrophysiological analysis of mitral cells in the isolated turtle olfactory bulb. *J Physiol (Lond)* **314,** 281–94.
32. Cicchetti, D. V., and Feinstein, A. R. (1990) High agreement but low kappa: II. Resolving the paradoxes. *Journal of Clin Epidemiol* **43,** 551–58.
33. Spitzer, R., and Fleiss, J. (1982) A design-independent method for measuring the reliability of psychiatric diagnosis. *J Psychiat Res* **17,** 335–42.
34. Agresti, A. (1990) *Categorical Data Analysis*, Wiley, New York.
35. Nadkarni, P. M., Marenco, L., Chen, R., Skoufos, E., Shepherd, G. M., and Miller, P. L. (1999) Organization of heterogeneous scientific data using the EAV/CR representation. *J Am Med Inform Assoc* **6,** 478–93.

2

Interoperability Across Neuroscience Databases

Luis Marenco, Prakash Nadkarni, Maryann Martone, and Amarnath Gupta

Summary

Data interoperability between well-defined domains is currently performed by leveraging Web services. In the biosciences, more specifically in neuroscience, robust data interoperability is more difficult to achieve due to data heterogeneity, continuous domain changes, and the constant creation of new semantic data models (Nadkarni et al., *J Am Med Inform Assoc* 6, 478–93, 1999; Miller et al., *J Am Med Inform Assoc* 8, 34–48, 2001; Gardner et al., *J Am Med Inform Assoc* 8, 17–33, 2001). Data heterogeneity in neurosciences is primarily due to its multidisciplinary nature. This results in a compelling need to integrate all available neuroscience information to improve our understanding of the brain. Researchers associated with neuroscience initiatives such as the human brain project (HBP) (Koslow and Huerta, *Neuroinformatics: An Overview of the Human Brain Project*, 1997), the Bioinformatics Research Network (BIRN), and the Neuroinformatics Information Framework (NIF) are exploring mechanisms to allow robust interoperability between these continuously evolving neuroscience databases. To accomplish this goal, it is crucial to orchestrate technologies such as database mediators, metadata repositories, semantic metadata annotations, and ontological services. This chapter introduces the importance of database interoperability in neurosciences. We also describe current data sharing and integration mechanisms in genera. We conclude with data integration in bioscience and present approaches on neuroscience data sharing.

Key Words: Data federation; data integration; data warehouses; mediator services; heterogeneous databases; ontology.

1. Importance of Data Interoperation in Neuroscience

The nervous system is one of the most complex of all biological systems, because it is involved in multiple, involuntary functions. These functions are intrinsically related in a hierarchical level of processes and interdependencies.

From: *Methods in Molecular Biology, vol. 401: Neuroinformatics*
Edited by: Chiquito Joaquim Crasto © Humana Press Inc., Totowa, NJ

They range from molecule interaction, to cell signaling, to cell processing, to cognition, and finally to organism behavior *(1)*. Unveiling each component at every one of these levels and understanding the interactions involved are crucial to formulating strategies to understand physiopathology of certain disorders of nervous system.

The constantly generated new data in neuroscience research are often used once, to achieve a primary purpose. Then these data, which are typically discarded, if properly shared, can still have great value for current and future research. There are many reasons impeding data sharing in sciences. Although legalities and authorship issues are worthy of consideration, the main reason still remains the lack of proper technology and standardized protocols to share those data.

The scientific community has been aware of the value that these data may still possess; they may hold information that can be used to conduct future research. But this can be facilitated if such data are properly shared, allowing other researchers to be able to discover those data first to later correlate that information with their own.

In neuroscience, as opposed to genomics or proteomics, data not only comprise sequences and annotations but other diverse data types such as cellular recordings, neural networks, neural systems, and brain images using different techniques and resolutions, all the way to behavioral information. Although neuroscience phenomena can ultimately be explained by molecular phenomena (e.g., gene expression and interactions), explanations of certain phenomena in terms of higher level concepts (taken, for example, from neurophysiology or psychology) are sometimes more appropriate and succinct. Olfaction, for example, can be studied at the biomolecular level of ligands interacting with olfactory receptors, concomitantly with neuronal interactions in the olfactory pathway, and with the reactions of the subject when exposed to the odor.

Multidisciplinary research is too difficult to be carried out by a small group. Sometimes a common goal investigation starts at different laboratories, each focusing on different aspects of a problem. Irrespective of the success or failure of these endeavors, it helps advance the science in the solution of the problem.

2. Databases and Web Databases

The term "database" refers to an organized collection of data. When the data contained in the database are made available via the Internet, the database is considered a Web database. Different databases differ with respect to the tools used to organize data. A "knowledge-base" is a special kind of database that not only contains low-level (primary, experimental) data but also higher level "rules"—assertions about the different types of data present in the system, for example. The knowledge information used to annotate raw data represents metadata. Some database systems provide mechanisms to annotate their data with different levels of expressiveness. The better the annotation, the easier automated agents can identify an extract their information without human intervention. A similar approach is used by the semantic Web by providing "smart data" *(2)* to annotate Web resources in general. Most database systems, particularly relational databases, lack built-in support for annotating the data in terms of their scientific domain. To improve the "smartness" in these databases, one must add those annotations at the database level, at an intermediary server, or by changing the underlying data model to a richer one. We will introduce samples of these techniques later in this chapter.

The heterogeneity of neurosciences databases in use today range across relational databases (Oracle, MS Access, SQL server, MySql, and PostgreSql), new semantic data models [Entity Attribute Value with Classes and Relationships (EAV/CR) *(3)*—Yale (NeuronDB and CellProp DB in mentioned in Chapter 1 are based on the EAV/CR schema, the architecture for which is described in Chapter 3), BrainML *(4)*—Cornel (Additional information about BrainML can be found in Chapter 6), Web Interfacing Repository Manager (WIRM) *(5)*—UW, and NeuroSys *(6)*—Montana SU among others], and files (MS Excel, text, and XML (*see* also, Chapter 4).

3. Data Sharing and Interoperability Standards

Interoperability refers to the ability of one system to share its information and resources with others. Database systems can interact in a variety of ways. These range from completely downloading all the database contents to accepting and responding to valid ad hoc queries. The following list describes different levels of interoperation approaches over the Web, categorized in order of increasing level of functionality:

1. Downloadable databases: This approach consists of making data downloadable via email, FTP or HTTP: this method is simple but nonscalable. The data recipient needs instructions on how to interpret the information within individual data sets. If the data format is nontextual, one must be familiar with the application used to create the set, in order to properly manipulate those data *(7)*. Furthermore, updates to the data will require re-download.
2. Web application wrappers: Some databases provide Internet-accessible mechanisms that access individual entries in the database. Examples of this are URLs that incorporate PubMed or GenBank IDs. If the content that is returned is a mixture of data and formatting/layout information (e.g., if the content is returned as a Web page), then the latter must be stripped off by a program. This strategy of "screen

scraping" is laborious and requires high maintenance, because changes to the layout can cause the stripping algorithm to fail. More sophisticated databases allow only data to be returned, using an XML-based format. A limitation of such an XML wrapper-based approach is that it is designed to return only one object at a time; this works for browsing but not when large data sets are to be extracted.

3. Interfacing via native application interfaces: Most modern database management systems (DBMSs) provide Web connectivity using specific interfaces, protocols, and languages built on top of TCP/IP. This approach is reasonable for Intranet scenarios (though difficult to use because of the expertise needed) but less appropriate for Internet situations because of security and scalability considerations.
4. Web services: With the emerging XML-derived technologies, the informatics community and W3C has adopted standard mechanisms for querying and representing basic data in a way that is hardware and software independent. This effort is represented in the Simple Object Access Protocol (SOAP) and derived technologies ***(8)*** to expose a list of interfaces and their definitions in a way that is machine understandable. This approach has clearly decreased the effort required to understand data types and modes of queries. For simple data, this may be appropriate, but for heterogeneous constantly changing domains as in neurosciences, this approach can have great maintenance overheads.
5. Semantic Web services: This latest effort explores the combination of semantic Web technologies and Web services. Theoretically, semantically annotated interfaces can be explored by cataloging engines and leveraged for data integration when part of the information they serve can be used to respond to a larger distributed query. Efforts to improve the self-description of Web services are being made in projects that aim to annotate Web services with semantic data ***(2)*** and the second generation of Web services ***(9)***.

Data interoperation between heterogeneous autonomous databases requires orchestration between components such as metadata repositories, metadata mappings, mapping rules, and language translation tools.

4. Semantic Data Models in Neuroscience

1. Evolving domain issues: Metadata is used to describe the information in the database. Because our understanding of neuroscience phenomena is continually evolving, new insights and/or hypothesis may require the structure of certain databases (which were designed to reflect the best available knowledge at a point in time) to be redesigned. This occurs more often than expected and there is no domain-independent relational schema that can be applied to neuroscience. One promising approach to address this issue is to create customized data modeling approaches that will facilitate schema adaptability to accommodate the new needs of the database. Among the new data modeling approaches are the EAV/CR (Nadkarni et al., *J Am Med Inform Assoc* 6, 478–93, 1999) created at Yale, the Common Data Model for

Neurosciences (CDM) (Gardner et al., *J Am Med Inform Assoc* 8, 17–33, 2001) at Cornell, and NeuroSys at Montana State University ***(6)***. Ontological systems such as Stanford's Protégé ***(10)*** system allow fast prototyping of knowledge bases but are not suitable for large-scale production use.

2. Security: Besides public data, some neuroscience databases contain private, incomplete, or noncurated data. Security is important to ensure that unregistered users do not tamper with the data or access incomplete or private data. If restricted data are available in certain tables, database security could be enough to enforce it. More often, access to some rows of data in a table is restricted, while information in other rows is not secured. The requirements of this type of security are called row-level security. The Oracle database system and some of the neuroscience-tailored data modeling approaches described above implement this type of security.
3. Motivation: The motivation for using semantic models comes from three different requirements. First, different databases use different terminology for the same concept; second, often, two databases use terms that are closely related (like "nerve cell" and "neuron"). A data integration system that uses the semantic model provided by the ontology can be used to related data objects that use two different concepts. The third requirement comes from the nature of queries scientists make with higher level terms. Very often, a query such as "find all patients who were given a 'working memory test' " should interpret the phrase "working memory test" by progressively transforming it into test names that are actually recorded in the databases.

5. Database and Resource Integration

Data integration between various sites in a database federation consisting of autonomous research groups can be achieved by using centralized data repositories or data mediator approaches. Computational resources (e.g., storage, processing) can be integrated using Web services. In both cases, metadata and interfaces must be standardized to allow heterogeneous systems to properly interact with each other.

The "centralized database integration approach" uses a global data repository to copy data from external databases. This approach proves convenient when the structure of the external databases remains constant, when the bandwidth to connect to those databases is low, or when public access to the remote DBMS is unlikely. Data repositories have the advantage of improved performance and facilitated query formulation. The major overhead is in the creation and maintenance of the global schema and data import programs. Data space requirements are significant because all query-able data from each of the databases must be copied over.

The "mediated database integration approach" maps several remote databases using a global schema to create a database federation ***(11,12)***. An integrated query is formulated in terms of the global schema that is later translated to the remote databases. The results are then integrated. Query translation and data integration process is easier in situations where the local database is modeled as a subset of the global schema using the metadata and DBMS. Unfortunately, in already existing databases, that scenario is difficult to achieve due to applications depending on the current schema and the preference of their owners for maintaining their autonomy. When remote databases use different schemas or other data models (RDBMS, Text, BrainML/XML, EAV/CR, etc.), an extra language and schema translation step is needed. Data mediation of this kind implies live connections to remote database systems and mechanisms to translate query structure and terms (semantics and syntax) every time a global query is formulated. This approach has advantages for databases whose owners want to preserve their freedom to evolve their structure and content as needed. In some systems, queries against distributed heterogeneous databases have performance disadvantages due to connectivity issues and the lack of query optimization mechanisms. Early efforts based on this approach, which incorporate enhanced semantic and evolvability features tailored for neuroscience, include EAV/CR and BrainML.

The major overhead on integrating heterogeneous databases consists of mapping their structures and semantics. Creating a $1 \times N$ mapping to a global schema simplifies the process but also limits the vision of the data to one "single normalized" perspective. Some database owners could see this single view of the domain as inconvenient when complementing their data with other data in the federation. They would like to see "complementing data" from the federation as an extension of their own database. An approach like this is being explored by the Query Integrator System and the Bioinformatics Research Network (BIRN) Ontology Server (OS) (BIRN-Bonfire) in a collaborative effort to integrate data using distributed database ontologies.

6. Evolvable Interoperability in Neurosciences (Approaches)

An important vision shared by several projects involved in neuroscience database research involves system adaptability to follow domain changes. One way to achieve this goal is to incorporate loosely coupled semantic mapping to database mediator systems that will allow local autonomy of both the domain ontology and the local databases in a multi-database system ***(13)***.

6.1. Query Integrator Server

Query Integrator Server (QIS) belongs to the class of mediator systems that limit themselves to metadata exchange. It uses a distributed architecture that is composed of three main functional units: Integrator Servers (ISs), Data Source Servers (DSSs), and the OS. These units form the system's middle tier, connecting "data consumers" (client applications requesting query execution) with "data providers" (back-end data sources containing the data), and knowledge sources (ontologies) (*see* **Fig. 1**). All servers use a DBMS (currently Microsoft SQL Server) in addition to a Web Server (Microsoft Internet Information Server).

- DSSs provide access to various data sources within a single group (or cooperating groups within an institution). In addition to traditional relational databases and EAV/CR databases that are built on top of relational technology, these source servers also access XML files and flat files (spreadsheets, text). DSS administrators add definitions of data sources to the DSS through a Web interface. *Schema capture* is the process of capturing metadata about a database (e.g., its table, column, and relationship definitions) into structured form. This process is partly automated

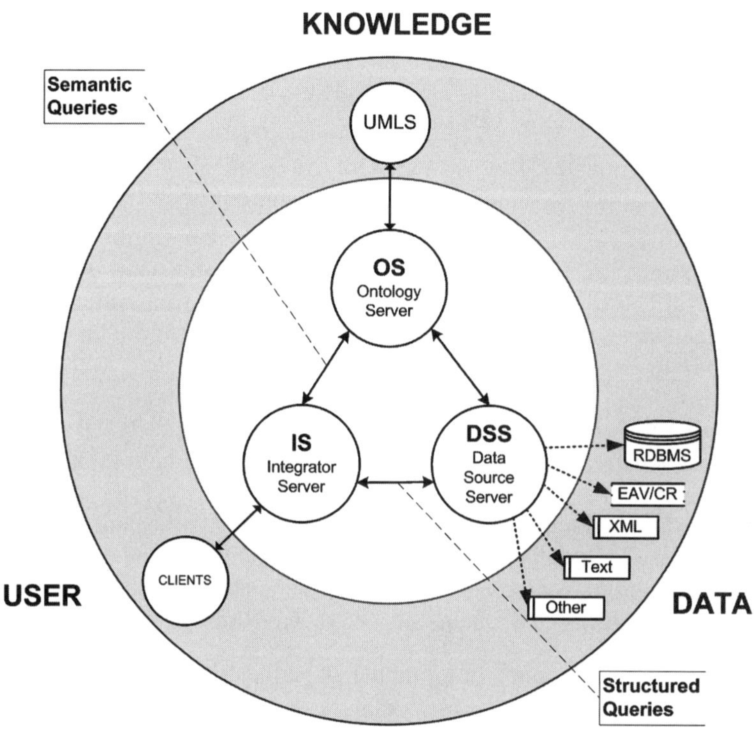

Fig. 1. Query Integration System—architectural overview. The Query Integration System architecture is based on three middle-tier servers. The Data Source Server (DSS) connects to disparate supported data sources. The Integrator Server (IS) stores, coordinates query execution, and returns query results to Web applications. The Ontology Server (OS) maps data sources' metadata and data elements to concepts in standard vocabularies. UMLS, Unified Medical Language System.

through connectivity technologies that query a database's system data dictionary, or an XML schema. However, the captured metadata need to be manually enriched by mapping schema elements, where possible, to elements in ontologies to assist automatic schema discovery, and by detailed textual annotations to provide semantic overview.

Each DSS itself is like a "meta-database" that accesses one or more individual databases at a site. The administrators of individual databases must define the subset of the data and metadata within their own schema that is "public." This is done through a three-step process for each data source.

- A special account with restricted privileges is created for the DSS. This account cannot alter data and can access only a limited set of tables or views.
- For tables containing both public and private data, an extra Boolean column is added to indicate whether the row in question is publicly accessible, with a default of "No/False," so rows that are to be made public must be manually set.
- The administrator creates views that define the subset of public columns/tables. Where the views use tables with both public and private data, the view must specify a filter that the "public" flag must be "True." The DSS account is now given permission to these views.

For non-account-oriented data, such as XML or text files, which must be accessed in their entirety, the DSS stores the URL of the source.

- ISs store "public" metadata from DSSs, as well as queries that access single or multiple data sources. They allow building of queries against the DSSs through a graphical user interface. QIS is primarily intended to allow other Web-based applications to execute pre-defined queries on the IS through Web service ***(14)*** mechanisms. That is, the IS operates "behind the scenes," and the federation's end-users connect to such an application rather than to the IS directly. IS administrators perform tasks such as registering new DSSs and registering individuals with domain expertise who can design queries.
- OSs: An OS maintains an integrated schema, plus content, of one or more controlled vocabularies used within the federation. Alternatively, it may provide a gateway to relate these vocabularies to standardized content maintained elsewhere, or it may replicate such content. Parts of the OS schema are recorded redundantly at the IS level.

The OS supports mapping of elements in individual data sources to vocabulary elements by curators who specialize in ontology development. More importantly, once this information is replicated on the DSS, mappings at both the metadata and data levels are also forwarded to the IS. The OS can now act as an information source map (ISM) ***(15)***. Such information makes it possible to ask questions of varying granularity, such as "which data sources contain information on neurons" (where mappings are likely to exist at the metadata level), or "which data sources contain information on cerebellar Purkinje cells" (where mappings will likely exist at the data level). The mapping to specific schema elements in a data source allows assisted query composition against that data source that would actually return the desired data.

Data and metadata elements from DSSs that are candidates for local concept creation are exported to the OS in a facilitated fashion, as discussed later. Other potential services envisaged for the OS are term translation and unit conversion. The OS also lets ontology curators collaborate with DSS curators to jointly define new "local" (federation-specific) concepts: this is necessary when existing ontologies offer insufficient coverage. This infrastructure can also be used to submit new, curated concepts to a standard vocabulary for inclusion in a new release.

Communication between the various QIS nodes is XML-encoded and HTTP-delivered to support communication through network firewalls. Asynchronous processing is supported using customized queuing services. Other software technologies used by the system are Microsoft Active Data Objects (ADO) for data and schema access, Extended Markup Language Document Object Model (XML DOM) and SAX (Simple Application Programming Interface for XML) for dataset manipulation, and Scalable Vector Graphics (SVG) ***(16)*** for standardized ER diagram generation.

6.2. BIRN Mediator and Tools

The BIRN-CC is charged with the development of an environment that enables researchers to execute queries across distributed heterogeneous data collections in complex ways that can generate new insights. To achieve this goal, the BIRN-CC has been developing a data integration system, referred to as the "mediator," that will enable researchers to perform these unified conceptual queries ***(17)***.

The BIRN mediation architecture builds on our work in knowledge-guided mediation for integration across heterogeneous data sources ***(12,18–21)***. In this approach, the mediator uses additional knowledge captured in the form of ontologies, spatial atlases, and thesauruses to provide the necessary bridges between heterogeneous data. Within a research collaboratory, each partner site creates their own database around the type of data acquired at that site and maps the content to one or more shared knowledge sources [e.g., the Unified Medical Language System (UMLS)]. The databases and knowledge sources are registered with the mediator to create a virtual data federation.

Figure 2 illustrates the utility of the semantic knowledge within the mediator for querying across multi-scale data. In a nonmediated query, a search for the term "cerebellar cortex" would return no results. In fact, the Cell Centred

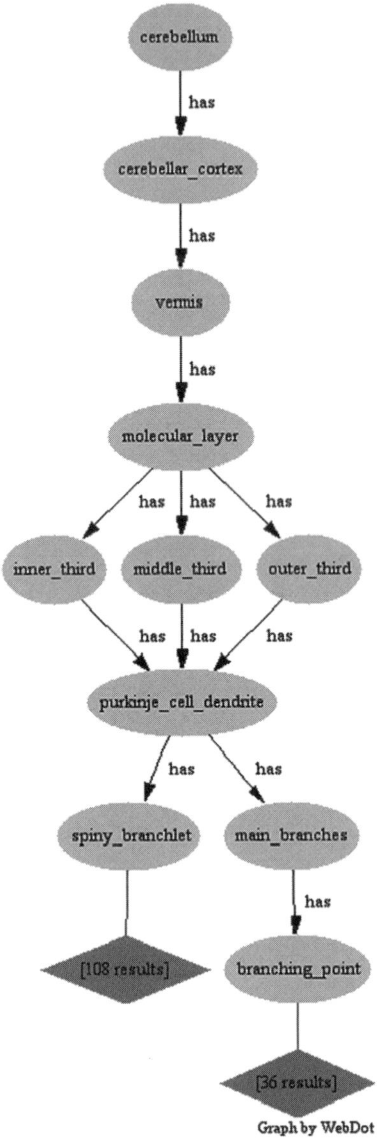

Fig. 2. Results of a query asking for images of protein labeling in cerebellar cortex. Clicking on the diamonds returns the raw data.

Database (CCDB) has many data sets that satisfy the query, because it has data on Purkinje neurons, Purkinje cell dendrites, and the neuropil of the molecular layer, all of which are components of the cerebellar cortex. However, in the CCDB, these data sets are indexed under the term "cerebellum," "Purkinje neuron," and "neuropil/molecular layer, cerebellum"; nowhere does the term "cerebellar cortex" appear. The problem can be solved by defining a virtual view at the mediator, which navigates "semantic links" in order to find all relevant data about the cerebellar cortex and uses this data to search registered sources.

In our present system, a relational source can declare some column to be "semantic," i.e., the content of the column can be related to an ontology X (e.g., UMLS, NeuroNames, Bonfire). For these columns, the mediator probes the source to construct an inverted term index whose terms are mapped to the specified ontology. One can further specify intersource join constraints containing both foreign key constraints and "joinability" statements. Once these two steps are performed for all sources, the mediator can take a set of keywords and automatically construct an integrated view on the fly and retrieve a user-specified query on the auto-constructed view. This greatly enhances the

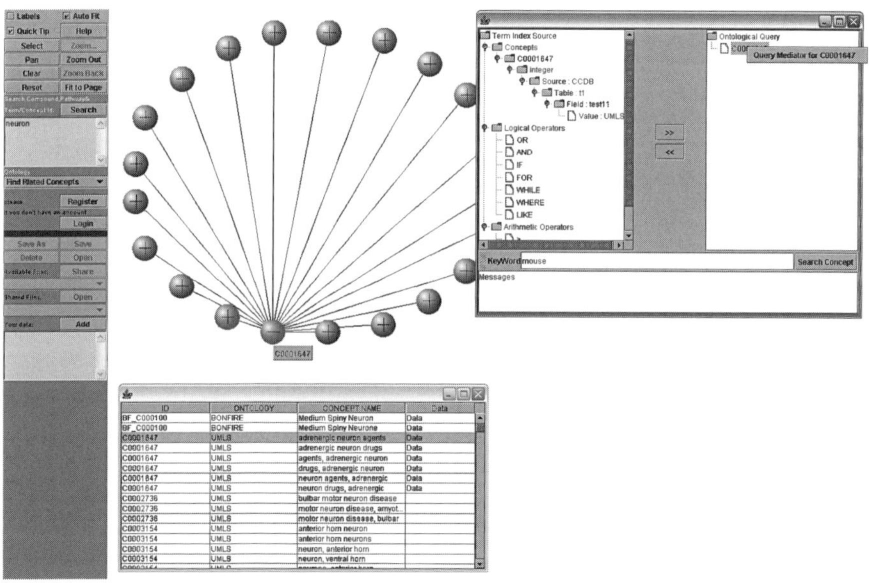

Fig. 3. Query results for the mediator are shown in context of the Unified Medical Language System (UMLS). Nodes that are referenced directly to data contained in one of the Bioinformatics Research Network (BIRN) databases are highlighted.

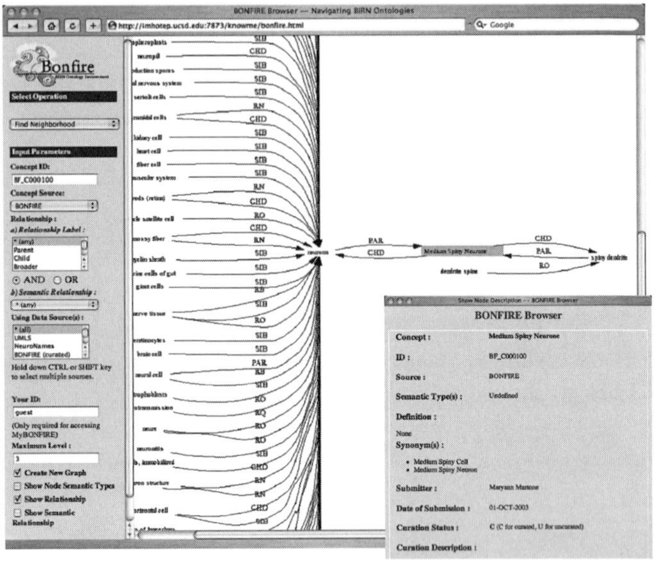

Fig. 4. Bonfire ontology tool: a graphical environment allowing users to query, browse, and edit ontologies. The browser above depicts the merging on a standard ontology (i.e., UMLS) with user-defined extensions (i.e., Medium Spiny Neuron that does not exist in any of the pre-defined source ontologies).

user's ability to "explore" the content of the federated data in case the user does not want to pose a query against the global schema. To facilitate such exploratory query, we have developed a new Mediator client called "VisANT-Builder" developed by integrating the Systems Biology platform VisANT *(22)* with the ontology Browser KnowMe (*see* **Fig. 3**) and View/Query Designer developed at BIRN.

The BIRN Data Integration Workspace provides researchers with the ability to browse data through ontological relations. The results of a query are displayed to the user as a graph of related entities and attributes. With this graph, the user can further search for related concepts by various graph operations such as "find neighbors," "find related," and "find shortest path." Color codes are used to mark those nodes that have data associated with the concept (*see* **Fig. 4**).

7. Conclusions

Database interoperability in the neurosciences poses a challenge. These challenges go hand in hand with the current need to create more "semantically-aware" applications. All the approaches described in this chapter along with text-mining and information extraction methodologies described in other chapters will be crucial to the future solutions of the interoperability issue.

References

1. Chicurel, M. (2000) Databasing the brain. *Nature* **406,** 822–5.
2. Daconta, M. C., Obrst, L. J., and Smith, K. T. (2003) *The Semantic Web: A Guide to the Future of XML, Web Services, and Knowledge Management*, Wiley Publishing, Inc., Indianapolis, Ind.
3. Marenco, L., Tosches, N., Crasto, C., Shepherd, G., Miller, P. L., and Nadkarni, P. M. (2003) Achieving evolvable web-database bioscience applications using the EAV/CR framework: recent advances. *J Am Med Inform Assoc* **10**(5) (abstract).
4. BrainML.org. Brain Markup Language. (2004) http://brainml.org. Last accessed on: June 1, 2006.
5. Jakobovits, R. M., Rosse, C., and Brinkley, J. F. (2002) WIRM: an open source toolkit for building biomedical web applications. *J Am Med Inform Assoc* **9,** 557–70.
6. Pittendrigh, S., and Jacobs, G. (2003) NeuroSys: a semistructured laboratory database. *Neuroinformatics* **1,** 167–76.
7. Gardner, E. Mouse Hunt. (2003) http://www.bio-itworld.com/archive/061503/mouse.html. Last accessed on: June 1, 2006.
8. Jepsen, T. (2001) SOAP cleans up interoperability problems on the Web. *IT Professional* **3**(1), 52–5.
9. Prescod, P. Second Generation Web Services. (2002) http://webservices.xml.com/pub/a/ws/2002/02/06/rest.html. Last accessed on: April 2007.
10. Noy, N. F., Crubezy, M., Fergerson, R. W., Knublauch, H., Tu, S. W., Vendetti, J., and Musen, M. A. (2003) Protege-2000: an open-source ontology-development and knowledge-acquisition environment. *AMIA Annu Symp Proc*, 953.
11. Hass, L. M., Schwarz, P. M., Kodali, P., Kotlar, E., Rice, J. E., and Swope, W. C. (2001) DiscoveryLink: a system for integrated access to life science data sources. *IBM System Journal* **40,** 489.
12. Martone, M. E., Gupta, A., and Ellisman, M. H. (2004) E-neuroscience: challenges and triumphs in integrating distributed data from molecules to brains. *Nat Neurosci* **7,** 467–72.
13. Elmagarmid, A. K., Rusinkiewicz, M., and Sheth, A. (1999) *Management of Heterogeneous and Autonomous Database Systems*, Morgan Kaufmann, San Francisco, Calif.
14. Kaye, D. (2003) *Loosely Coupled: The Missing Pieces of Web Services*, RDS Press.
15. Masys, D. (1992) An evaluation of the source selection elements of the prototype UMLS Information Sources Map. *Proc Annu Symp Comput Appl Med Care*, 295–8.

16. Andersson, O., Armstrong, P., Axelsson, H., Berjon, R., Bezaire, B., Bowler, J., Brown, C., Blultrowicz, M., and Capin, T. (2003) World Wide Web Consortium. http://www.w3.org/TR/2003/REC-SVG11-20030114/.
17. Astakhov, V., Gupta, A., Santini, S., and Grethe, J. S. (2005) Data Integration in the Biomedical Informatics Research Network (BIRN). *DILS 2005*, pp. 317–20.
18. Ludäscher, B., Gupta, A., and Martone, M. E. (2003) A model based mediator system for scientific data management. In *Bioinformatics: Managing Scientific Data* (Lacroix, Z., and Critchlow, T., Eds.), pp. 335–70, Morgan Kaufmann Publishers.
19. Ludäscher, B., Gupta, A., and Martone, M. (2000) A Mediator System for Model-based Information Integration. *Int. Conf. VLDB*,
20. Gupta, A., Ludäscher, B., Martone, M. E., Rajasekar, A., Ross, E., Qian, X., Santini, S., He, H., and Zaslavsky, I. (2003) BIRN-M: a semantic mediator for solving real-world neuroscience problems. *ACM Conf. on Management of Data (SIGMOD) (Demonstration)*, 678.
21. Ludäscher, B., Gupta, A., and Martone, M. E. (2001) Model-Based Mediation with Domain Maps. 17th Intl. Conference on Data Engineering (ICDE), Heidelberg, Germany, IEEE Computer Society.
22. Hu, Z., Mellor, J., Wu, J., and DeLisi, C. (2004) VisANT: an online visualization and analysis tool for biological interaction data. *BMC Bioinformatics* **5**, 17.

3

Database Architectures for Neuroscience Applications

Prakash Nadkarni and Luis Marenco

Summary

To determine effective database architecture for a specific neuroscience application, one must consider the distinguishing features of research databases and the requirements that the particular application must meet. Research databases manage diverse types of data, and their schemas evolve fairly steadily as domain knowledge advances. Database search and controlled-vocabulary access across the breadth of the data must be supported. We provide examples of design principles employed by our group as well as others that have proven successful and also introduce the appropriate use of entity–attribute–value (EAV) modeling. Most important, a robust architecture requires a significant metadata component, which serves to describe the individual types of data in terms of function and purpose. Recording validation constraints on individual items, as well as information on how they are to be presented, facilitates automatic or semi-automatic generation of robust user interfaces.

Key Words: Relational databases; entity–attribute–value; metadata-driven database architectures.

1. Introduction

Sound data management is a fundamental aspect of laboratory and clinical research, and the field of neuroscience is no exception. When such research is performed on a medium to large scale, the use of robust database engines becomes more or less mandated. We provide here a practical guide to the informaticians who must support such research efforts. We contend that, to avoid wasted effort, it is worth considering the distinguishing features of research databases in general, and neuroscience databases in particular, and the requirements that they must satisfy. It is useful to consider the application of techniques

From: *Methods in Molecular Biology, vol. 401: Neuroinformatics*
Edited by: Chiquito Joaquim Crasto © Humana Press Inc., Totowa, NJ

and designs that have proven successful in this area. Rather than describe a single monolithic system that would claim to solve all problems, we describe individual components that can be applied if they are a good fit to the particular problem that the database designer is trying to solve.

2. Characteristics and Requirements of Medium- to Large-Scale Neuroscience Research Databases

- Diversity/complexity of data: Multiple types of experiments centered on a broad theme typically address a research problem. The larger the research group and the more varied the specific research interests, the more diverse the data that are being generated. In terms of database design, schemas with 50–100 tables are not uncommon.
- Frequent enhancements to the data model: As one learns more about the phenomena that are being studied, the data model requires to be continually enhanced. From the design perspective, new tables get added to the schema, and additional columns are added to existing tables.

  For complex and continually evolving schemas, it is highly desirable for the databases to contain self-describing content, which can double as online documentation.
- User-interface requirements: revisions to the data model necessitate revisions to the user interface, because one must provide a means for the additional data elements to be captured from end-users (or external applications) as well as a means for displaying the new content. Although the tools to facilitate user-interface building are continually improving, some degree of automation—particularly, enabling motivated power users who do not necessarily possess programming skills to generate aspects of the user interface—is highly desirable.
- Support of search across various classes/tables in the model: When end-users first access a database with a highly complex data model, there are two basic paradigms for organizing the user interface. One widely used approach is to provide navigational shortcuts to functional categories of data, typically via menu choices. This is appropriate for high-frequency (e.g., daily) users of the system, who want to carry out a specific task—e.g., recording of details of a particular type of experiment.

  However, many databases also have the responsibility of publishing stable subsets of their data on the Web. Users such as scientific collaborators or unaffiliated researchers in the field, who are unlikely to be intimately familiar with the data sets' contents and organization, will need to browse the content for items of interest. For the robust support of Web-based browsing, it is important to remember that users have been conditioned by the ubiquitous Web search engine to search for data of interest through the use of one or more keywords. They will not necessarily remember to, or care to, limit the search to a specific category of data, which implies a search across all tables.

  To complicate the situation, neuroscience, like most of biomedicine, is characterized by the significant presence of synonyms (e.g., "norepinephrine" and "noradrenaline" refer to the same molecule) as well as homonyms (phrases with the same spelling but different meanings)—the partial phrase "olivary nucleus" is ambiguous and could refer to one of the several structures in the brain with distinct functions.

It is also necessary to consider whether to facilitate search through use of separate hardware. When the publishable data subset is large (e.g., >500 gigabytes) and/or external users access the data heavily (e.g., 300+ hits/day), it makes sense to migrate the browsable subset to a separate machine rather than use the same system used for data entry, so as not to interfere with the system's responsiveness for the internal users. However, the design features that support cross-table searches are applicable no matter what hardware architecture is employed.

- Support for controlled vocabularies: The importance of using standard terminologies to codify and describe clinical observations has long been known, and the biosciences are also realizing their importance, notably in the description of observations related to gene function as exemplified by the Gene Ontology *(1)*. Databases for clinical and basic neuroscience are therefore likely to contain a significant reference component, which hold the contents of vocabularies that are likely to be needed. While individual vocabularies tend to vary modestly from each other, standard sub-schema designs that can handle an unlimited number of vocabularies without the need to proliferate the number of tables in the database continually have long been known.

  Controlled vocabularies also help with the search-by-synonym requirement described above. Having to curate the database contents and manually specify synonymous relationships between numerous neuroscience terms is labor-intensive, but to a significant extent it may not be necessary, because this information is already available electronically, through reference sources (controlled vocabularies) that are often freely available. The world's largest controlled vocabulary, the National Library of Medicine's Unified Medical Language System (UMLS) *(2)*, is a compendium of vocabularies from several sources. Although experimental neuroscience is underrepresented in the UMLS (except at an introductory/undergraduate level of knowledge), the UMLS provides a useful starting point for building useful vocabularies for neuroscience as well as for neuroscience synonym content. We will consider the schema of the UMLS, which is actually fairly small and simple and can readily be replicated within neuroscience databases that need to support access to vocabularies.

We now discuss the architectural means by which realization of the above requirements is facilitated. We should note that certain designs impact multiple requirements favorably. Furthermore, some design decisions may virtually mandate other decisions, in that the former are so greatly facilitated by the latter that it becomes inconceivable to implement them in isolation.

3. Dealing with Diversity, Complexity, and Change: Simplifying Complex Database Schemas

By simplification, we mean the process of reducing the number of physical tables in a database schema. Database engines usually have fairly generous limits on the number of tables in a schema (e.g., in Microsoft SQL Server 2005, for example, this limit is close to 2 billion). The practical limits on schema complexity are human rather than electronic—specifically, the challenge of fully understanding schemas and maintaining them expeditiously as the domain evolves. If fewer tables can do the work of many, without loss of information, the schema is simpler to understand. More important, the program code that operates the system (and accesses these tables) also becomes more modular and generic and consequently easier to maintain in the long term. We consider a few of schema-simplification strategies below.

3.1. Consolidating Lookup Tables

In any schema, there are columns in tables whose values are constrained to a limited set of choices, which range from 2 to 100. For conciseness, these choices may be stored internally as numeric or character codes, each of which corresponds to a descriptive phrase. In the user interface, these phrases are displayed rather than the codes, and the column itself is presented as a pull-down list. When representing the codes and descriptive phrase information, the natural tendency is to designate a lookup table to contain them. As the schema evolves, such lookup tables can easily run into the hundreds, while each table may contain less than 10 rows on average. Apart from the visual clutter introduced in the database schema diagram, these lists are not always static, and allowing designated end-users to maintain their contents becomes a major user-interface chore.

A technique to simplify the maintenance of such lists is used in several systems, but it is esoteric enough that it bears repetition here. It consists of consolidating all potential tables into two. The first table (designated Choice_Sets) contains three columns—an auto-incrementing integer primary key, a brief name of the list, and a more extended description. The second table (designated Choice_Set_Values) contains three columns—a link to the Choice_Sets table, a code column, and a descriptive phrase column (used for the user interface)—and optionally a fourth column containing a more detailed description of the item. Such a design allows centralized administration of all lookup lists, which can be located by name or description. Once the desired list is located, items can be added or edited.

3.2. Consolidating "Bridge" Tables: Capturing Arbitrary Many-to-Many Relationships

In any schema that captures information on numerous categories of data, several of these categories are related to each other in various ways. For example, in a neuronal database, a neuron contains different functional compartments, each compartment has specific channels and receptors, and neurons themselves are connected to other (afferent/efferent) neurons. The traditional method of recording associations between classes is to create "bridge" (many-to-many) classes/tables, which are linked to each of the parent classes. It can easily be seen that such bridge classes can become physically very numerous.

The solution here involves two design decisions.

- Maintaining a central "Objects" table: This table contains names and nutshell descriptions for every top-level item in the database, irrespective of which table it lies. The primary key of this table is a sequentially generated ID. In most of the other tables of the system, each row has an Object ID column, which points back to this table. (This column may double as primary key, because it is unique across the system.) Originally pioneered in the early 1990s by Tom Slezak's team at Lawrence Livermore for the chromosome 19 mapping effort *(3)*, the Objects table approach is well known to the curators of large systems such as Entrez, maintained at the National Center for Biotechnology Information (NCBI) *(4)*. The UMLS also uses this approach, where the "objects" are concepts in the meta-thesaurus.
- Creating a Binary_Relationships table: For binary relationships (between a pair of objects, which turn out to be the most common in practice), the standard design approach is to maintain two columns for the IDs of the two objects involved in the relationship, with additional columns that record the type of the relationship or that qualify the relationship. The permissible sets of relationships or relationship qualifiers constitute choice sets, as described above. This approach is used in the UMLS.
  Note that this approach does not work for relationships that involve more than two objects. In such a case, one has to resort to the traditional bridge table. (An alternative approach, which requires a significantly different data modeling perspective, is possible through the EAV/CR (EAV with Classes and Relationships) approach discussed below.)

3.3. "Generic" Means of Capturing Information: Entity–Attribute–Value Modeling

Recording information as attribute–value pairs associated with an object (or "entity") is a venerable approach, first employed in the "Association List" data structure of the List Processing (LISP) language *(5)*. An "attribute" is an aspect or property of the object that we wish to describe, and the "value" is the value of that attribute. The entity–attribute–value (EAV) modeling of data has long been used to capture clinical data *(6–8)*, where the number of potential attributes that describe an entity (i.e., a patient at a given point in time) can run into the hundreds of thousands, even though a relatively modest number of attributes actually apply to a given entity—that is, the attributes are *sparse*. An example of an EAV triple is <Patient XYZ, 7/1/2005>, <serum Calcium level>, and <10.5 mg/dl>. EAV is also used in general-purpose repositories such as the

Windows registry ***(9)*** and also provides the conceptual underpinnings of the Resource Description Framework (used to describe Web resources) ***(10)***, which uses the term "property" as a synonym for attribute.

In many complex database schemas, the number of categories (classes) of data, along with the associations between categories, runs into several dozens or even into hundreds, whereas the number of items per class is relatively modest (e.g., a couple of hundred). In neuroscience, for example, brain structures, neurotransmitters, channels, and receptors are categories of data that exhibit this characteristic. Attribute–value pairs can be used to describe information on such categories of data.

3.3.1. Extending EAV: The EAV/CR Data Model

It can be seen that ensuring data validity is problematic with a basic EAV approach, because one may erroneously apply properties that are inapplicable to an object. For example, neurons have a location property but not an odor property. It is highly desirable to constrain the permissible attributes of an object based on the category (class) that to which it belongs.

The EAV/CR model, first described in 1999 ***(11)***, is intended to support the rapid evolution of schemas in fields such as neuroscience. (It is employed extensively in SenseLab ***(12)***, one of the components of the Human Brain Project ***(13)*** that focuses on data related to the olfactory system.) This model superimposes the principles of object orientation, such as classes, inheritance, and constraints, on the basic EAV model.

The way this is done is through tables that record rigorous and detailed descriptions of each class of data in the system. These tables comprise *metadata*, "data that describe and define other data" ***(14)***. While metadata is an important component of large databases, the metadata in an EAV/CR system is "active" in that the system software consults it continually to perform a variety of tasks. The most important of these tasks is data validation, but other tasks include data presentation and automatic user-interface generation. (To illustrate with very simple examples, if the metadata records that a particular attribute's values are based on a choice set, a Web page can be generated to present that attribute as a pull-down "Select" list. Similarly, a field that is likely to contain a significant amount of text can be shown as a multi-line scrolling text area rather than a single-line text box.)

EAV/CR relies on an "Objects table" (described above). All Entities/Objects in the system are entries in this table, and every entity must belong to a Class. In some cases, the Values of certain attributes can themselves be other objects. (This is useful to record, for instance, that a particular cell type possesses specific receptors, where a receptor itself is an object defined by other attributes, such as sequence and affinity for particular molecules.)

The ability to treat objects as values where needed allows us to model inter-class relationships where more than two objects are involved. An example of an N-ary relationship is the encoding of the following information: "benzaldehyde (an odor molecule) increases (effect) cyclic AMP (a second messenger molecule) in Catfish (a species) melanophores (a tissue)." Such a relationship is essentially modeled as a class of data (as it would be in a conventional database design), but the individual "fields" that describe the relationship—odor, second messenger, species, tissue, effect—are stored in attribute–value form, where the "entity" refers to the fact being described.

3.3.2. Caveats Regarding EAV Data Storage

Attribute–value data modeling pays off only when the attributes are both numerous and sparse, and/or the number of objects per class is relatively small. When these criteria are met, as for the millions of data points in a collection of microarray experiments, conventional relational modeling of a class of data (as a table) is a more rational decision. In actual practice, therefore, no production data model is completely attribute–value based.

From our own experience and that of collaborators, we have realized that the task of writing tools to manage data that are stored in attribute–value form is not trivial when attempted from first principles. The major challenge comes in dealing with user-interface issues—the data must be presented to the user, in most circumstances, as though it were conventionally structured (i.e., as rows and columns)—this is effected through *pivoting*, a procedure that transforms rows into columns and vice versa.

Although recent versions of some databases (e.g., Microsoft SQL Server 2005) now provide built-in support for pivoting within the data manipulation language [Structured Query Language (SQL)], there are numerous complications, such as dealing with null/missing values. In general, if one is to build a system with reasonable rapidity, it is best to start with a body of source code that one can modify (our own code is available freely, see http://ycmi.med.yale.edu/TrialDB/). Even then, we have found that for many developers, the task is too formidable and/or error-prone. Fortunately, the metadata infrastructure alluded to above can be used even with conventional data organization.

The topic of metadata is important enough to merit its own subsection below. EAV data modeling is an example of a decision design that mandates another design decision (the use of metadata), but metadata use itself is independent of the data model.

4. Self-Describing Content: The Use of Metadata

All databases contain metadata in one form or another; it is only a matter of extent. Database Management Systems (DBMSs) themselves use internal metadata—the contents of "system tables"—to maintain internal consistency and check the validity of queries against the database before executing them. This metadata is created automatically by the DBMS through the developer's actions (e.g., executing CREATE/ALTER TABLE statements in SQL) and cannot be edited by the developer. The Choice_Sets and Choice_Set_Values tables described in **Subheading 3.1** are examples of developer-defined metadata.

The idea of metadata can be extended to include detailed online documentation on the tables, and columns within a table, which is integrated into the schema. Such documentation provides details from a semantic/domain perspective rather than only a structural one. The advantages of doing so become clear in a schema with numerous tables/classes, where integrated documentation not only facilitates in-depth understanding of the schema but can be utilized in interfaces that support composition of queries against the system by advanced users and developers. (It is well known that the hardest part of devising queries against large systems is locating the schema elements that are needed for a given query.)

Synchronizing the online descriptions with the current status of the schema has traditionally been difficult, because the documentation of the schema tends to lag behind its current structure. Recently, this chore has become considerably simpler in DBMSs such as Microsoft SQL Server 2005, which support Data Definition Language Triggers *(15)*—code that executes whenever one or more elements of the schema change—such as the addition or deletion of a table or of columns to/from a table.

4.1. User-Interface Generation: Presentation Metadata

Historically, the internal metadata of high-end DBMSs is concerned only with data and query correctness, because, in the client–server architecture of systems such as Sybase and Oracle, data presentation was supposed to be the responsibility of applications software. However, presentation information about individual data elements is an important aspect of the overall system functionality. For example, the order in which the columns in a table are presented to the user is not necessarily the same as their physical sequence within the table, and the explanatory caption presented for the column is typically more descriptive than the physical column name. In the absence of a centralized approach to organizing and storing presentation information, there is a very high likelihood of the user interface being inconsistent in large projects when the same elements are presented in different parts of the application, because such decisions are left to individual programmers.

Automating user-interface generation requires significant initial effort, because the specifications embodied in the presentation metadata must be transformed by code into a functioning interface object such as a Web page. In the long term, however, such effort pays for itself, especially when small developer teams are maintaining large systems. In this circumstance, the user-interface development can be the responsibility of designated super-users who can design interfaces for their own data subsets without having to be expert programmers. It is fairly straightforward to build metadata definition/editing interfaces using mature two-tier client–server technology, as described by us ***(16)***. (These interfaces do not need to be Web based because they can be limited to a handful of users within the intranet.)

Although microcomputer DBMSs (notably Microsoft Access) capture some presentation metadata ***(16)***, such as caption and pull-down information, the developers of client–server development frameworks such as PowerBuilder first realized the importance of centralizing presentation metadata much more extensively, through the use of "extended attributes" stored in specially designated server tables, which were accessed uniformly by the user-interface generation framework. Such techniques, however, can readily be duplicated without needing to purchase special development frameworks. We illustrate below the details of the EAV/CR metadata.

4.2. Active Metadata and EAV Systems

We have already stated that EAV-modeled systems are workable only in the presence of a significant metadata component. Such systems, in fact, trade a simplified data schema for a complex metadata schema. This trade-off becomes worthwhile for systems with numerous classes. Although this may seem to be a disadvantage, it turns out to be of long-term benefit, because the metadata sub-schema can be used both for internal documentation and for user-interface generation. Its use for the latter also forces the system's maintainers to ensure that it is specified to a reasonable degree of completeness. Furthermore, the metadata allows support of a *mixed schema*—that is, certain data classes can be stored in conventional relational tables rather than as EAV triplets. In fact, a so-called "EAV/CR" database must hold the vast majority of its data in conventional tables rather than in EAV form: it is the metadata component that distinguishes such schemas.

It is best to describe a metadata schema that supports both documentation and presentation through example, and the example we have chosen is that of our own EAV/CR model. It should be noted that the elements described below constitute a minimal set—that is, more elements may be added for a particular purpose. (The current version of the EAV/CR metadata schema can be accessed at the URL http://senselab.med.yale.edu/senselab/site/dsArch/.) In the SenseLab schema, for example, it was found necessary to segregate the overall user interface into a set of conceptual portals to organize the numerous classes into logical categories based on the interests of browser-users who accessed their data, and so we have a higher order data category, the "database," to which a class belongs. (A database is virtual—we have a single physical database, and in retrospect, the term "portal" might have been more appropriate, but this work was commenced before "Web portal" became a commonly used phrase.) Among the fields that define a portal are several that constitute a "theme"—foreground and background colors, logos, and so on. The idea is that, for a user browsing across several portals, the individual portals are visually distinguished.

4.3. Details of the EAV/CR Metadata Schema: Classes and Attributes

In a database's logical schema, Classes are analogous to tables and Attributes to fields within those tables. In EAV/CR, the Classes and Attributes tables are the most important part of the metadata and contain not only the logical schema description but also information essential to a Web-based user interface. Their structure is described in **Fig. 1**, which illustrates the Choice_Set table discussed in **Subheading 2.1**.

4.3.1. The Structure of the Classes Table

This table has the following fields, at a minimum.

- Class name, Description (both short and detailed descriptions can be stored), and a unique Class ID.
- EAV_Flag (Boolean)—whether the class is stored in EAV form or conventional.

In our own schemas, we record much more extensive detail, with the idea that the system should contain self-descriptive information. Thus, we also record information about every logical or physical class in the system, including the Classes table itself. We have found such structured, functional description of a class very useful when we have to generate a schema for a new vendor database engine. (For example, we record the primary key expression for a table as well as theme/cosmetic information for form generation. Furthermore, because different DBMSs deal with the issue of handling auto-incrementing primary keys differently, it is useful to record the fact that a particular table's primary key shows such behavior.)

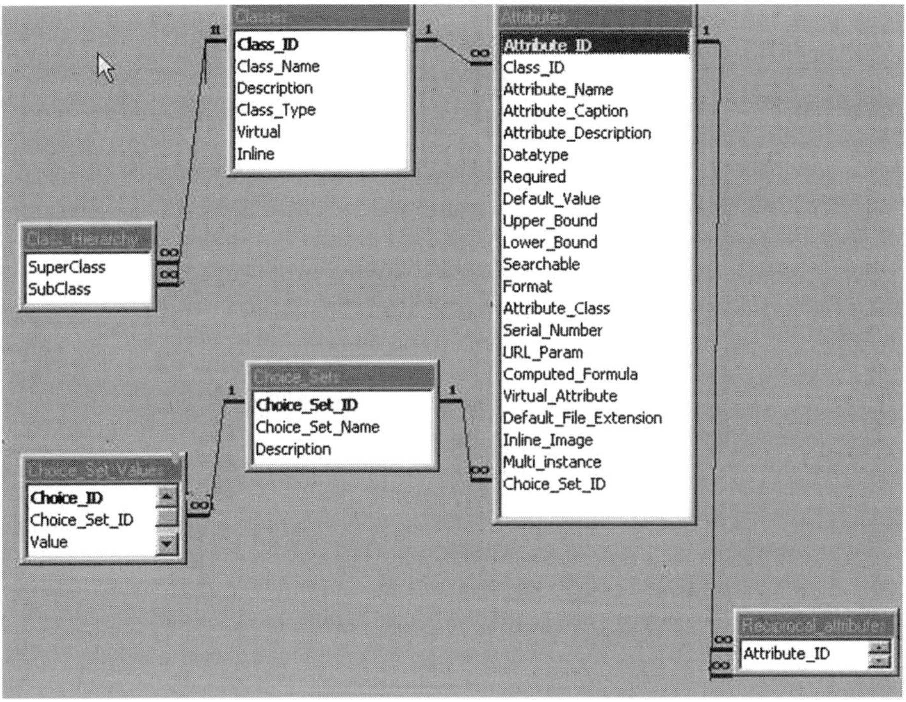

Fig. 1. The EAV/CR metadata sub-schema.

4.3.2. The Class_Hierarchy Table

This table has two fields, Parent class and Child class, both linked to the Classes table. It records parent–child relationships between class definitions. It is consulted when the user specifies a query based on a super-class that might encompass sub-classes as well. (In pharmacology, for example, the class "Drug Family" is a parent of the class "Drug.")

4.3.3. The Attributes Table

This table has the most detail of all the metadata tables. Among the most important fields in this table are

- Attribute ID (a unique ID), Class ID (points to the Class this attribute belongs to), Internal Name, a Caption (seen by the user), Description (for documentation), and Serial Number (order of presentation in a generic Web form).
- Datatype: It is one of integer, real, date, string, long string, binary, enumerated, ordinal, or Boolean. The last three are subtypes of integer. Enumerated and ordinal

attributes are based on choice sets (see above); ordinals are a special case of enumerated where the items in a set are conceptually ordered in some way, allowing greater than/less than comparisons.

- Attribute Class: For "Object" attributes only. Indicates the class of the attribute itself.
- Choice Set ID: For enumerated and ordinal attributes only. Points to the choice set.
- Default Value: (Applicable only for integer, real, string, and date/time datatypes.) If specified, and the actual value of this attribute does not happen to be stored for a particular instance, then this value is presented to the user. Upper Bound and Lower Bound, if applicable, are used for data entry validation along with datatype.
- Required (Boolean): If true, this value must be supplied for a new record.
- Width and Height: These numbers, applicable to strings and images, indicate how the attribute is to be displayed in a Web form. (A short string may be displayed either as an INPUT field, if the height is zero, or a TEXTAREA field, where the height indicates the number of columns. A long string will always be displayed as a TEXTAREA.) For numbers and dates, the width is computed based on defaults or on the format (see below) if specified.
- Format (picture): A datatype-specific string indicating how a value is to be formatted when displayed, e.g., dates may be shown with date and time or date alone, and real numbers can be displayed with a certain number of decimal places.
- Searchable (Boolean): If true, it indicates that a field for this attribute should be included in the search form generated by the system to let the user search for objects within a class on complex Boolean criteria.
- Computed Formula: Certain attributes may be computed based on the value of other attributes (if they are non-null). This field holds a Javascript template—an expression with placeholders that are replaced by the values of the appropriate attributes during runtime.

4.4. Specific Examples of Classes and Attributes: The NeuronDB Database

NeuronDB is one of the databases in SenseLab, which contains information on individual types of neurons in human and other species. This may be accessed via the URL http://senselab.med.yale.edu/senselab/NeuronDB/default.asp, whose contents are very briefly summarized in the account below.

The following are some of the NeuronDB classes: Brain Regions; Neurons; Channels; Receptors; Neurotransmitters/Neuromodulators; Currents; Canonical Neuron Forms; Bibliographic References.

We now describe the attributes of an individual class (Neuron). Many of these attributes, indicated by asterisks, reference other classes in the database. A neuron consists of one or more canonical compartments. Each compartment is associated with one or more receptors* (inputs from other neurons*), intrinsic currents*, and output neurotransmitters* that act efferently on other neurons*.

In addition, the entire neuron may be associated with one or more computational models as well as bibliographic citations*. Finally, there may be experimental data on individual neurons (accessed by hyperlinking to the Cornell database) as well as microscopy data [accessed by hyperlinking to the Cell-Centered Database (CCDB)] at UCSD.

The description of individual classes and attributes is accessible via the URL http://senselab.med.yale.edu/senselab/site/dsArch/?lb=tree, which provides a hierarchical view of the schema of NeuronDB and other databases in SenseLab.

4.4.1. Managing Reciprocal Semantics

A Reciprocal Attributes table is used to store reciprocal semantic information when semantic tags are assigned to objects in a relationship. For example, in a neuronal pathway class, if we record, for a pair of neuronal types, that one set of neurons is "afferent" to another, then we can alternatively record that the second set is "efferent" to the first. "Afferent" and "efferent" are examples of semantic inverse/reciprocal relationships. In neuroanatomy, another common pair of relationships that records associations between super-structures and sub-structures is "contains/is-part-of." Maintaining such a table facilitates navigation from any object to any related object within the database and allows the content to be used as the basis of a semantic network, as discussed in **ref. *17***.

5. Support for Vocabularies and Search

Originally, the term "controlled vocabulary" simply meant a collection of key phrases (terms) used to describe a domain. It is clear that terms in isolation are not very useful, unless one can relate them to each other. *Thesauri* and *ontologies* represent evolved forms of vocabularies. A *thesaurus* defines individual *concepts*, *synonyms* for concepts (e.g., "acetylsalicylic acid" for "aspirin"), and *antonyms* if applicable (e.g., "agonists" versus "antagonists"), as well as hierarchical *relationships* between specific concepts and more general concepts (e.g., between aspirin and non-steroidal anti-inflammatory drugs). The hierarchy need not be strict: a specific concept can be related to more than one general concept. Thus, chlorpromazine is simultaneously a dopamine antagonist, an anticholinergic, and a phenothiazine. (The last relationship is important in considering specific side effects such as skin and ocular pigmentation.)

Ontologies ***(18)*** represent refinements of thesauri, in that

- In an ontology, one can segregate concepts into *categories* (classes), where each class is described by *attributes* or properties that describe individual concepts belonging to that class. Examples of classes in the realm of therapeutics include therapeutic agents, diseases/ailments, definitions of adverse effects, and genes and gene variants, both common (polymorphisms) and rare. Examples of attributes include chemical structure, which applies to drugs, or sequence, which applies to genes/gene variants.

- The relationships between concepts and concept classes can be more general (e.g., related not just to categorization). Thus, a "used-in-treatment-of" relationship can exist between individual therapeutic agents and diseases.
- Because concepts are segregated into categories, the permissible properties/attributes for a given object/concept can be appropriately constrained. Additional means of constraining what the ontology may express include description logics *(19)*.

Ontologies are traditionally modeled using the sub-schema illustrated in **Fig. 2.** The Objects/Entities table has been described earlier: this table is usually called "Concepts," because an ontology deals with concepts in a domain. Each concept has associated Properties, which are simply attribute–value pairs: for more voluminous data, the properties of each class could be modeled as traditional relational tables for each class. (By combining this with the EAV/CR approach, using the Classes and Attributes tables, we can constrain the attributes that are applicable to individual objects/concepts appropriately.) Each concept has one or more synonyms, or Terms, which allow location of a concept by alternative keywords.

Because the Objects table records information on all objects in the database irrespective of the physical locations of their details, it follows that the Terms/Synonyms table also supports cross-table keyword search of objects. For efficiency purposes, the individual words in the phrase constituting a term can be recorded in a table of Keywords. The Keyword table has an auto-generated integer primary key, and the Bridge table (Keyword_Terms) supports fast Boolean search when multiple keywords are specified—the UMLS

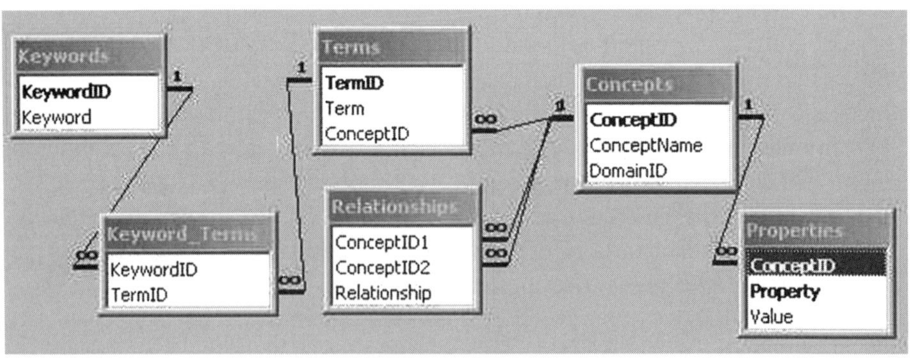

Fig. 2. A sub-schema for ontology support.

includes such a bridge table, which is called a *concordance* table. The (Binary) Relationships table is used to record relationships between pairs of concepts, as described in **Subheading 3.2**.

6. Conclusions

It can be seen that certain design decisions, such as the use of an Objects table, Classes/Attributes table to record descriptions of the data elements, and Relationships tables, are applied productively in different contexts. Indeed, these design elements prove to be so versatile that any non-trivial research database, such as used in medium- to large-scale neuroscience efforts, can benefit from incorporating them into the schema. The use of EAV modeling, on the contrary, is optional—it resembles a loaded gun in that it requires significant experience and programming expertise to get comfortable with.

References

1. Gene Ontology Consortium. (2004) An Introduction to the Gene Ontology; http://www.geneontology.org/GO.doc.html. Last accessed: 11/26/04.
2. Lindberg, D. A. B., Humphreys, B. L. and McCray, A. T. (1993) The Unified Medical Language System. *Methods Inf. Med.* **32**, 281–91.
3. Slezak, T., Wagner, M., Yeh, M., Ashworth, L., Nelson, D., Ow, D., et al. (1995) A Database System for Constructing, Integrating, and Displaying Physical Maps of Chromosome 19. In: Hunter, L. and Shriver, B. D., editors. *Proceedings of the 28th Hawaii International Conference on System Sciences*, Wialea, Hawaii. IEEE Computer Society Press, Los Alamitos, CA, p. 14–23.
4. National Center for Biotechnology Information. (1998) *NCBI Software Development Toolkit*. National Center for Biotechnology Information, National Library of Medicine, Bethesda, MD.
5. Winston, P. H. (1984) *Artificial Intelligence*. 2nd ed. Addison-Wesley, Reading, MA.
6. Stead, W. W. and Hammond, W. E. (1988) Computer-Based Medical Records: The Centerpiece of TMR. *MD Comput.* **5(5)**, 48–62.
7. Huff, S. M., Haug, D. J., Stevens, L. E., Dupont, C. C. and Pryor, T. A. (1994) HELP the Next Generation: A New Client-Server Architecture. In: *Proceedings of the 18th Symposium on Computer Applications in Medical Care*, Washington, DC. IEEE Computer Press, Los Alamitos, CA, p. 271–5.
8. Friedman, C., Hripcsak, G., Johnson, S., Cimino, J. and Clayton, P. (1990) A Generalized Relational Schema for an Integrated Clinical Patient Database. In: *Proceedings of the 14th Symposium on Computer Applications in Medical Care*, Washington, DC. IEEE Computer Press, Los Alamitos, CA, p. 335–9.
9. Petrusha, R. (1996) *Inside the Windows 95 Registry*. O'Reilly Associates, Sebastopol, CA.

10. World Wide Web Consortium. (2002) Resource Description Framework (RDF); http://www.w3c.org/RDF/. Last accessed: 02/23/02.
11. Nadkarni, P. M., Marenco, L., Chen, R., Skoufos, E., Shepherd, G. and Miller, P. (1999) Organization of Heterogeneous Scientific Data Using the EAV/CR Representation. *J Am Med Inform Assoc* **6(6)**, 478–93.
12. Shepherd, G. M., Healy, M. D., Singer, M. S., Peterson, B. E., Mirsky, J. S., Wright, L., et al. (1997) Senselab: A Project in Multidisciplinary, Multilevel Sensory Integration. In: Koslow, S. H. and Huerta, M. F., editors. *Neuroinformatics: An Overview of the Human Brain Project*. Lawrence Erlbaum Associates, Mahwah, NJ, p. 21–56.
13. Koslow, S. H. and Huerta, M. F. (1997) *Neuroinformatics: An Overview of the Human Brain Projects*. Lawrence Erlbaum Associates, Mahwah, NJ.
14. Marco, D. (2000) *Building and Managing the Metadata Repository*. Wiley, New York.
15. Microsoft Corporation. (2005) *Microsoft SQL Server 2005*. Microsoft Corporation, Redmond, WA.
16. Nadkarni, P. M., Brandt, C. A. and Marenco, L. (2000) WebEAV: Automatic Metadata-Driven Generation of Web Interfaces to Entity-Attribute-Value Databases. *J Am Med Inform Assoc* **7(7)**, 343–56.
17. Microsoft Corporation. (2003) *Microsoft Access for Office*. Microsoft Corporation, Redmond, WA.
18. McCray, A., Aronson, A., Browne, A., Rindflesch, T., Razi, A. and Srinivasan, S. (1993) UMLS Knowledge for Biomedical Language Processing. *Bull Med Libr Assoc* **81(2)**, 184–94.
19. Pidcock, W. and Uschold, M. (2003) What are the Differences Between a Vocabulary, a Taxonomy, a Thesaurus, an Ontology, and a Meta-Model? http://www.metamodel.com/article.php?story=20030115211223271. Last Accessed: 03/24/06.
20. Ceusters, W., Smith, B. and Flanagan, J. (2003) Ontology and Medical Terminology: Why Description Logics are Not Enough. In: *Towards an Electronic Patient Record (TEPR)*, San Antonio, TX. Medical Records Institute, Boston, MA.

4

XML for Data Representation and Model Specification in Neuroscience

Sharon M. Crook and Fred W. Howell

Summary

EXtensible Markup Language (XML) technology provides an ideal representation for the complex structure of models and neuroscience data, as it is an open file format and provides a language-independent method for storing arbitrarily complex structured information. XML is composed of text and tags that explicitly describe the structure and semantics of the content of the document. In this chapter, we describe some of the common uses of XML in neuroscience, with case studies in representing neuroscience data and defining model descriptions based on examples from NeuroML. The specific methods that we discuss include (1) reading and writing XML from applications, (2) exporting XML from databases, (3) using XML standards to represent neuronal morphology data, (4) using XML to represent experimental metadata, and (5) creating new XML specifications for models.

Key Words: XML; MorphML; NeuroML; neuronal morphology; neuroinformatics.

1. Introduction

The complexity of problems in neuroscience requires that research from multiple groups across many disciplines be combined. Such collaborations require an infrastructure for sharing data and exchanging specifications for computational models, and the information published by one group must be in a form that others can use *(1)*. EXtensible Markup Language (XML) technology provides an ideal representation for the complex structure of models and neuroscience data, as it is an open file format and provides a language-independent method for storing arbitrarily complex structured information. Software and

From: *Methods in Molecular Biology, vol. 401: Neuroinformatics*
Edited by: Chiquito Joaquim Crasto © Humana Press Inc., Totowa, NJ

database developers in many fields, including neuroscience, have enthusiastically adopted XML due to its simplicity, flexibility, and its relation to the Hypertext Markup Language (HTML) standard for Web pages. Like HTML, XML is composed of text and tags that explicitly describe the structure and semantics of the content of the document. Unlike HTML, developers are free to define the tags that are appropriate for their application. For example, an electrophysiology database might use tags such as <species>, <age>, and <recording-site>, whereas an anatomy application might define tags such as <brain-region> and <staining-method>. The self-describing structure of XML ensures that it can be processed easily within a computer program, but a user can also easily read and understand an XML document as demonstrated in **Example 1**.

Example 1

```
<?xml version=“1.0” encoding=“utf-8”?>
<experiment>
<description>This is an example of an <emph>XML file</emph> with
   information about an experiment.</description>
<sample species=“rat” strain=“a100” age=“17 days”/>
<recordings>
<rec startTime=“0” dataFile=“rec1.dat”/>
<rec startTime=“1 min” dataFile=“rec2.dat”/>
</recordings>
</experiment>
```

In this example, the first line indicates the version of XML used (1.0) and the encoding scheme used for the text. The standard is UTF-8 (universal text format, 8 bits), which is a coding scheme for Unicode text. This allows most characters to be stored in a single byte but also supports international character sets such as Chinese or Korean in multiple bytes. The <description> element has *mixed content*, which includes both text and subtags (e.g., <emph>XML file</emph>). The notation <rec . . . /> is a shorthand notation for the pair <rec> . . . </rec> and is often used where an element has no required subelements.

Because XML is extremely flexible, many applications use only a subset of the features available. The most general approach allows a free mixing of tags and text, and many word processors save documents in this style of XML. Neuroinformatics applications typically use a restricted subset of XML without any document-style mixed content, where XML is used simply as an exchange format for structured data within databases or programs. In this case, the tags that provide structure to the data stream in an XML document are sometimes considered to have an equivalent representation in an XML schema. An XML schema defines the hierarchy of XML elements and their attributes and specifies the structures that are syntactically correct in a compliant stream of data. The use of an XML schema makes it easy to validate documents and create database tables for data element storage and access. The XML schema elements are also equivalent to what object-oriented programmers refer to as an *object class*, which facilitates the generation of code for reading and writing data documents or for processing and analyzing data. All these aspects of XML lead to its greatest benefit—its use facilitates large-scale, interdisciplinary communication and collaboration *(2)* (*see* **Notes 1** and **2**).

2. Materials

This section includes information on some of the projects, software, and materials relevant to the use of XML in neuroscience. NeuroML is an open source project for XML in neuroscience, which was initiated due to growing interest in defining a common data format for describing and exchanging neuronal models, data, literature, methods, and other aspects of neuroscience *(3)*. In **Subheading 3**, we use examples from NeuroML to demonstrate the use of XML in neuroscience. Contributions to NeuroML are made through a Web site that serves as a repository for information on the continuing development of NeuroML standards *(4)* and includes discussion boards, mailing lists, schema distributions, documentation, and the NeuroML Development Kit *(5)*, which is described in **Subheading 3.5** (*see* **Note 3**).

Creating standards for the description of neuroscience data and models is difficult for several reasons. Experimental and theoretical studies are done at a variety of levels of scale, from protein interactions to large-scale networks of neurons. In addition, different software packages for data analysis and simulation have different and changing capabilities that create a moving target and make it impossible to address every aspect of data and model specification. However, it is sometimes possible to restrict the scope of a particular XML standard to converge on an agreed notation for one limited aspect of neuroscience modeling or data. For this *intersection* approach, the emphasis is on defining the XML schema for that particular standard. This approach is attractive because it makes the resulting model or data description useable by many tools with overlapping functions. However, it may only encompass a subset of desired functionality, and implementation for a particular tool will sometimes require extensions to the XML standard. In **Subheading 3.3**,

MorphML is used as an example of an XML standard. Under the umbrella of NeuroML, MorphML has been developed for representing three-dimensional reconstructions of neuronal morphology in both data analysis and model simulation software ***(6,7)***.

Owing to the limitations in developing standards and for flexibility in storing their own data, software developers often choose to create their own custom XML formats. For data or parts of a model for which no standards exist, rather than developing arbitrary XML, NeuroML provides a common method for serializing object models. For this *union* approach, the emphasis is on making it easy to define any object model and to serialize it in XML using a systematic coding of an object tree. An example of this approach is given in **Subheading 3.5**, where the NeuroML Development Kit is used to define a new XML language.

Several XML applications other than NeuroML have been developed that are relevant to problems in neuroscience. One of these is the Systems Biology Markup Language (SBML) ***(8)***, which is a standard for specifying models of biochemical reaction networks and thus provides a common format for differential equation models of intracellular pathways ***(9,10)***. However, owing to this limited scope, it provides no formalism for the spatial aspects of three-dimensional interactions and does not include support for modeling the electrical properties of receptors, a critical aspect of most neuronal models. Another is BrainML, which provides formalisms for representing time series data, spike trains, experimental protocols, and other data relevant to neurophysiology experiments ***(11,12)***. The BrainML project at Cornell University School of Medicine also includes efforts to develop ontologies for neuroscience data. Some of these XML applications use MathML ***(13)*** for representing mathematical equations when needed.

3. Methods

Here, we describe some of the common uses of XML in neuroscience, with case studies in representing neuroscience data and defining model descriptions. The methods that we will discuss include (1) reading and writing XML from applications, (2) exporting XML from databases, (3) using XML standards to represent neuronal morphology data, (4) using XML to represent experimental metadata, and (5) creating new XML specifications for models.

3.1. Reading and Writing XML from an Application

One of the advantages of XML is that it is a simple text format and can be read and written by any programming language using standard text I/O (*see* **Note 4**). However, XML is not simple to parse with ad hoc code, and as a result, most software developers choose to use one of the standard XML libraries supplied for all major programming languages. The main alternatives for parsing are Simple Application Programming Interface (API) for XML (SAX), Document Object Model (DOM), the XML Pull Parser (XPP), and data binding.

SAX is a standard low-level event-based interface for parsing XML for most languages *(14)*. The developer provides a small number of method callbacks. Then the parser tokenizes the XML source and invokes the user's methods in order. An example is shown in **Note 5**. Using SAX can be inconvenient, as the programmer must develop a stack to maintain the position in the XML tree, but it is a fast and memory efficient choice for large XML files.

The DOM interface loads an XML file into memory and provides functions for traversing the tree in any order *(15)*. It is much simpler for programmers to use than SAX but has a cost for large XML files and does not work for XML files larger than memory.

The XPP interface *(16)* is less well known than SAX but provides a superior streaming interface where control remains with the user's program. The idea is that the user's program calls a nextToken() method to fetch the next available token, whereas in SAX, the parser has control of the thread of execution and invokes the user's code. Thus, XPP combines the efficiency of SAX for large XML files with much of the convenience of DOM.

An alternative to parser interfaces is to use data binding where there is a direct high-level mapping between classes and fields in code and the elements and attributes of an XML file. This is the most convenient style of parser for programmers to use, as it regards XML as simply a language-independent object serialization framework, and reading and writing of XML files can be performed with a single method call. An additional advantage is that the parser need not be rewritten when the XML schema changes, which is usually the case for the low-level SAX, DOM, and XPP-based parsers. **Subheading 3.5** illustrates one library, the NeuroML Development Kit, which supports data binding for Java *(5)* (*see* **Note 6** for other libraries).

3.2. Exporting XML From Databases

Example 2 shows how hypertext preprocessor (PHP) *(17)* can be used to generate XML from the contents of a relational database, in this case containing experimental recordings. The advantage for a remote user in having the information exported as XML rather than as an HTML Web table is that the information is much simpler to data mine. This example is intended to illustrate the simplicity of generating XML documents using code that is similar to existing code for creating the HTML views for Web browsers.

Example 2

```
<?php
$db = mysql_connect("localhost", "dbmgr", "passwd")
mysql_select_db("neuro",$db)
$results = mysql_query("select * from recordings",$db)
print "<recordings>\n"
while ($myrow = mysql_fetch_row($results)){
print "<rec startTime=".$myrow[4]." file=".$myrow[5]."/>\n"
}
print "</recordings>\n"
?>
```

Most bioscience databases on the Web now support the export of information in XML as well as the HTML format suitable for Web browsers. This makes it simpler to use the databases from programs that issue "http:" requests for data mining. An interesting example is **ref.** ***18***. It provides XML export from the object-oriented ACeDB database for gene-related information in *Caenorhabditis elegans*.

3.3. Using XML Data Standards: MorphML

Neuronal morphology data are important for the study of many areas of neuroscience including neural development, aging, pathology, and neural computation *(19)*. Generally, the axonal and dendritic arborizations of a neuron are represented using a collection of points, diameters, and connections in three dimensions, but there are many different data formats. Some examples are those used by different neuron-tracing systems such as Eutectic's Neuron Tracing System, Microbrightfield's Neurolucida, the Nevin binary branch tree (BBT) syntax, and the Douglas syntax *(20)* as well as the formats required by software such as NEURON (*see* **Chapter 6**) *(21,22)*, GENESIS (*see* **Chapter 7**) *(23)*, and cvapp *(24)*. To address the problems associated with multiple data formats, MorphML was developed for use as an XML standard for quantitative neuroanatomical data *(7)* or as a representation for cell morphology in biologically realistic modeling efforts as part of NeuroML. **Example 3** provides a fragment of a MorphML document that represents the simple branching structure illustrated in **Fig. 1**.

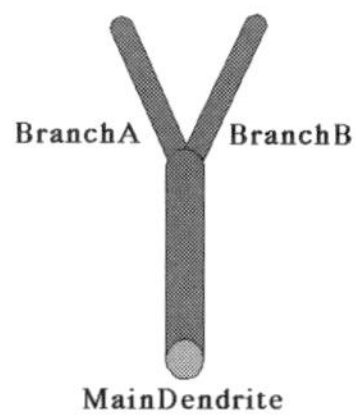

Fig. 1. Schematic representation of a branching dendrite.

Example 3

```
&lt;segments&gt;
&lt;!– Segment: MainDendrite, ID: 1–&gt;
&lt;segment id=”1” name=”MainDendrite”&gt;
&lt;proximal x=”0” y=”0” z=”0” diameter=”4”/&gt;
&lt;distal x=”0” y=”16” z=”0” diameter=”4”/&gt;
&lt;/segment&gt;
&lt;!– Segment: BranchA, ID: 2–&gt;
&lt;segment id=”2” name=”BranchA” parent=”1”&gt;
&lt;proximal x=”–1” y=”16” z=”0” diameter=”2”/&gt;
&lt;distal x=”–6” y=”25” z=”0” d=”2”/&gt;
&lt;/segment&gt;
&lt;!– Segment: BranchB, ID: 3–&gt;
&lt;segment id=”3” name=”BranchB” parent=”1”&gt;
&lt;proximal x=”1” y=”16” z=”0” diameter=”2”/&gt;
&lt;distal x=”6” y=”25” z=”0” diameter=”2”/&gt;
&lt;/segment&gt;
&lt;/segments&gt;
```

The level 1 MorphML schema describes only the anatomical aspects of neural data such as cell morphology and fiducials used for registration and alignment; the level 2 schema also includes a standard for describing the dynamics of ion channels and their densities throughout the neuron. The MorphML Web site ***(6)*** also includes a validation tool that will validate a particular XML file against the various XML schema documents that define the NeuroML and MorphML specifications.

One example of the use of the MorphML standard is the Virtual RatBrain Project, which centers around a repository of peer-reviewed three-dimensional cellular anatomical data from multiple brain regions of the rat ***(25)***. All neuroanatomical data are stored and exchanged using the MorphML standard.

The database can be accessed by tools, which allow users to visualize and analyze this evolving, three-dimensional data set. For example, the MorphML Viewer allows a user to view multiple data sets in the context of a reference brain and zoom in to view particular data. To view images of this brain atlas and associated tools, see **ref. *25***.

3.4. XML for Experimental Metadata

One of the most valuable uses for XML in neuroscience is the annotation of experimental data with structured *metadata* that describes experimental conditions in a machine readable format as demonstrated in **Example 1**. This makes it possible for researchers to create laboratory databases of experimental data, with the potential for data mining and automated analyses across multiple data sets such as images and electrophysiology recordings. XML is also an ideal format for supporting publication of neuroscience data on the Web. Routine publication of experimental data along with structured metadata in XML describing experimental conditions would simplify the task of building computational models, which rely on both electrophysiology and neuroanatomy data.

A small number of applications and databases have been developed, which begin to address this need. The NeuroSys database *(26)* is an open-source laboratory data management system that stores data in XML. The Catalyzer *(27)* program allows users to import experimental data files and automatically extracts metadata from the headers. Users can add extra fields and notes on experimental conditions, and the resulting catalogs are saved in custom XML and can be published as a Web site. Other examples are provided in **Note 7**.

3.5. Using the NeuroML Development Kit to Define New XML Languages

The NeuroML Development Kit *(5)* provides one method for defining an XML language that represents the complex structures used by a particular program. **Figure 2** shows a state transition diagram for a simple model, which we use to illustrate the development of a new XML format for representing the model in **Example 4**.

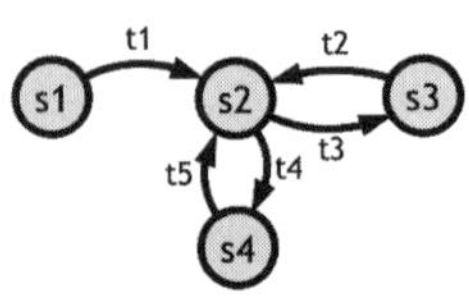

Fig. 2. A state transition diagram for a sample model.

Example 4

Step 1: Define the classes for the model.

```
class State {
public String name;
public State(String n) {name=n; }
}
class Transition {
public String name;
public String from;
public String to;
public Transition(String n, String f, String t) {name=n; from=f; to=t; }
}
class Diagram {
public List states = new List(“State”);
public List transitions = new List(“Transition”);
}
```

Step 2: Build the model in XML.

```
<neuroml class=“Diagram”>
<states>
<state name=“s1”/>
<state name=“s2”/>
<state name=“s3”/>
<state name=“s4”/>
</states>
<transitions>
<transition name=“t1” from=“s1” to=“s2”/>
<transition name=“t2” from=“s3” to=“s2”/>
<transition name=“t3” from=“s2” to=“s3”/>
<transition name=“t4” from=“s2” to=“s4”/>
<transition name=“t5” from=“s4” to=“s2”/>
</transitions>
</neuroml>
```

Step 3: Load the model into software for manipulation.

```
class Test {
public static void main(String[] args) {
//Reading an arbitrary file takes a single line of code
Diagram d = (Diagram) XMLIn.loadFile(“test.xml”);
//at this point we have the model in memory and can
//simulate /manipulate it.
```

```
}
}
```

Step 4: Write the modified model out to an XML file.

```
XMLOut.toFile(d, "test2.xml");
```

Step 5: Create extensions.

It is the nature of modeling to extend the features of model descriptions. For this example, one might wish to add a rate or probability to each transition or add annotations to each of the states. We add a field to the class as follows.

```
class Transition {
public String name;
public String from;
public String to;
public double probability;
}
```

Now existing code will read XML entries of the format:

```
<transition name="t1" from="s1" to="s2" probability="0.5"/>
```

Further documentation and additional examples supplied with the NeuroML Development Kit illustrate support for other features, including class inheritance and references to other objects within the XML file. Note that although this example is provided in Java, any other programming language can read XML files by using an appropriate data-binding tool kit (*see* **Note 6** for references to data-binding kits for other languages).

4. Notes

1. XML is not always the most appropriate format for data. The major liability of XML is that the format is verbose, so that file sizes can be significantly larger than binary formats. Thus, XML files will take longer to parse than custom binary formats. For high volumes of data, such as images, a binary format is more appropriate, but when file size is not a concern, the advantages of clarity and self-description outweigh the time and space overheads due to the use of XML. XML files compress well using standard compression tools such as zip and gzip. There are also XML-specific compression tools such as Xmill ***(28)***, which achieve slightly better performance.
2. The following example of an XML file includes tags from two different *namespaces* available at mydoc.org and MathML.org. It is common to bind a short prefix to a URL, which is used to distinguish the tags in the file. In this case, the prefix m: is assigned to the namespace from the MathML standard.

```
<mydoc xmlns="http://www.mydoc.org"
```

```
xmlns:m="http://www.mathml.org">
<name>bob</name> <!– from the mydoc.org namespace –>
<m:math> <!– ab + c in MathML –>
<m:apply>
<m:plus/>
<m:apply>
<m:times/>
<m:ci> a <m:/ci>
<m:ci> b <m:/ci>
</m:apply>
<m:ci> c <m:/ci>
</m:apply>
</m:math>
</mydoc>
```

The advantage of using namespaces is that it allows unambiguous mixing of several XML languages; the disadvantage is that the resulting XML loses some of its readability to people and is slightly more difficult to read in software.

3. Currently, many of the most widely used neural simulators are adopting XML formats for serializing model descriptions. The GENESIS simulator is undergoing a major redevelopment effort and will use an XML representation for model specifications ***(29)***. Other software packages for model development and simulation also use XML representations including the Neuron Simulator group ***(21,22)*** and the developers of neuroConstruct ***(30)*** and Catacomb ***(31)***. All these groups are contributors to the ongoing effort to create standards under the umbrella of NeuroML.
4. One drawback of using basic text I/O is that the programmer must allow for special cases, such as quoting illegal characters. There are some restrictions on the characters that can be used in XML text as summarized in **Table 1** .
5. The example below uses the SAX XML parser and demonstrates an implementation of the startElement method, which is called when the parser encounters an opening tag. This method echoes the element name and attributes; a real application would need to respond differently to different element tags.

Table 1
Character Replacements in EXtensible Markup Language (XML) Text

Character	<	>	&	"	'
Replacement	<	>	&	&quote;	'

```
public void startElement(String namespaceURI,
String sName, String qName,
Attributes attrs) throws SAXException {
System.out.println(" element: "+qName);
if (attrs != null) {
for (int i=0; i<attrs.getLength(); i++) {
String aName = attrs.getLocalName(i);
if ("".equals(aName)) aName = attrs.getQName(i);
System.out.println(" attr "+aName+"=\""+attrs.getValue(i)+"\"");
 }
 }
 System.out.println();
 }
```

6. In **Subheading 3.5**, we provide an example of the use of the NeuroML Development Kit, which supports data binding for Java ***(5)***. Other libraries for Java include Castor ***(32)*** and JAXB ***(33)***. There are additional libraries for other software such as PHP ***(17)***, Python ***(34)***, Perl ***(35)***, and C++ ***(36)***.
7. Many disciplines in biology are starting to produce XML descriptions of metadata; the microarray community has been active in developing the MIAME requirements for metadata and the MAGE XML standard ***(37)***. In imaging, the Openmicroscopy group is developing an object model for confocal imaging ***(38)***.

Acknowledgments

The authors thank Joseph Svitak, Padraig Gleeson, and Robert Cannon for helpful discussions of issues relating to NeuroML and MorphML as well as contributions to schema and tool development. S.C. gratefully acknowledges the support of the National Science Foundation through IBN 0091117. F.H. is supported by a UK MRC eScience grant in Neuroinformatics.

References

1. Martone, M. E., Gupta, A. and Ellisman, M. H. (2004) e-Neuroscience: challenges and triumphs in integrating distributed data from molecules to brains. *Nature Neuroscience* **7**, 467–472.
2. Bourret, R. XML and Databases, http://www.rpbourret.com/xml/XMLAndDatabases.htm.
3. Goddard, N., Hucka, M., Howell, F., Cornelis, H., Shankar, K. and Beeman, D. (2001) Towards NeuroML: model description methods for collaborative modelling in neuroscience. *Philosophical Transactions of the Royal Society of London. Series B, Biological Sciences* **356**, 1209–1228.
4. Crook, S., Howell, F. and Svitak, J. NeuroML Sourceforge Site, http://neuroml.sourceforge.net.

5. Howell, F. NeuroML Development Kit, http://www.neuroml.org/ndk.html.
6. Crook, S. MorphML: An XML Application for Neuronal Morphology Data, http:/morphml.org.
7. Qi, W. and Crook, S. M. (2004) Tools for neuroinformatic data exchange: an XML application for neuronal morphology data. *Neurocomputing* **58–60C**, 1091–1095.
8. SBML Team. Systems Biology Markup Language, http://www.sbml.org.
9. Hucka, M., Finney, A., Sauro, H. M., Bolouri, H., Doyle, J. C., Kitano, H., Arkin, A. P., Bornstein, B. J., Bray, D., Cornish-Bowden, A., Cuellar, A. A., Dronov, S., Gilles, E. D., Ginkel, M., Gor, V., Goryanin, I. I., Hedley, W. J., Hodgman, T. C., Hofmeyr, J.-H., Hunter, P. J., Juty, N. S., Kasberger, J. L., Kremling, A., Kummer, U., Le Novere, N., Loew, L. M., Lucio, D., Mendes, P., Minch, E., Mjolsness, E. D., Nakayama, Y., Nelson, M. R., Nielsen, P. R., Sakurada, T., Schaff, J. C., Shapiro, B. E., Shimizu, T. S., Spence, H. D., Stelling, J., Takahashi, K., Tomita, M., Wagner, J. and Wang, J. (2003) The systems biology markup language (SBML): a medium for representation and exchange of biochemical network models. *Bioinformatics* **19**, 524–531.
10. Finney, A. and Hucka, M. (2003) Systems biology markup language: level 2 and beyond. *Biochemical Society Transactions* **31**, 1472–1473.
11. Gardner, D., Knuth, K. H., Abato, M., Erde, S. M., White, T., DeBellis, R. and Gardner, E. P. (2001) Common data model for neuroscience data and data model exchange. *Journal of the American Medical Informatics Association: JAMIA* **8**, 17–33.
12. BrainML Team. BrainML Site, http://www.brainml.org.
13. W3C Math Working Group. MathML, http://www.w3org/TR/MathML2/.
14. SAX, http://sax.sourceforge.net.
15. Dom4J, http://www.dom4j.org.
16. XML Pull Parser Homepage, http://www.xmlpull.org.
17. SimpleXML, http://uk.php.net/manual/en/ref.simplexml.php.
18. Linking to WormBase, http://www.wormbase.org/about/linking.html.
19. Cannon, R. C., Howell, F., Goddard, N. H. and De Shutter, E. (2002) Non-curated distributed databases for experimental data and models in neuroscience. *Network: Computation in Neural Systems* **13**, 415–428.
20. Capowski, J. (1989) *Computer Techniques in Neuroanatomy*, Plenum Press, New York.
21. Hines, M. L. and Carnevale, N. T. (1997) The NEURON simulation environment. *Neural Computation* **9**, 1179–1209.
22. Hines, M. L. and Carnevale, N. T. (2001) Neuron: a tool for neuroscientists. *The Neuroscientist* **7**, 123–135.
23. Bower, J. M. and Beeman, D. (1998) *The Book of GENESIS: Exploring Realistic Neural Models With the GEneral NEural SImulation System*, second edition, Springer-Verlag, New York.

24. Cannon, R. C., Turner, D. A., Pyapali, G. and Wheal, H. V. (1998) An on-line archive of reconstructed hippocampal neurons. *Journal of Neuroscience Methods* **84**, 49–54.
25. Zaborsky, L., McDonnell, N., Vasanyi, P. and Howell, F. The Virtual Ratbrain Project, http://www.ratbrain.org.
26. Pittendrigh, S. and Jacobs, G. (2004) NeuroSys: a semi-structured laboratory database. *Neuroinformatics* **1**, 167–176.
27. Axiope Catalyzer, http://www.axiope.com.
28. XMILL: An Efficient Compressor for XML, http://www.cs.washington.edu/homes/suciu/XMLTK/xmill/www/index.html.
29. Crook, S., Beeman, D., Gleeson, P. and Howell, F. (2005) XML for model specification in neuroscience: an introduction and workshop summary. *Brains, Minds, and Media* 1:bm228(urn:nbn:de:009-3-2282).
30. Gleeson, P., Silver, R. A. and Steuber, V. neuroConstruct, http://www.neuroconstruct.org.
31. Cannon, R. Catacomb, http://www.catcmb.org/.
32. Castor. Castor XML Binding for Java, http://www.castor.org.
33. Sun Microsystems. JAXB, http://java.sun.com/xml/jaxb.
34. Ogbuji, U. XML Data Bindings in Python, http://www.xml.com/pub/a/2003/06/11/py-xml.html.
35. Perl-XML Frequently Asked Questions, http://perl-xml.sourceforge.net/faq/.
36. The Apache XML Project, http://xml.apache.org/xerces-c/.
37. Microarray and Gene Expression Data Society, http://www.mged.org.
38. Open Microscopy Environment, http://www.openmicroscopy.org.

5

Creating Neuroscience Ontologies

Douglas M. Bowden, Mark Dubach, and Jack Park

Summary

The insufficiency of terminological standards in neuroscience is increasingly recognized as a serious obstacle to interoperability. Adoption of a controlled vocabulary is a successful solution for small numbers of groups that work closely together but is impractical for large numbers of groups who represent diverse areas of research, index information by various legitimate nomenclatures, or publish in different languages. Interoperability among such disparate databases requires a translation mechanism, or "mediator," to enable communication and data sharing among databases. Shared ontologies are essential components of a mediator. An ontology codifies the relations between terms of multiple nomenclatures and the concepts they represent. Neuroanatomy is central to neuroscience, and neuroanatomical terminology represents a core portion of the vocabulary of neuroscience. We have created in NeuroNames an ontology of 2500 neuroanatomical concepts referenced by 15,000 terms in seven languages. NeuroNames is the mediator for BrainInfo, a portal to neuroanatomy on the Web. We hope that a description of our experience in establishing interoperability between BrainInfo and other neuroscience Web sites may be useful to others engaged in the development of ontologies for neuroscience.

Key Words: Ontology; neuroinformatics; information technology; interoperability; mediator; data mining; brain atlases; Topic Maps; NeuroNames; BrainInfo; spatial database.

1. Introduction

Ontology, in the broadest sense, is the "study of being; a doctrine of the being and relations of all reality" *(1)*. A *scientific ontology* is a codification of the relations among concepts in a specific domain of scientific research and the words that represent them. An essential feature of scientific concepts is that they are defined operationally, that is, on the basis of well-specified operations

From: *Methods in Molecular Biology, vol. 401: Neuroinformatics*
Edited by: Chiquito Joaquim Crasto © Humana Press Inc., Totowa, NJ

that lead to a consistent result that can be observed by any reasonable person (*see* **Subheading 3.3**). A *neuroscientific ontology* defines the relations among concepts in the particular domain of neuroscience and links the concepts to the terms that represent them in data repositories and scientific reports.

Historically, ontology was largely the concern of philosophers, linguists, and authors of dictionaries. With the emergence of informatics as an applied science, the practical significance of ontologies has become apparent to a broader segment of the academic community. They are essential tools if one is to fully exploit computer technology for locating and acquiring information from an ever-expanding network of knowledge resources.

First-generation Web search engines, such as AltaVista and Lycos, indexed knowledge sources on the basis of character strings and syntax, e.g., word location in a sentence, sentence location in a paragraph, etc. They made great strides toward informing users of possible locations of information, but the *string-based indexation* is subject to huge false-positive error. Google *(2)* and other second-generation Web search systems greatly enhanced info-retrieval accuracy by adding a semantic component to their indexes. In Google, the enhancement consisted of extracting and applying the implied semantic relatedness of words appearing in close proximity within Web documents, a form of citation indexing. The resulting, empirically derived map of semantic relatedness greatly improved the retrieval precision of Web search, but it still falls short of the retrieval accuracy achievable by ontology-mapped information.

If one searches Google for "connections of ventromedial nucleus of the hypothalamus," one obtains a list of 15,400 possible locations with no way of knowing which if any contains a complete listing of the information desired. *Concept-based indexation* of information allows search engines to respond with greater precision than indexation based on character-string or statistical-context *(3)* (also discussed in Chapter 1). Ontologies, as codifications of the meanings of terms used in particular domains of knowledge, are essential for full exploitation of the search power of information technology. The development of ontology-based search engines to rival the precision of traditional search engines requires the investment of considerable effort to create a thorough and accurate ontological framework and to map concepts occurring in documents to appropriate nodes in the ontology. We describe here our extensive experience with the performance of these critical tasks in the neuroscience domain.

We are building in BrainInfo, *(4)* a portal to neuroscientific information on the Web. The goal is to create a Web site from which visitors can navigate with the fewest possible clicks to comprehensive information about specific brain structures located deep in other Web sites. We found that, while we can hyperlink to relevant pages in Web sites anywhere in the world, the terms used in text and labels on images at other sites are often different from the terminology of queries posed by our users. Recognizing the impracticality of inducing other sites to revise their terminology, we concluded that BrainInfo needed the capability of translating the terms used at target sites into terms known to BrainInfo's visitors. For example, if BrainInfo directed a visitor to a table listing connections of the hippocampus in the Brain Information Management System (BIMS) at the University of Southern California ***(5)***, it should post a message telling the visitor the relevant nomenclature used at the site. The message appearing in BrainInfo sidebar reads: "HIPPOCAMPUS: Look for ammon horn, CA1, CA2, or CA3." For this, we needed an ontology containing the information that "ammon horn" is a synonym of "hippocampus," BrainInfo's default name for the structural concept, and that CA1, CA2, and CA3 are parts thereof. To meet that kind of need, we developed a nomenclature, NeuroNames ***(6–8)***.

The complexity of a scientific ontology depends upon the size of the knowledge domain it encompasses and the degree to which workers in the area have adopted a standardized terminology. As we developed NeuroNames as a codification of neuroanatomic nomenclature, we gradually recognized that neuroanatomists agree for the most part as to the primary structures that comprise the mammalian brain. Comparing our macaque brain atlas ***(9,10)*** with two standard rodent atlases ***(11,12)***, we found that, if one excluded cortical structures, about 90% of primary structures in the macaque brain have morphologic homologues in the rodent ***(13)***. But neuroanatomists lack anything resembling a standard terminology. Although the number of primary structures (*see* **Subheading 3.1**) in the macaque atlas is about 350, our early survey of major human and nonhuman primate neuroanatomy texts and atlases identified more than 2000 legitimate English and Latin names for them.

This chapter describes our approach to the development of NeuroNames as an ontology of neuroanatomy, its application to the development of BrainInfo as a portal to neuroanatomy on the Web, and current efforts to extend the ontology to concepts of neuroscience in general.

2. Materials

The creation of a scientific ontology requires specialized personnel, hardware, software, and sources of authoritative information.

2.1. Personnel

A *curator* performs the scholarly, low-technology work of identifying the concepts and nomenclature of the knowledge domain, establishing the relations among concepts and recording the pertinent information in computerized forms that transmit it to appropriate locations in the database. A *database manager/programmer* performs the high-technology work of (1) creating input forms that allow the curator to enter information with maximal accuracy, completeness, and efficiency; (2) developing and maintaining a concept-oriented database; and (3) programming retrieval routines that respond to Web queries by the display of information that is precise, accurate, and complete. An *image database manager/programmer* develops software for the creation of digital canonical atlases, overlays, and Web-display routines necessary for the illustration of anatomical concepts that define the structural names in the ontology.

2.2. Software and Hardware

The NeuroNames ontology and the BrainInfo portal that it subserves have been developed in a configuration that includes a Symbolic Database, a Spatial Database, a Map Server, and an extensive Java (5.0) applet and user interface run by the visitor's Web browser on the visitor's computer. The Symbolic Database contains NeuroNames, the input forms for populating the databases, and most of the text and numeric information about concepts. The Spatial Database provides visual illustrations of concepts. The Map Server receives queries from the Java applet over the Web and uses the databases to produce and return appropriate Web displays or to send the visitor to pages in other Web sites to answer the query.

The Symbolic Database is an MS-SQL database on a Dell PowerEdge 1800 server (Intel Xeon 3.4-GHz CPU, 4 GB RAM, 136 GB storage, running MSWindows Server 2003; the database occupies less than 1 GB storage). The Spatial Database consists of ESRI GIS shapefiles (a standard Geographic Information Systems format) on a Dell PowerEdge SC-1420 server (Intel Xeon 3-GHz CPU, 2 GB RAM, 136 GB storage, running MSWindows Server 2003; the database occupies less than 5 GB storage). The BIMS server (Visual Basic 6.0, integrated with ESRI MapObjects APIs) runs on the SC-1420. Assorted .aspx, .jsp, .gif, .jpg, and .html files also reside in the front end of the Web site. Another SC-1420 is used as a development platform, and several other workstations are networked with the servers to enable curators to populate the Symbolic and Spatial Databases.

2.3. Sources of Authoritative Information

A scientific ontology must contain a comprehensive set of term definitions. NeuroNames currently consists of about 2500 core concepts represented by 15,000 neuroanatomical terms. Each term is accompanied by citation of at least one source that uses the term in the sense recorded. Most of the terms in the current version of NeuroNames were obtained from neuroanatomical textbooks, atlases, reviews, research articles, and translations thereof. In the BrainInfo database, the source is recorded in standard reference format and linked to the individual term in the database. It is also linked to a text description of the publication which can be an abstract in PubMed, a description written by the curator, or a page in the publisher's Web site that describes the publication. Some sources are other neuroscience Web sites, in which case there is a hyperlink to the page where the term is defined or used. The citations for the terms representing a concept provide the operational definition for that concept.

In a few cases, NeuroNames itself is cited as the source of a term. This occurs when (1) in order to complete a Latin nomenclature to parallel the English key names for structures in the classical brain hierarchy, it is necessary to construct a Latin name; or (2) it is necessary to standardize terminology across sources that use the same term to refer to different entities. In NeuroNames, we adopted a convention of referring to cortical areas as "area [author's acronym] of [author]-[year of publication]," e.g., "area 9 of Brodmann-1909." Although NeuroNames is thus occasionally cited as the source of a term, it is never the source of a concept.

The Neuroscience Information Framework, supported by the NIH Blueprint for Neuroscience Research and endorsed by the Society for Neuroscience, is conducting a series of workshops in which expert scientists representing different subdomains of neuroscience assemble and organize the terminology used in different areas of interest. As we extend NeuroNames from neuroanatomy into other areas of neuroscience, we plan to incorporate those terminologies into NeuroNames citing products of the workshops as the authoritative sources.

3. Methods

3.1. Define the Domain

The first step in creating an ontology is to define the *knowledge domain* and the set of concepts to which the vocabulary belongs. In our case, the first domain was neuroanatomy; the concepts were brain structures from the level of the whole brain down to the level of primary structures. *Primary structures* were defined as the smallest structures in the brain that are commonly designated by names unique within the domain of neuroanatomy. Lobes, gyri, tracts, and nuclei have, for the most part, unique names and are handled as primary structures. Their substructures, layers and subnuclei, tend to have names like "molecular layer" or "densocellular part." Within the domain of neuroanatomy, these are ambiguous terms whose meanings are only clear if the reader or listener knows the structure where they are located. The location is communicated either by extension of the name to include the parent structure or by context. Such terms include, for example, the "molecular layer of the dentate gyrus" or "densocellular part of the medial dorsal nucleus." Excluding these substructures, the initial domain captured by NeuroNames was the terminology of gross and microscopic neuroanatomy of the brain down to the level of gyri, tracts, and nuclei.

3.2. Collect the Terms

The set of names that represent concepts in a domain is the nomenclature of the domain. The structure names in the first version of NeuroNames ***(6)*** were assembled by going through the Encephalon section of *Nomina Anatomica* ***(14)***, the most authoritative compilation of classical neuroanatomical terminology. Terms from the *Nomina* were supplemented extensively from several of the most comprehensive neuroanatomical textbooks ***(15–17)***, atlases of the human ***(18–20)*** and macaque brain ***(21–28)***, and Anthoney's explication of ambiguities in neurological terminology ***(29)***. Because of synonyms, this survey yielded roughly six times as many names as the number of volumetric structures of the primate brain as well as additional names for about 390 superficial structures. Subsequently, we have integrated terms from more than 100 other sources in seven languages and expanded species to include the rat and mouse to reach the current numbers of 2500 concepts and 15,000 terms.

3.3. Define the Relations of Terms to Concepts

The feature of a scientific ontology that distinguishes it from others, such as the ontologies underlying some psychiatric models, metaphysical philosophies, and religions, is that in a scientific ontology all terms represent operationally defined concepts. For a neuroanatomic ontology, the operation is usually visual examination of a portion of the brain prepared in some replicable manner. Most structures of classical neuroanatomy are defined by visual observation of the dissected brain or by microscopic examination of formalin-fixed, Nissl-stained histological sections.

The definition of scientific concepts is the job of the research scientist who grapples with the problems of identifying significant entities and causal relationships among them. The scientist conceptualizes, defines, and communicates concepts of physical, neuroanatomic entities in text and images. With the advent of computerized knowledge bases, the text and images are increasingly stored in digital information management systems. An important job of the research informaticist is to codify in an ontology the relations of words to concepts, and of concepts to each other, for the indexing, storage, and retrieval of information from digital repositories. An ontology is the product of the informaticist's codification. We have adopted a definition of a *neuroanatomical ontology* as "human-readable text and human-recognizable visual images specifying relationships between the words of a nomenclature and the operationally defined structural concepts they represent" ***(3)***, modified from Gruber ***(30)***.

In neuroanatomy, the words that represent structural concepts are nouns or adjective-noun phrases defined by visual images of the structures that the eyes and brains of neuroanatomists lead them to believe exist physically in three-dimensional space. Thus, the next step in the development of the ontology based on NeuroNames was to link the names of structures to images. For neuroanatomical concepts, the most informative and authoritative source of images is a brain atlas in which structures are outlined (segmented), so that every point in the brain is identified unambiguously with one and only one named structure ***(13)***. In a comprehensive brain atlas, the index, which links the names of structures to the pages of the atlas that show each structure, is a hard-copy neuroanatomical ontology of the species and developmental stage the atlas represents. Drawing terms from NeuroNames (the nomenclature) and images from Primate Brain Maps (our atlas of the macaque brain), we constructed the ontology of adult macaque neuroanatomy that is the basic reference tool for macaque brain data in BrainInfo. The first version of the NeuroNames/BrainInfo ontology was stored in a database where each structure name was linked to images of the structure for display on BrainInfo ***(3,24)***.

3.4. Create a Default (Controlled) Vocabulary

Most concepts have several names, one of which is selected to represent the concept in a given knowledge repository. The set of names selected for use in a given repository is a controlled vocabulary. A *controlled vocabulary* is a limited ontology in which each concept is represented by one name. The NeuroNames ontology includes multiple nomenclatures, many of which are used by others as controlled vocabularies for specific atlases and databases. A controlled vocabulary is a practical necessity for systems, including BrainInfo, that take an ontologic approach to indexing information. Thus, we have extracted from the total NeuroNames vocabulary a set of terms, one name per concept that is used to compose text definitions and label images in BrainInfo and to index information at other Web sites, to which BrainInfo serves as a portal. The most important criterion for selection of terms for a controlled vocabulary is that each term be familiar to users of the system where it is to be applied. To avoid the misconception that we regard the set of terms extracted from NeuroNames for BrainInfo as superior to vocabularies used in other systems, we call it the *default vocabulary* of BrainInfo.

The BrainInfo default vocabulary has two major functions. First, it provides a unique, human-recognizable label by which the information about a given concept can be indexed. In this respect, terms in the default vocabulary are the equivalent, for the human user, of the unique numeric IDs assigned to concepts for the benefit of the computer. They are essential for unambiguous storage and retrieval of information about particular concepts. Second, the default vocabulary provides a complete, nonredundant set of terms for the curator to use in writing neuroanatomical text for BrainInfo visitors who seek information about the concept. It minimizes the size of the vocabulary that visitors must know to use the Web site.

The default vocabulary of BrainInfo is a set of key names, one per concept, that curators use to reference concepts within BrainInfo. The key name is selected from the set of English names accumulated in NeuroNames for the concept (*see* **Subheading 3.2**). Briefly, the criteria for the selection of a key name are (1) avoid eponyms, e.g., use "hippocampus," not "Ammon's horn"; (2) avoid awkward Latin spellings, e.g., use "locus ceruleus," not "locus coeruleus"; (3) avoid duplication of terms–if the same name is used for two different structures, select a different name for one of them; (4) avoid numeric designations, e.g., use "superior frontal gyrus," not "Gyrus F1"; (5) short is better than long, e.g., use "amygdala," not "amygdaloid nuclear complex" ***(7,8)***; and (6) recently, consider the relative frequency of use as measured by the number of abstracts retrieved when the term is submitted to PubMed ***(31)***. Every English and Latin term was submitted to PubMed for the 25-year interval (1976–2000), and the number of abstracts retrieved was used as an additional criterion in the selection of default terms. For example, "sixth cranial nerve" is a term used in 104 abstracts retrieved by PubMed; the synonym "abducent nerve" retrieves 64, and the synonym "abducens nerve" retrieves 1270; so, "abducens nerve" was selected as the key name for BrainInfo's default vocabulary.

Actually, BrainInfo has four default vocabularies. (1) The default vocabulary described above is a set of Default Long Names (*see* **Table 1**, point 6) that

Table 1
Coding Attributes of a NeuroNames Topic (Paraventricular Nucleus of the Thalamus) in XTM

Concept or attribute	Description and example	Representative extract of XTM code
1—A structural concept is a *Topic* in the Topic Map of Neuro Names.	A unique character string within NeuroNames' subdomain "neuroanatomy"; in most cases, the topic id is based on the Default Short Name (*see* point 7) with a suffix, if necessary, to make it unique; the *topic id* contains no spaces, for use in Web addresses	<topic id="ParaventricularNucleusTh">
2—The location of defining information about the structure is the *subjectIdentity* of the Topic	The address of an authoritative definition of the structure is represented as a *subjectIdentity*, which may refer to a page in a conventional publication or the URL of a page in a Web site	<subjectIdentity> <subjectIndicatorRef xlink:href="http://braininfo.rprc.washington.edu/Scripts/hiercentraldirectory.aspx?ID=291&toback=0"> </subjectIdentity>
3—Species and theoretical models in which the anatomical structure is found are *Scopes* of the Topic	Species, such as "human" and "mouse," and models, such as "classical brain hierarchy" and "stress circuit" are represented as *scopes* of the Topic	<scope> <topicRef xlink:href=#human/> </scope> <scope> <topicRef xlink:href=#classical brain hierarchy/> </scope>

(Continued)

Table 1
(*Continued*)

Concept or attribute	Description and example	Representative extract of XTM code
4—The relation of the structure to a model is specified by an *association id*	Within the scope of the model "classical hierarchy," the relation "paraventricular nucleus of the thalamus" is a substructure *member* of an *association* in which the "median nuclear group" is the superstructure *member*	<association id= "ParaventricularNucleusTh_MedianNuclearGroup_ association"> <member> <roleSpec> <topicRef xlin:href="substructure" </roleSpec> <topicRef xlink:href="# Paraventricular NucleusTh"/> <roleSpec> <topicRef xlin:href="superstructure" </roleSpec> <topicRef xlink:href="# MedianNuclear Group"/> <member> </association>
5—Each name of the structure is a *baseName* scoped for its language	A baseName is any word, multi-word term, acronym, or other character string used by authors in any language to refer the concept in text or images in any medium. The example is "Nucleus paraventricularis thalami," scoped as a Latin name for the structure	<baseName> <baseNameString>Nucleus paraventricularis thalami</baseNameString> <scope> <topicRef xlink:href=#la/> </scope> </baseName>

6—One of the names is scoped as the *Default Long Name* for use in BrainInfo	The Default Long Name is a name selected from English names for the concept that is used in situations where a term unique within the NeuroNames domain "neuroanatomy" is needed, for example, in writing a template for natural language definitions. For this structure, it is "paraventricular nucleus of the thalamus"	<baseName> <baseNameString>paraventricular nucleus of the thalamus</baseNameString> <scope> <topicRef xlink:href=#en/> </scope> <scope> <topicRef xlink:href=#DLoName/> </scope> </baseName>
7—Another name is scoped as the *Default Short Name* for use in BrainInfo	The Default Short Name is a shorter term that is not necessarily unique within the domain of "neuroanatomy" but is unambiguous in situations where the context is known, for example, in a natural language definition where the long name (*see* point 6) has appeared earlier. If the reader already knows that the structure is in the thalamus, not the hypothalamus, "paraventricular nucleus" suffices	<baseName> <baseNameString>paraventricular nucleus </baseNameString> <scope> <topicRef xlink:href=#DShName/> </scope> </baseName>

(Continued)

Table 1
(*Continued*)

Concept or attribute	Description and example	Representative extract of XTM code
8—A third "name" is scoped as the Default Acronym for use in BrainInfo	The Default Acronym is a very short character string used to label illustrations and as shorthand for the concept in text. It is unique within NeuroNames domain "neuroanatomy." (Acronyms are treated as a language in NeuroNames.) Thalamic nuclei are commonly referred to by their acronyms, in this case "PV"	<baseName> <baseNameString>PV</baseNameString> <scope> <topicRef xlink:href=#ac/> </scope> <scope> <topicRef xlink:href=#DAcronym/> </scope> </baseName> </topic>

meet most needs within the Web site for unique symbolic representation of structures in text definitions and tables. The key names in BrainInfo are the Default Long Names of the structures. As the Default Long Names must be unique within the domain of neuroanatomy, some are too long to be used repeatedly in a given passage of text. (2) To avoid awkward repetition of such terms, we have designated a Default Short Name (*see* **Table 1**, point 7) for each concept. These are not necessarily unique. For example, the Short Name of "paraventricular nucleus of the hypothalamus" is "paraventricular nucleus." "Paraventricular nucleus" by itself is ambiguous, because there is also a paraventricular nucleus in the hypothalamus. The Long Name is used the first time the concept is mentioned in a passage. Subsequent references to the concept in the same context use the Default Short Name. (3) BrainInfo also has a default vocabulary of unique acronyms (*see* **Table 1**, point 8) for labeling atlas images and text references to structures whose Short Names are themselves long. Many of these are common and easily recognized. For example, "VPL" is more commonly used than "ventroposterolateral nucleus." (4) Finally, as we convert NeuroNames to a concept-oriented database (*see* **Subheading 3.7**), we must create a unique name, a "topic id" for each concept in Topic Maps format. The distinguishing feature of the topic id, which may otherwise be identical to the default Long Name, is that spaces between words are eliminated to allow use in Web addresses (*see* **Table 1**, point 1).

3.5. Link Concepts to Operational Definitions

Whereas, in any ontology, relating names to concepts is essential, in a scientific ontology relating the concepts to operational definitions is equally so. Neuroanatomists define structural concepts in two ways, symbolically by text definitions and spatially by reference to location in canonical atlases. Thus, the defining element of the NeuroNames anatomic ontology is the link between the key name for the concept and its illustration in a canonical atlas. Just as there exist, in addition to the key name, a number of other legitimate names for a given structural concept, there exist, in addition to the defining images of a canonical atlas, a number of illustrations (instances) of the structure. Currently, BrainInfo illustrates hundreds of structures by linking key names to images, either within BrainInfo or at other Web sites. The images illustrate homologous structures in several species, including the human–Digital Anatomist ***(32)***, Atlas of the Human Brain ***(33)***, Medical Neuroscience ***(34)***, and Whole Brain Atlas ***(35)***; the macaque–BrainInfo ***(4)***, BrainMaps.org ***(36)***, and LONI ***(37)***; and the mouse–Mouse Brain Library ***(38)*** and Allen Brain Atlas ***(39)***.

The features of a brain atlas that make it suitable as a canonical reference are described elsewhere ***(3)***. Most important is that structures be defined by boundaries in a comprehensively segmented representation of the brain that allows unambiguous mapping of data in relation to classical neuroanatomic landmarks. Several brain atlases suitable for use as canonical atlases are available on compact disk ***(10,33,40–42)***. In addition, templates for mapping macaque data for publication can be downloaded from the BrainInfo Web site ***(43)***.

3.6. Link Concepts to Concepts

Relations among concepts are defined in *models*. Two kinds of model predominate in neuroanatomy: hierarchies and systems ***(3)***. The simpler model is the *hierarchy*. In the classical neuroanatomical hierarchy, for example, the brain divides into forebrain, midbrain, and hindbrain, which, in turn, subdivide through as many as nine levels down to the primary structures that constitute the basic uniquely named structures of the brain (*see* **Subheading 3.1**). In NeuroNames, data regarding hierarchical relationships are stored by specifying each hierarchy to which the concept belongs, such as the "classical brain hierarchy," and its parent within that hierarchy. For example, the parent of "superior frontal gyrus" in the classical hierarchy is "frontal lobe."

We distinguish two kinds of hierarchy, partitive hierarchies and categorical hierarchies. In a *partitive hierarchy*, the "children" of a given "parent" constitute a closed set; the sum of the parts equals the whole. The Hierarchical Index in Primate Brain Maps ***(10)*** is an example. There every point in the brain belongs to one and only one primary structure, and the primary structures are successively grouped into superstructures such that the whole brain is represented at every level of the hierarchy. Partitive hierarchies are most useful in quantitative neuroanatomy. When concepts are related in a partitive hierarchy, the value of an attribute of a higher level concept is the sum of values of that attribute across the lower level concepts; for example, the number of pyramidal cells in the superstructure hippocampus is the sum of the numbers in its substructures, areas CA1, CA2, and CA3.

In a *categorical hierarchy*, the children of a given parent constitute an open set; that is, any number of concepts can be accommodated. An example of a categorical hierarchy is a model that categorizes cells into neurons and glia and neurons into bipolar cells, pyramidal cells, and so on. If one discovers a new kind of neuron, one simply adds it to the concept of neurons. There is no need, as in the partitive hierarchy, to redefine the components of the superordinate concept to accommodate the new concept. By the same token, one cannot assume a categorical hierarchy to be comprehensive.

The concepts in hierarchical models are unirelational; the entire hierarchy can be reconstructed by specification of a single relation for each entity. This is achieved in the brain hierarchy by specification of the structure to which the given structure is subordinate. *Systemic models* differ from hierarchical models in that multiple kinds of relations can be specified among concepts in the system. The most common systemic model in neuroanatomy is a system of neuronal pathways. Combinations of pathways represent the anatomically overlapping circuits and networks of structures that mediate a physiological or behavioral function. The basic unit of such *connectivity (hodological) models* is a connection between two structures. Connections are specified by a structure of origin, a pathway, and a target. CoCoMac ***(44,45)***, a Web database of connectivity in the macaque brain, and BAMS ***(5)***, a database of connectivity in the rat, are based on systemic, connectivity models of brain structure.

3.7. Create a Concept-Oriented Database

The six steps above (*see* **Subheadings** 3.1–3.6) must be implemented in a database of some kind. The development of NeuroNames has proceeded through three types of database. The first was populated in Apple/Macintosh HyperCard, the first pre-Web software for desktop computers that enabled a computer-naïve neuroscientist to create data-entry forms and mixed text/image displays with navigation by hypertext ***(6)***; the database was distributed in the form of HyperCard "stacks" on floppy disks. In 1998, NeuroNames was ported to a Microsoft ACCESS database that was interfaced to GIS software (*see* **Subheading 2.2.**) to allow dissemination over the Web ***(24)***. In 2004, it was ported to an SQL database, which offered increased capacity, enhanced security, and greater ease of creating and modifying Web pages.

The databases supporting most Web portals to neuroscientific information are organized according to a "library model" in which the entities are the locations of database resources ***(48)*** rather than concepts. Metadata as to the kinds of data stored at the sites are coded as attributes of the resource. Typically, a user addresses a query in the form of a character string to the portal, and the portal provides a hyperlinked list of resource Web sites where the data may exist. Users go to a Web site, consult site maps, or navigate series of menus in search of information about the entity of interest. In such a file structure, concepts are treated as attributes of locations in the database. Users must know, or learn through interaction, first with the portal and then with each Web site, the local terminology by which information is indexed. The user must then navigate to files where it may be. If it is not there, the user must back out and navigate elsewhere.

To establish interoperability among databases on the Web requires the development of concept-oriented mediators to translate among the controlled vocabularies of the sites that need to communicate. Currently, we are developing software for translation from our SQL database into Topic Maps (XTM), an XML-based formal standard ***(46)*** particularly suited to indexing information by concept rather than by resource. In a concept-oriented database, locations are coded, along with definitions, illustrations, names, and so on, as attributes of concepts. Some of the XTM tags for attributes of concepts in NeuroNames are described in **Table 1** (column 3). Information is organized by concept (topic) rather than by character string (name) or resource (location) ***(47,48)***.

For our purposes, Topic Maps promises three major advantages over other XML derivatives ***(49)***. (1) By straightforward accommodation of multiple names for the same concept, XTM enables a user to query a knowledge source using any legitimate name (synonym) for the concept; users of the system do not have to explore a controlled vocabulary for a relevant term to initiate a search. (2) By coding for context, XTM eliminates the "homonym problem." Terms that have more than one meaning, such as "paraventricular nucleus," are coded for a concept, or meaning, within a context ("topic" within a "scope"). Thus, again, in contrast to systems where search is based on a controlled vocabulary, the user can post a query using a term that has multiple legitimate meanings, such as "paraventricular nucleus"; the system notes the ambiguity and questions the user as to which concept he or she has in mind. This resolves, before the search begins, an ambiguity that accounts for large numbers of false-positive errors in string-based searches. (3) Ontologies stored in XTM databases are more readily merged than ontologies coded in other systems that are currently available ***(48)***. XTM can incorporate an ontology from a new subdomain into a more comprehensive, existing ontology without requiring content experts from both systems to meet and negotiate relations among terms and concepts. Concepts in the domain ontology are defined operationally with sufficient specificity that a content expert from a new database can determine individually the concepts that have equivalents in the existing domain ontology, add concepts to the domain ontology as necessary, and merge the new ontology into the domain ontology with minimal necessity for personal interaction. We believe that XTM will remove the single greatest sociological obstacle to interoperability among large databases, namely, the difficulty of assembling individuals with the necessary expertise to accomplish a task that, for most experts, is inherently onerous.

3.8. Dissemination

Neuroinformatics is an applied science. Like the products of engineering or medicine, the products of a neuroinformatics effort are of little value if they are not used. Thus, dissemination of the product is a critical aspect of methodology in ontology development. The NeuroNames ontology aids dissemination of neuroanatomical information in several ways. The most powerful mode of dissemination is through BrainInfo, both as a primary knowledge base of information about the macaque brain and as a portal to neuroanatomical information at some 50 other sites on the Web. Visitors to BrainInfo look up the meanings of thousands of neuroanatomical terms in several languages and explore an atlas of the macaque brain for the names of structures unfamiliar to them (*see* **Fig. 1**).

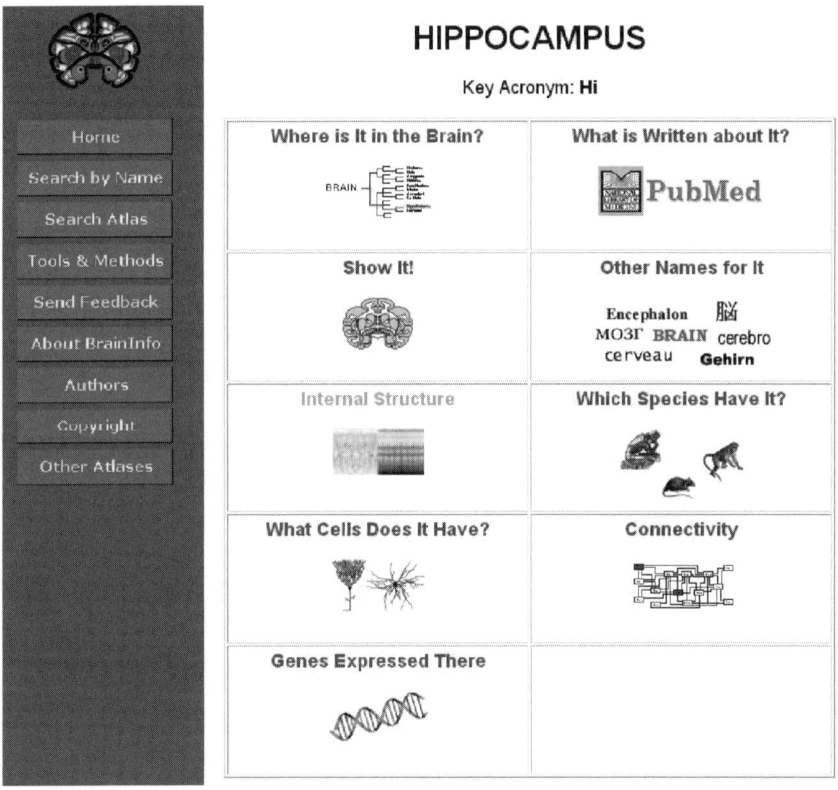

Fig. 1. The Central Directory for the hippocampus offers visitors the opportunity to obtain many kinds of information about the structure.

They find synonyms and text displays of more than 2500 concepts, many of which are illustrated by images at the BrainInfo site or by hyperlink to pages in other Web sites that show the structure. In 2006, more than 120,000 unique visitors, 550 per day, visited an average of 25,000 pages a day.

In addition to looking up the definitions of specific structures, visitors to BrainInfo can download the neuroanatomic subontologies of particular species based on the brain atlases most commonly used for mapping data in the mouse ***(39,41,50)*** and the macaque ***(8)***. Managers of Web sites with controlled vocabularies based on one of those atlases use the subontologies to establish interoperability with BrainInfo. They obtain a listing of neuroanatomical structures in the nomenclature of the atlas, the key names of those structures in NeuroNames, and the Web addresses of the central directories to information about those structures in BrainInfo. (See NeuroNames Ontologies ***(43)***/Home Page.) We anticipate that translation of NeuroNames into XTM will enhance its algorithmic accessibility as a shared, canonical neuroanatomical knowledge source.

A final mode of distribution of the classical NeuroNames brain hierarchy is through the Unified Medical Language System (UMLS) of the U.S. National Library of Medicine ***(51)***. The UMLS is an amalgamation of vocabularies used in various areas of biology and medicine. By codifying relationships among concepts in different "source vocabularies," developers of the Metathesaurus of the UMLS are creating a master, unified lexicon of the biomedical sciences. The Metathesaurus of the UMLS establishes synonymy and new relationships between concepts and terms from the different source vocabularies. It strives to do so in a way that preserves the meanings, attributes, hierarchical connections, and other relationships among terms. An early version of NeuroNames was incorporated into the Metathesaurus in 1999. It is available for download by research groups and institutions wishing to use it as a medium for indexing and exchanging neuroanatomical data.

Acknowledgments

The authors express their thanks to William Bug, Senior Analyst and Ontological Engineer at Drexel University, for a generous and constructive critique of this chapter in its penultimate form; to Evan Song who has contributed most to development of the computerized forms and other routines necessary to populate the NeuroNames ontology in the SQL database; to Ioulia Kocheleva, curator, for accurate and efficient establishment of hyperlinks between BrainInfo and other neuroscience Web sites; and to Erik McArthur for skilled processing of neuroanatomic images to populate BrainInfo itself. This work is supported by grants LM-008247, MH69259, and RR-00166 from the U.S. National Institutes of Health to the University of Washington, Seattle.

References

1. OED (2005) *Oxford English Dictionary* (http://dictionary.oed.com).
2. Google (2003) Google Website (http://google.com); General Review (http://www.searchengineshowdown.com/features/google/review.html); Technical Review (http://www.google.com/technology/pigeonrank.html).
3. Bowden, D. M. and Dubach, M. (2005) Neuroanatomical nomenclature and ontology, in *Databasing the Brain* (Koslow, S. H. and Subramaniam, S., eds.), John Wiley and Sons, Hoboken, NJ, pp. 27–45.
4. BrainInfo (2002) BrainInfo, National Primate Research Center, University of Washington, Seattle, WA (http://braininfo.rprc.washington.edu).
5. BAMS (2003) Brain Architecture Management System, University of Southern California, Los Angeles, CA (http://brancusi.usc.edu/bkms).
6. Martin, R. F., Dubach, J., and Bowden, D. M. (1990) NeuroNames: human/macaque neuroanatomical nomenclature, in *Fourteenth Annual Symposium on Computer Applications in Medical Care: Standards in Medical Informatics* (Miller, R. A., ed.), American Medical Informatics Association, Washington, DC, pp. 1018–1019.
7. Bowden, D. M. and Martin, R. F. (1995) NeuroNames brain hierarchy. *Neuroimage* 2:63–83.
8. Bowden, D. M. and Dubach, M. F. (2002) NeuroNames 2002. *Neuroinformatics* 1:43–60.
9. Martin, R. F. and Bowden, D. M. (1996) A stereotaxic template atlas of the macaque brain for digital imaging and quantitative neuroanatomy. *Neuroimage* 4:119–150.
10. Martin, R. F. and Bowden, D. M. (2000) *Primate Brain Maps: Structure of the Macaque Brain*, Elsevier Science, Amsterdam, The Netherlands.
11. Paxinos, G. and Watson, C. (1998) *The Rat Brain in Stereotaxic Coordinates*, 4th Edition, Academic Press, San Diego, CA.
12. Swanson, L. W. (1998) *Brain Maps: Structure of the Rat Brain – A Laboratory Guide with Printed and Electronic Templates for Data, Models and Schematics*, Elsevier, Amsterdam, The Netherlands.
13. Bowden, D. M. and Martin, R. F. (1997) A digital rosetta stone for primate brain terminology, in *Handbook of Chemical Neuroanatomy: The Primate Nervous System*, Part 1, Vol. 13 (Bloom, F. E., Bjorklund, A., and Hokfelt, T., eds.), Elsevier Science B.V., Amsterdam, The Netherlands, pp. 1–35.
14. International Anatomical Nomenclature Committee (1983) *Nomina Anatomica*, 5th Edition, Williams & Wilkins, Baltimore, MD.
15. Carpenter, M. B. and Sutin, J. (1983) *Human Neuroanatomy*, Williams & Wilkins, Baltimore, MD.

16. Crosby, E. C., Humphrey, T., and Lauer, E. W. (1962) *Correlative Anatomy of the Nervous System*, Macmillan, New York.
17. Paxinos, G. (ed.) (1990) *The Human Nervous System*, Academic Press, San Diego, CA.
18. Riley, H.A. (1943) *An Atlas of the Basal Ganglia, Brain Stem and Spinal Cord*, Williams & Wilkins, Baltimore, MD.
19. Roberts, M. P., Hanaway, J., and Morest, D. K. (1987) *Atlas of the Human Brain in Section*, Lea & Febiger, Philadelphia, PA.
20. Ono, M., Kubik, S., and Abernathey, C. D. (1990) *Atlas of the Cerebral Sulci*, Georg Thieme Verlag, Stuttgart, Germany.
21. von Bonin, G. and Bailey, P. (1947) *The Neocortex of Macaca mulatta*, University of Illinois Press, Urbana, IL.
22. Krieg, W. J. S. (1975) *Interpretive Atlas of the Monkey's Brain*, Brain Books, Evanston, IL.
23. Kusama, T. and Mabuchi, M. (1970) *Stereotaxic Atlas of the Brain of Macaca fuscata*, University Park Press, Baltimore, MD.
24. Bowden, D. M. and Dubach, M. F. (2002) BrainInfo, in *A Practical Guide to Neuroscience Databases and Associated Tools* (Koetter, R., ed.), Kluwer Academic Publishers, Norwell, MA, pp. 259–273.
25. Oertel, G. (1969) Zur Zyto und Myeloarchitektonik des Rhombencephalon des Rhesusaffen (Macaca mulatta Zimmerman). *J. Hirnforsch.* 11:377–405.
26. Shantha, T. R., Manocha, S. L., and Bourne, G. H. (1968). *A Stereotaxic Atlas of the Java Monkey Brain (Macaca irus)*, S. Karger, Basel, Switzerland.
27. Szabo, J. and Cowan, W. M. (1984) A stereotaxic atlas of the brain of the cynomolgus monkey (Macaca fascicularis). *J. Comp. Neurol.* 222:265–300.
28. Olszewski, J. (1952) *The Thalamus of the Macaca mulatta. An Atlas for Use With the Stereotaxic Instruments*, S. Karger, Basel, Switzerland.
29. Anthoney, T. R. (1994) *Neuroanatomy and the Neurologic Exam: A Thesaurus of Synonyms, Similar-Sounding Non-Synonyms, and Terms of Variable Meaning*, CRC Press, Boca Raton, FL.
30. Gruber, T. (1993) What is an ontology? (http://www-ksl.stanford.edu/kst/what-is-an-ontology.html).
31. PubMed (2003) PubMed, National Library of Medicine, Bethesda, MD (http://www.ncbi.nlm.nih.gov/PubMed).
32. DA (2002) Digital Anatomist Project, University of Washington, Seattle, WA (http://www9.biostr.washington.edu/da.html).
33. Mai, J., Assheuer, J., and Paxinos, G. (1998) *Atlas of the Human Brain*, Book With CD-ROM, Academic Press, Elsevier Science, Amsterdam.
34. Medical Neuroscience (2001) Medical Neuroscience, Loyola University, Chicago, IL (http://www.meddean.luc.edu/lumen/MedEd/Neuro/Neuro.html).
35. WBA (1999) Whole Brain Atlas, Harvard Medical School, Boston, MA (http://www.med.harvard.edu/AANLIB/cases/caseM/case.html).

36. BrainMaps.org (2005) Brain Atlas, University of California, Davis, CA (http://www.brainmaps.org).
37. LONI (2002) Laboratory of Neuroimaging, University of California, Los Angeles, CA (http://www.loni.ucla.edu/).
38. MBL (2002) Mouse Brain Atlas, Mouse Brain Library, University of Tennessee, Memphis, TN (http://www.nervenet.org/mbl/atlas170/atlas170_frame.html).
39. AIBS (2005) Allen Reference Atlas, Allen Institute of Brain Science, Seattle, WA (http://www.brainatlas.org).
40. Nowinski, W. L., Thirunavuukarasuu, A., and Benabid, A. L. (2005) *The Cerefy Clinical Brain Atlas on CD-ROM*, 2nd Edition, Thieme New York, New York.
41. Paxinos, G. and Franklin, B. J. (2000) *The Mouse Brain in Stereotaxic Coordinates*, 2nd Edition With CD-ROM, Elsevier Science, Amsterdam, The Netherlands.
42. Swanson, L. W. (2004) *Brain Maps III: Structure of the Rat Brain*, 3rd Edition With Electronic Templates for Data, Models and Schematics, Elsevier Science, Amsterdam, The Netherlands.
43. BrainInfo (2002) BrainInfo, Neuroscience Division, National Primate Research Center, University of Washington, Seattle, WA (http://braininfo.rprc.washington.edu).
44. Kamper, L., Bozkurt, A., Rybacki, K., Geissler, A., Gerken, I., Stephan, K. E., and Kötter, R. (2002). An introduction to CoCoMac-Online, in *Neuroscience Databases: A Practical Guide* (Kötter, R., ed.), Kluwer Academic Publishers, Boston, MA, pp 155–169.
45. CoCoMac (2003a) Collations of Connectivity Data on the Macaque Brain, Heinrich Heine University, Düsseldorf, Germany (http://www.mon-kunden.de/cocomac).
46. Park, J. and Hunting, S. (eds.) (2003) *XML Topic Maps: Creating and Using Topic Maps for the Web*, Addison-Wesley, Boston, MA.
47. Newcomb, S. R. (2003) A perspective on the quest for global knowledge interchange, in *XML Topic Maps: Creating and Using Topic Maps for the Web* (Park, J. and Hunting, S., eds.), Addison-Wesley, Boston, MA, pp. 31–50.
48. Freese, E. (2003) Topic Maps and RDF, in *XML Topic Maps: Creating and Using Topic Maps for the Web* (Park, J. and Hunting, S., eds.), Addison-Wesley, Boston, MA, pp. 283–326.
49. Hunting, S. (2003) How to start topic mapping right away with the XTM specification, in *XML Topic Maps: Creating and Using Topic Maps for the Web* (Park, J. and Hunting, S., eds.), Addison-Wesley, Boston, MA, pp. 81–100.
50. Hof, P. R., Young, W. G., Bloom, F. E., Belichenko, P. V., and Celio, M. R. (2000) *Comparative Cytoarchitectonic Atlas of the C57BL/6 and 129/Sv Mouse Brains*, Elsevier, Amsterdam, The Netherlands.
51. Hole, W. T. and Srinivasan, S. (2002) Adding NeuroNames to the UMLS metathesaurus. *Neuroinformatics* 1:61–64.

II

Computational Neuronal Modeling and Simulation

6

Model Structure Analysis in NEURON

Toward Interoperability Among Neural Simulators

Michael L. Hines, Tom M. Morse, and N. T. Carnevale

Summary

One of the more important recent additions to the NEURON simulation environment is a tool called ModelView, which simplifies the task of understanding exactly what biological attributes are represented in a computational model. Here, we illustrate how ModelView contributes to the understanding of models and discuss its utility as a neuroinformatics tool for analyzing models in online databases and as a means for facilitating interoperability among simulators in computational neuroscience.

Key Words: NEURON; ModelView; ModelDB; interoperability; XML; BrainML; CellML; MorphML; NeuroML.

1. Introduction

Sometimes working on a solution for one problem seems to create a new problem or at least reveals a problem that was waiting to be uncovered. Such was the case with the availability of computational models in neuroscience. For many years, a common lament among computational neuroscientists was the difficulty of obtaining the source code used in published models. This constituted a serious barrier to a key element of scientific method: reproducibility.

We and our collaborators have been addressing this problem by developing ModelDB, a WWW-accessible resource designed to serve as a mechanism for attributed reuse of published models *(1,2)*. Based on the EAV/CR schema described in Chapter 3, ModelDB is still very much a neuroinformatics research project, but it is already seeing heavy use by modelers from labs around the world. In the short time since its inception, it has grown to more than 290 entries, most of which are executable.

However, the presence of a large number of runnable, published models in ModelDB brought to light a problem that seems obvious in retrospect: even when source code is available, it can be hard to determine exactly what biological properties are represented in a model. Source code is generally hard to read and understand, because model developers tend to have their own individual programming styles and generally do not place a high priority on presentation when writing code. Furthermore, programs are often organized in such a way that parameters are assigned values at run time that differ from what was specified at the top of the code.

Because most of the entries in ModelDB were written for the NEURON simulation environment *(3,4)*, we decided to make it easier to deal with the problem of understandability by adding a tool to NEURON that analyzes a model and presents a concise, yet complete, graphical and textual summary of its anatomical and biophysical properties. This tool, which is called ModelView, is part of NEURON's graphical user interface, and in this chapter, we describe its main features and present an example of its use by analyzing one of the entries in ModelDB.

2. Materials

1. The NEURON simulation environment, which runs under OS X, UNIX/Linux, and MSWindows *(3)*. Installers, source code, and documentation are available from http://www.neuron.yale.edu/ *(4)*.
2. ModelDB *(1)*, whose home page is http://senselab.med.yale.edu/senselab/modeldb/*(2)*.
3. To illustrate the features of the ModelView tool, we selected a very complex hippocampal Ca1 pyramidal cell model that was developed by Poirazi et al. *(5)* for a study of synaptic integration. We retrieved this model following instructions provided at ModelDB's WWW site. For simplicity, we used ModelDB's "Auto-launch" feature, which downloads the model code and then makes NEURON automatically compile the biophysical mechanisms and execute the model code. This brought up a graphical window that contained three buttons, including one that was labeled "hyperpolarization-current," which we clicked on to start a simulation that showed how the model is affected by an injected hyperpolarizing current. After the end of this simulation, we clicked on the "Init" button in NEURON's RunControl panel to restore the model to its initial condition. At this point, the model cell was ready for analysis.

3. Methods

3.1. ModelView in NEURON

Clicking on NEURON Main Menu/Tools/Model View (*see* **Fig. 1**) brings up a ModelView tool (*see* **Fig. 2**), which initially shows a top-level summary of the model. In NEURON, a cell is represented by one or more sections, each of which consists of one or more computational nodes called segments *(3)*. The first item in the summary ("183 sections; 331 segments") shows the total number of sections and segments for all the cells in the model. The second and final items in the summary are the numbers of real and artificial cells, and density mechanisms and point processes; the latter two include ion channels, synaptic receptors, and electrodes *(3)*. The Poirazi model contains one real cell, several density mechanisms, and one point process.

The main window of the ModelView tool presents a tree-structured, hierarchical textual outline of the properties of a model that can be explored by clicking on any line of text. Clicking on a line that starts with an asterisk will expand or contract the next level of the tree. Clicking on "* 1 real cells," and then on the newly appeared "* root soma[0]," brings up a "shape plot" that shows the cell's anatomical configuration and expands the outline to display

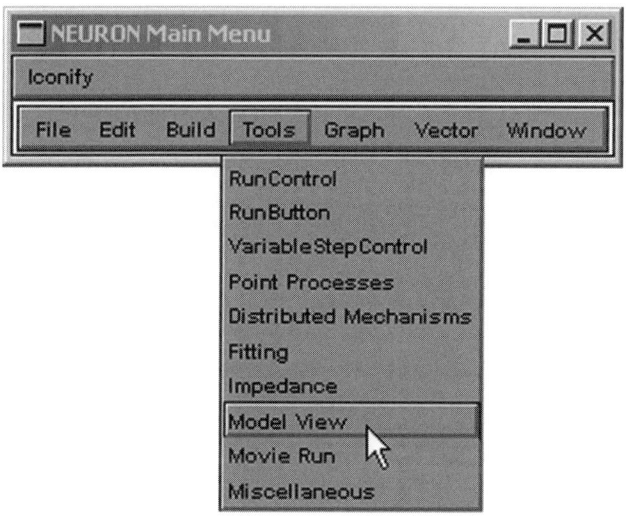

Fig. 1. Click on Tools/Model View in the NEURON Main Menu toolbar to bring up a ModelView tool.

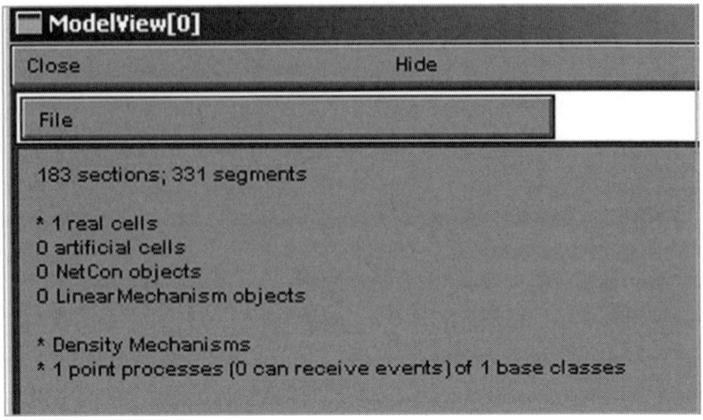

Fig. 2. A newly created ModelView tool displays a top-level summary of the model. Note: This and many of the following images are cropped to focus on regions of particular interest.

a list that reports how many duplicate and unique parameters are distributed throughout the cell (*see* **Fig. 3**). The first three lines of this list are the total number of sections and segments in this particular cell, the related nseg (spatial discretization) parameter distribution, and the number of inserted mechanisms.

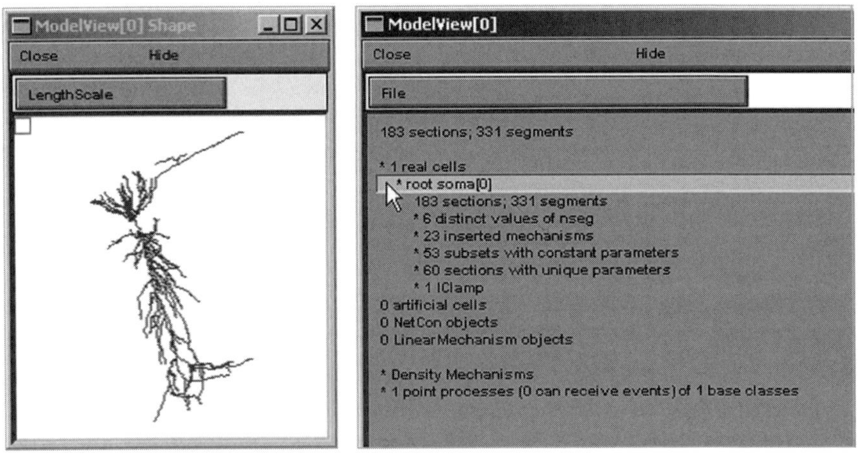

Fig. 3. Clicking on "* 1 real cells" and then on "* root soma[0]" brings up a shape plot of the model cell (left) and expands the outline.

Expanding "* 23 inserted mechanisms" reveals which ion channels are used in the model (*see* **Fig. 4**). Clicking on "* cat" expands ModelView's outline, disclosing that the cat mechanism's conductance density, gcatbar_cat, varies over the cell. It also makes the shape plot redraw the cell so that only sections with the cat mechanism are in red (*see* **Fig. 5**). To see exactly how gcatbar_cat varies with location, we just click on "* 36 gcatbar_cat" to expand the outline; this also brings up a "range graph" window with a top panel that uses a color scale to indicate the spatial variation of gcatbar_cat, and a bottom panel that plots gcatbar_cat as a function of path distance in microns from the soma (*see* **Fig. 6**).

ModelView analyzes the model to find subsets of sections over which one or more parameters have uniform values. In this example, there are 53

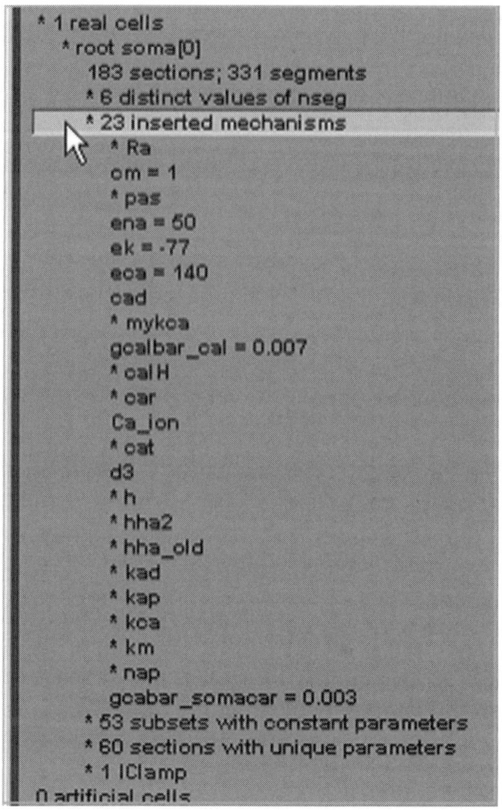

Fig. 4. Clicking on "* 23 inserted mechanisms" expands the outline to reveal which ion channels the model uses.

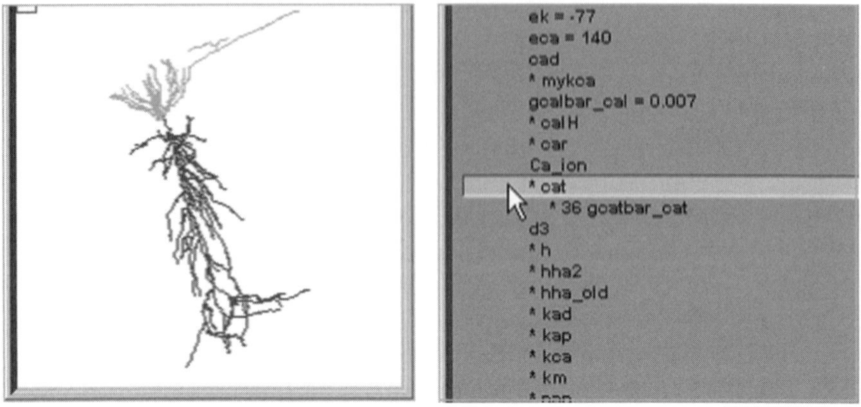

Fig. 5. Selecting "* cat" makes the shape plot draw the sections that contain this current in red (dark here), and shows that its conductance density is nonuniform.

subsets with constant parameters (*see* **Fig. 3**, right). The number of sections that have unique parameter values is also calculated and displayed (60 here). The fact that these numbers are so large indicates that some parameters have heterogeneous values. For a simpler model, the number of subsets with constant parameters, and the number of sections with unique parameters, would help identify the combinations of values that are used and also their spatial distribution. Selecting a parameter from these lists, e.g., "eca = 140," highlights the part of the shape plot where the parameter has that value (*see* **Fig. 7**).

The final top-level groups are Density Mechanisms and Point Processes, which expand into text trees and spawn graphs that show the distributions of these items. Homogeneous Parameters (*see* **Fig. 8**) lists those parameters that occur with just a single value, whereas Heterogenous Parameters lists those that take multiple values. Selecting a homogeneous parameter, such as el_hha2, highlights that part of the shape plot where this parameter exists (*see* **Fig. 9**). If a heterogeneous parameter is selected, such as gkabar_kad, the range graph shows the overall complexity (*see* **Fig. 10**), whereas the ModelView's text and the shape plot can be used to explore the details of the distribution (*see* **Fig. 11**).

The Point Processes group reveals that this model has only one point process: a current clamp. The ModelView browser shows us that this is located at the soma (*see* **Fig. 12**).

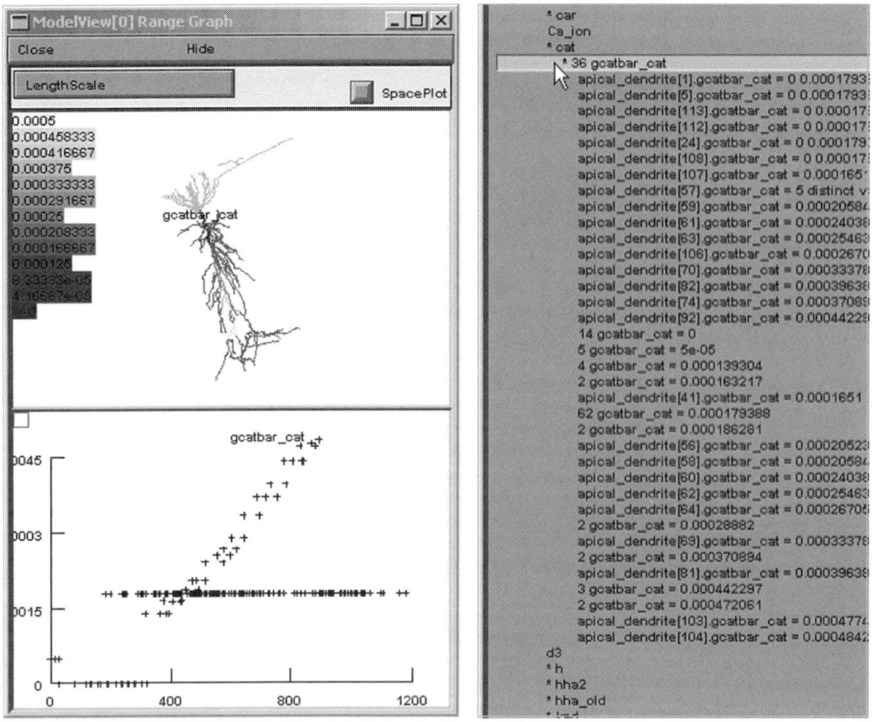

Fig. 6. Clicking on "* 36 gcatbar_cat" brings up a range graph (left) that uses a color scale and graph to convey a sense of the spatial distribution of parameter values. Detailed parameter values are provided in ModelView's expanded outline.

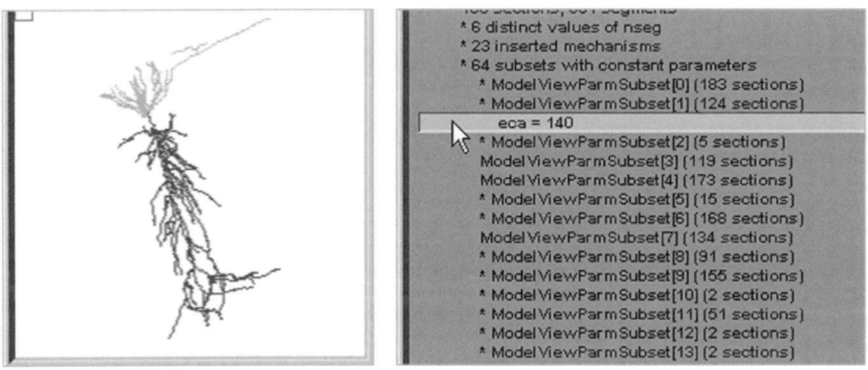

Fig. 7. Browsing the list of subsets with constant parameters allows the discovery of sections in which a parameter has a particular value, e.g., "eca = 140" in this case.

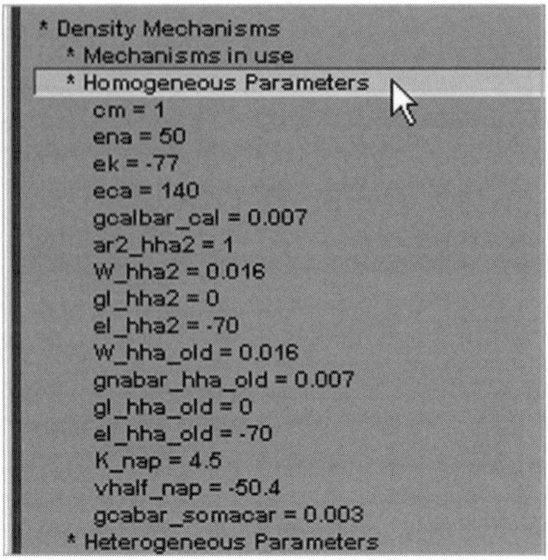

Fig. 8. Clicking on "Homogeneous Parameters" shows which parameters have a single value over the entire model cell.

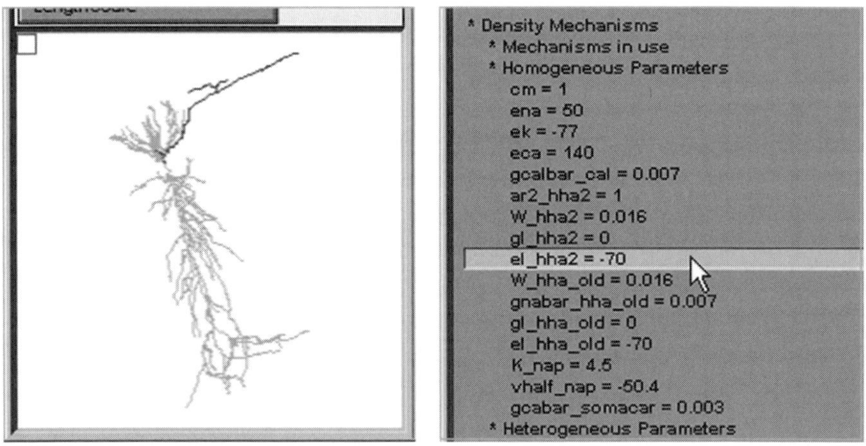

Fig. 9. Selecting "el_hha2 = –70" reveals that this parameter (and the hha2 mechanism) is restricted to what looks like axonal branches of the model cell.

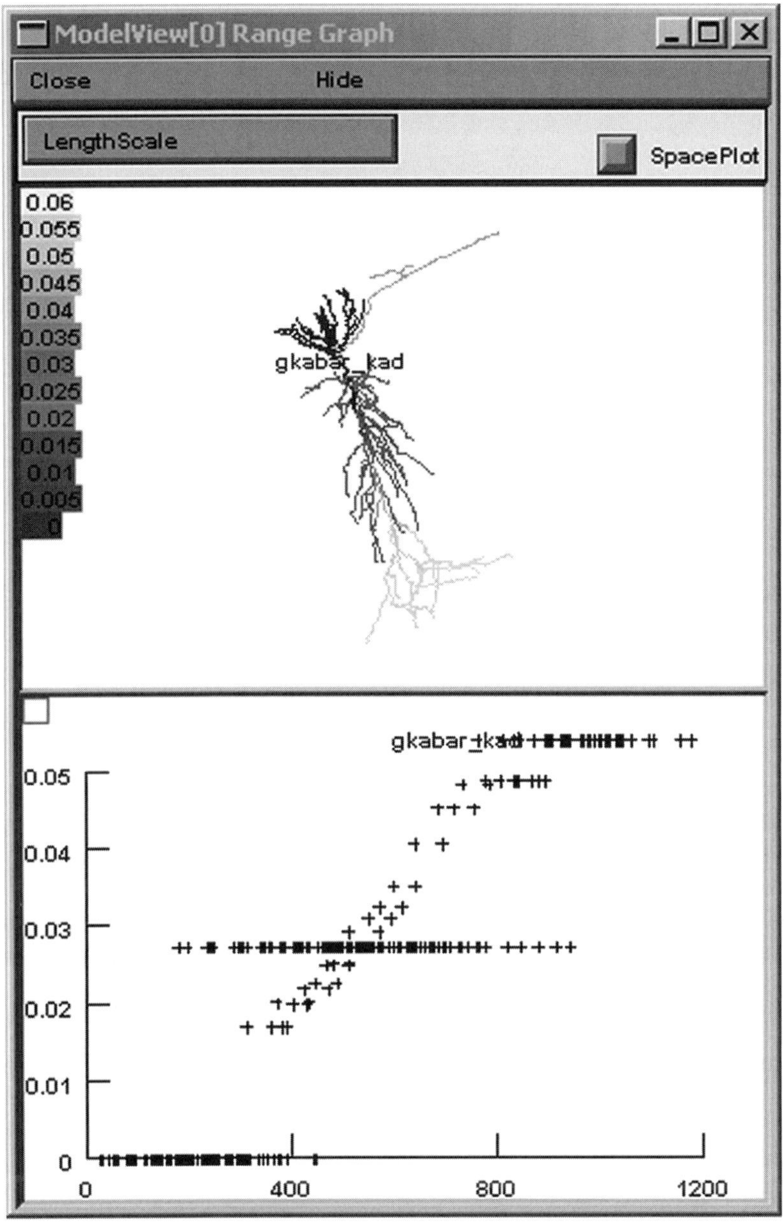

Fig. 10. Selecting "* gkabar_kad" from the ModelView's list of Heterogeneous Parameters brings up a range graph that shows the spatial distribution of this parameter's values.

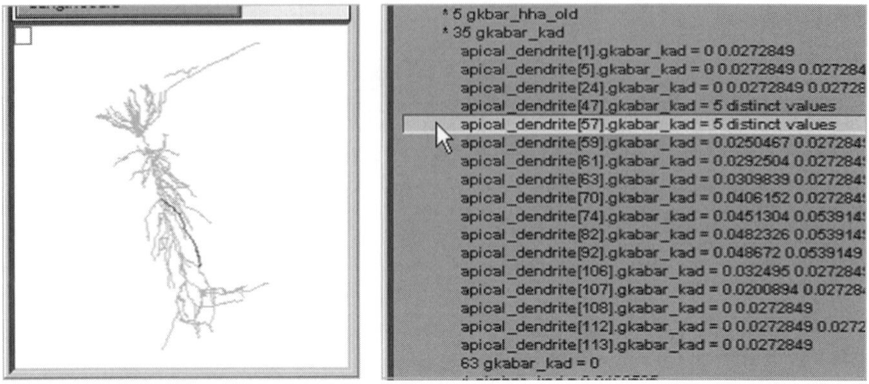

Fig. 11. Details of the distribution of inhomogeneous parameters are easily examined by clicking on individual lines in the ModelView tool. This figure shows that apical_dendrite[57] (dark branch near middle of shape) has five different values of gkabar_kad along its length.

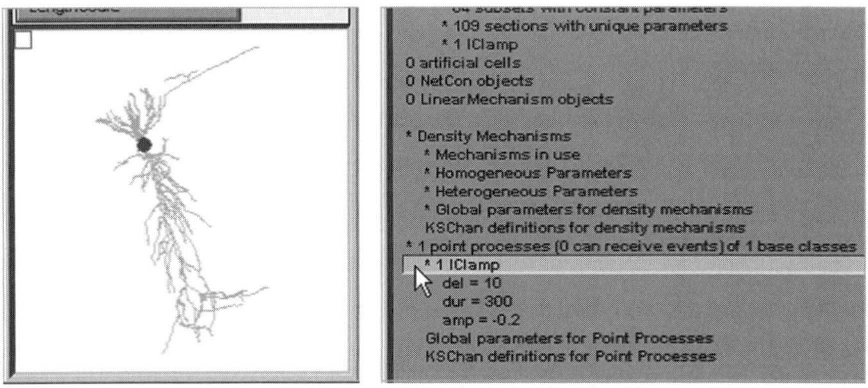

Fig. 12. The dot in the shape plot marks the location of the IClamp.

3.2. ModelView in ModelDB

We are developing a ModelView tool that will be incorporated into ModelDB. This will allow investigators to browse the morphology and biophysical specification for NEURON models stored in ModelDB.

3.3. ModelView and Interoperability: Future Goals

ModelView can output an XML description of the model's morphology and biophysical specification. ModelView uses MorphML *(6)* to describe the morphology of the cell. We are collaborating with the MorphML group to evolve the MorphML specification for interoperability. The biophysical XML specification is under development, and we have plans to provide translation tools to previous and developing standards. We will continue to refine the XML output to make it suitable for communication with neuroinformatics tools as well as other modeling software such as Catacomb *(7)* and GENESIS *(8)* (*see* **Chapter 7**). We will adopt, develop through informal collaboration and communication, and promote NeuroML *(9)* (whose goal is to establish standard descriptions of computational models that are simulator independent) and BrainML *(10)* (whose goal is to develop standards in the general description of neuroscience data). We will also attempt to implement the tools required to import and export models with a third specification, CellML *(11)* (whose purpose is to store and exchange computer-based biological models).

4. Conclusions

ModelView is a powerful model exploration tool that quickly and conveniently summarizes and displays information about NEURON models that can be hard to find if one has only the model code to search through. By presenting this information in an intuitive, browseable textual and graphical format that is easily understood, ModelView can help an investigator answer questions about a model that would otherwise have been very tedious to resolve. This tool is capable of further enhancements to match the evolving capabilities of NEURON and needs of computational neuroscientists. For example, we are developing XML interoperability tools that will extend ModelView's functionality to further aid investigators in understanding models and that will help translate models to simulator platforms different than the one in which they were originally written.

Acknowledgments

We are grateful for the support from NIH grants NS11613 and 5P01DC004732 (Human Brain Project).

References

1. Migliore, M., Morse, T.M., Davison, A.P., Marenco, L., Shepherd, G.M., Hines, M.L. (2003) ModelDB: making models publicly accessible to support computational neuroscience. *Neuroinformatics* **1**, 131–134.
2. ModelDB http://senselab.med.yale.edu/senselab/modeldb/.
3. Hines, M.L. and Carnevale, N.T. (2001) NEURON: a tool for neuroscientists. *The Neuroscientist* **7**, 123–135.

4. NEURON http://www.neuron.yale.edu/.
5. Poirazi, P., Brannon, T., Mel, B.W. (2003) Arithmetic of subthreshold synaptic summation in a model CA1 pyramidal cell. *Neuron* **37**, 977–987.
6. MorphML http://www.morphml.org/.
7. Catacomb http://www.catcmb.org/.
8. GENESIS http://www.genesis-sim.org/GENESIS/.
9. NeuroML http://www.neuroml.org/.
10. BrainML http://brainml.org/.
11. CellML http://www.cellml.org/.

7

Constructing Realistic Neural Simulations with GENESIS

James M. Bower and David Beeman

Summary

The GEneral NEural SImulation System (GENESIS) is an open source simulation platform for realistic modeling of systems ranging from subcellular components and biochemical reactions to detailed models of single neurons, simulations of large networks of realistic neurons, and systems-level models. The graphical interface XODUS permits the construction of a wide variety of interfaces for the control and visualization of simulations. The object-oriented scripting language allows one to easily construct and modify simulations built from the GENESIS libraries of simulation components. Here, we present procedures for installing GENESIS and its supplementary tutorials, running GENESIS simulations, and creating new neural simulations.

Key Words: Computational neuroscience; realistic neural modeling; simulations; GENESIS; tutorials.

1. Introduction

The GEneral NEural SImulation System (GENESIS) is a software platform for realistic modeling of biological systems. In this case, "realistic models" are defined as those models that are based on the known anatomical and physiological organization of neurons, circuits, and networks *(1)*. Thus, for example, realistic single-cell models usually include dendritic morphology *(2,3)*, whereas realistic network models attempt to duplicate known efferent and afferent projection patterns *(4,5)*. Models of this type at all levels of analysis require that finer scale single components be linked together into larger structures whose emergent behavior is then predicted numerically. GENESIS supports the simulation of neural systems ranging from subcellular components *(6,7)*

From: *Methods in Molecular Biology, vol. 401: Neuroinformatics*
Edited by: Chiquito Joaquim Crasto © Humana Press Inc., Totowa, NJ

and biochemical reactions ***(8,9)*** to complex models of single neurons ***(2,10)***, simulations of large networks ***(4,5,11)***, and systems-level models ***(12)***. An extensive list of publications describing models implemented with GENESIS is available at the GENESIS Web site (http://www.genesis-sim.org/GENESIS/). GENESIS, with its graphical interface XODUS, is written in C and currently runs with the X Window System on most UNIX platforms including Linux, MS Windows under the Cygwin environment (http://www.cygwin.com/), and Mac OS/X. Efforts are well underway to build a platform-independent, Java-based interface for the system, to be available in late 2006 as GENESIS 3.0.

GENESIS, and the version for parallel computers, PGENESIS, may be obtained from the GENESIS Web site. The GENESIS distribution comes with full source code, documentation, and example simulation scripts. It is licensed under the GNU General Public License and is distributed at no cost. The following sections present a brief guide to installing GENESIS, running GENESIS simulations, and creating new neural simulations with GENESIS.

2. Materials—Acquiring and Installing GENESIS and the Documentation

The first step is to download the latest GENESIS distribution from the GENESIS Web site http://www.genesis-sim.org/GENESIS. If you are using Linux, Microsoft Windows with Cygwin (a UNIX-like environment for Windows), or Mac OS/X, you may download the appropriate GENESIS distribution that includes a precompiled executable file named *genesis*. This will save you the trouble of compiling GENESIS from the source code that is included with the GENESIS distribution.

The GENESIS distribution may be installed in whatever directory is most convenient. You do not need "root" privileges to install within your home directory. For a system-wide installation, */usr/local* is a good place to install GENESIS. The distributed compressed tar file unpacks to produce the directory named *genesis* and its subdirectories. Use a terminal window to change to the directory in which you wish the GENESIS directory tree to reside, and (if the file you downloaded is called *genesis.tar.gz*) give the UNIX command (*see* **Note 1**):

```
gunzip -c genesis.tar.gz | tar xvf -
```

Begin by reading the *README* file in the *genesis* directory, and then follow the compilation instructions in *genesis/src/README*. If you are using a package with a precompiled executable, there will be installation instructions in a *README* file within the unpacked *genesis* directory.

After you have installed GENESIS, the hypertext GENESIS Reference Manual can be found in *genesis/Hyperdoc/Manual.html*. The Reference Manual is primarily a summary of the syntax used by the GENESIS Script Language Interpreter, the commands that it recognizes, and the GENESIS "objects" that are used as "building blocks" for constructing simulations. However, it will not tell you very much about the process of using GENESIS to construct neural simulations. For a "Users Guide" to GENESIS, we recommend *The Book of GENESIS* ***(13)***. The free Internet edition may be downloaded from http://www.genesis-sim.org/GENESIS/bog/bog.html. Part I of the book teaches concepts in neuroscience and neural modeling by means of interactive computer tutorials on subjects ranging from neuronal membrane properties to cortical networks. These chapters may be used without any prior knowledge of the GENESIS simulator or computer programming. Part II is intended to teach the use of the GENESIS scripting language for the construction of one's own simulations. The reader may modify these simulations and use them as a starting point for the development of original simulations.

2.1. Installing the GENESIS Neural Modeling Tutorials

The GENESIS Neural Modeling Tutorials package is an evolving set of HTML tutorials intended to introduce realistic neural modeling with GENESIS. They present an efficient path to the creation of new GENESIS simulations, recognizing that modelers generally start with existing simulations. The current version available from the GENESIS Web site is derived from earlier versions used at the first annual meeting of the World Association of Modelers Biologically Accurate Modeling Meeting (http://www.WAM-BAMM.org) and the annual EU Advanced Course in Computational Neuroscience (http://www.neuroinf.org/courses/EUCOURSE).

The main tutorial in the set is a multi-part GENESIS Modeling Tutorial ***(14)*** intended as a "quick start" to creating simulations with GENESIS. It provides sufficient tools and background information to rapidly begin creating cells and networks with GENESIS. The description of the methods for constructing GENESIS simulations in **Subheading 3.2**, below, draws heavily upon the background provided in this tutorial.

The GENESIS Neural Modeling Tutorials package contains several directories of sample simulations, as well as supplementary hypertext material to provide the background in computational neuroscience that is assumed by the GENESIS Modeling Tutorial. There are hyperlinks to this material throughout the GENESIS Modeling Tutorial. To install the Tutorials package, download the Tutorials archive from the GENESIS Web site http://www.genesis-sim.org/GENESIS. It is available as either *Tutorials.tar.gz* or *Tutorials.zip*, which may be extracted with "unzip" or one of the tools used with Windows for extracting zip files. In either case, it extracts to a directory (folder) called *Tutorials*. We recommend that you extract the archive in the *genesis* directory, alongside the *Scripts*, *Doc*, *Hyperdoc*, and other GENESIS directories. This will allow you to treat it as an extension of the documentation in the GENESIS distribution.

The GENESIS tutorials refer to this *Tutorials* directory, which contains the tutorials themselves, and other directories such as the *cells* directory, which contains cell models used by the individual tutorials. The *index.html* file in the *Tutorials* directory is the master index to the tutorials. It would be a good idea to set a bookmark to this file in your Web browser, so that you will have the links to the tutorials and to the GENESIS documentation easily available.

3. Methods

3.1. Running GENESIS Simulations

Although GENESIS has a rich variety of graphical interfaces, GENESIS simulations are started from the command line in a terminal window. After logging into a machine on which GENESIS is installed, open a terminal window (*see* **Note 1**). To allow room for the graphical display, move the terminal window into the lower left hand corner of your screen. (This is normally done by using the left mouse button to click and drag on the title bar of the window.)

The demonstration and tutorial simulations provided with GENESIS are located in directories ("folders" in Windows terminology) that are beneath the *genesis/Scripts* directory. Many of these simulations were designed as interactive tutorials for teaching concepts in neurobiology and neural modeling and are described in Part I of *The Book of GENESIS* ***(13)***. Each of these has online help and a number of suggested exercises or "experiments" that may be either assigned as homework in a course or used for self-study. Several of them are derived from existing research simulations and may be taken apart and modified to create your own simulations.

A typical GENESIS simulation using XODUS will display one or more control panels—windows that contain labels, buttons, toggle buttons, and labeled text fields (called dialog boxes in GENESIS) for display or entry of data. In addition, there will typically be graphs and other displays that allow you to visualize results of a simulation or to interact with it through clicks of the mouse. Here, we will illustrate the use of the XODUS GUI with the *traub91* tutorial, located in *Scripts/traub91*. This is a tutorial based on a GENESIS simulation of the hippocampal pyramidal cell model of Traub et al. (***15***). A description of the properties of the various types of ionic channels used in the model may be found in Chapter 7 of *The Book of GENESIS*.

The tutorial explores some of the properties of this model and also serves as an introduction to the "Neurokit" simulation environment. Neurokit is a very flexible and general GENESIS program that reads in the details of a single-cell model and its channels from supplemental files. Various experiments may then be performed by clicking with the mouse or by entering values in dialog boxes. If you wish to learn more about the advanced features of Neurokit, you may run Neurokit from the *Scripts/neurokit* directory and try some of the examples that are suggested in the online help. To begin, change into the *Scripts/traub91* directory and give the command "`genesis Neurokit`" to begin the simulation. For example, if GENESIS has been installed in */usr/local/genesis*, you would type the commands

```
cd /usr/local/genesis/Scripts/traub91
genesis Neurokit
```

Note that UNIX is case sensitive, so that you will have to capitalize "Neurokit." If the files have been properly installed and the necessary paths have been set, you should be rewarded with messages indicating that GENESIS is loading. Eventually, the Neurokit menu bar will appear at the top of the screen, as shown in **Fig. 1**. The menu bar and the other windows that will later appear are examples of XODUS "forms." As with the terminal window and the other windows that are used with X Windows, they may be moved and resized by clicking and dragging with the mouse. This is particularly useful if the resolution of your display does not allow everything to be visible on your screen. As with most GENESIS tutorial simulations, there is a "help" button that can be used to bring up a window with scrollable text, graphical images, or additional control forms.

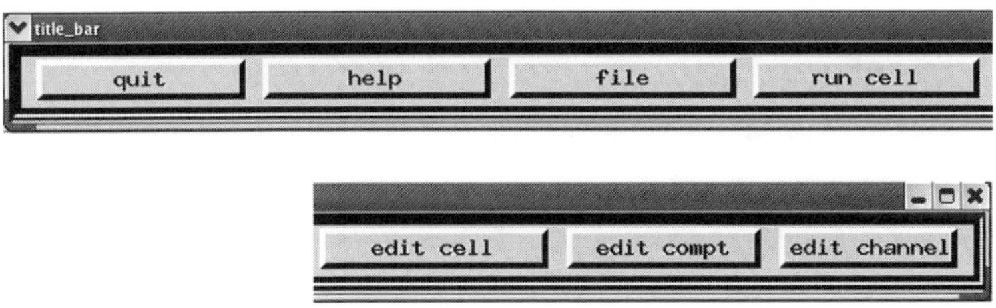

Fig. 1. The Neurokit menu bar. For legibility, the menu bar, which extends across the top of the screen, has been split into a left part (above) and a right part (below).

Click on the "help" button, and a detailed description of the tutorial will be displayed. (As usual, when we say "click on...," we mean "click the left mouse button on..." unless another button is specified.) When you feel that you are ready to begin, click on the "CANCEL" button at the bottom of the help window.

Click on the "file" option on the title bar. The "file" menu will appear, showing various dialog boxes indicating that we are running the CA3 cell model simulation. As the default values in the boxes are all correct, click the "Load from file" button on the "file" menu. A few messages will appear in your terminal window. Then click "run cell" on the menu bar. You should see the "SIMULATION CONTROL PANEL" (*see* **Fig. 2**) and a "Cell Window" containing a representation of the cell with a graph below (*see* **Fig. 3**).

Initially, the "Recording" button under the "ELECTROPHYSIOLOGY" heading will be highlighted, indicating that you are ready to plant virtual recording electrodes in the cell. The help file describes a number of experiments that

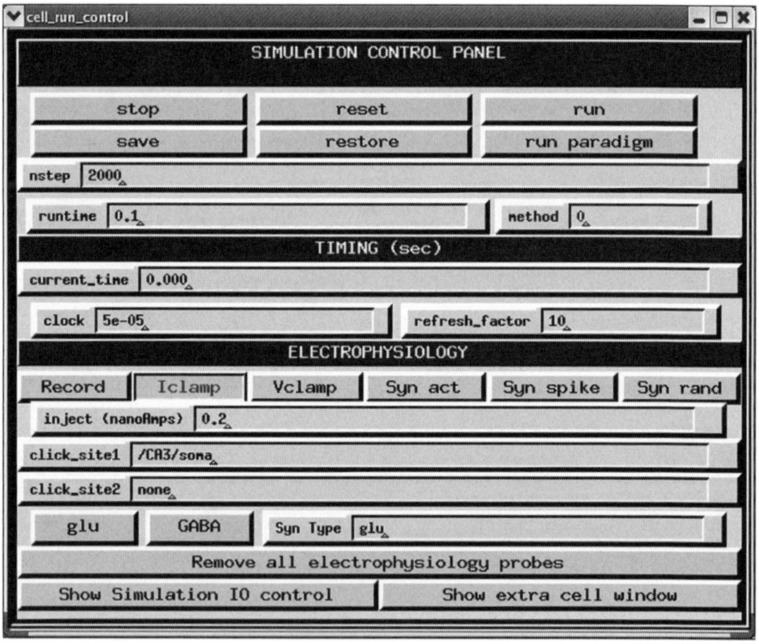

Fig. 2. The Neurokit SIMULATION CONTROL PANEL. The ELECTROPHYSIOLOGY dialog boxes reflect the choice of the "Iclamp" toggle to provide a steady 0.2 nA current injection to the soma.

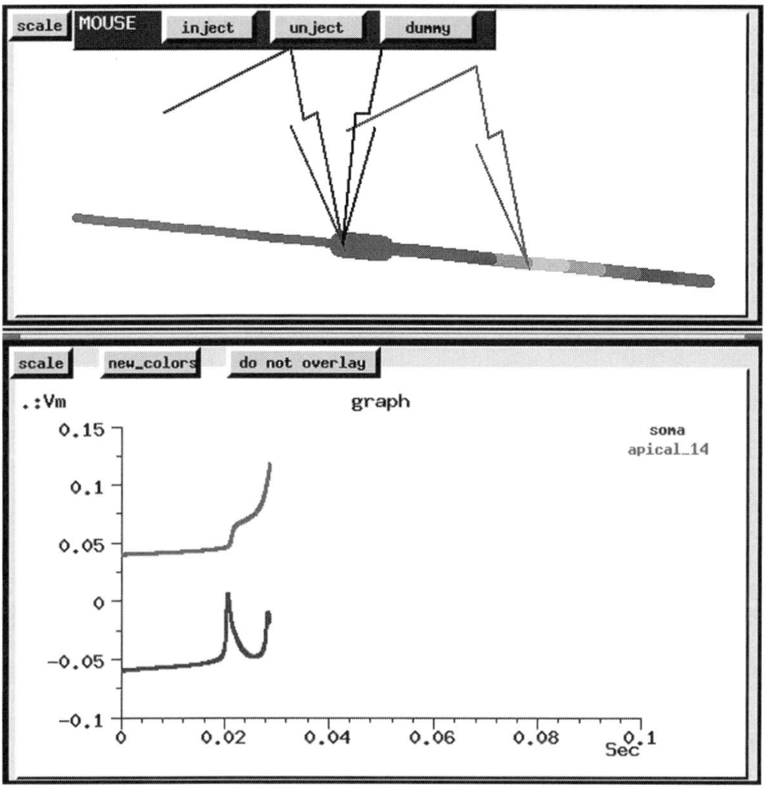

Fig. 3. The Neurokit Cell Window. The cell representation (top) shows the location of the recording and injection electrodes and a color-coded representation of the membrane potential in each compartment. The graph (below) plots the membrane potential in the two compartments selected for recording: the soma (lower plot) and apical dendrite compartment 14 (upper plot).

may be performed on the cell. The default one is to reproduce the bursting behavior observed in this cell with an injection current of 0.2 nA to the soma.

Start by clicking the left button in the round soma. A recording electrode should appear, and the box labeled "click_site1" should display "/CA3/soma." To plant a recording electrode in compartment 14 of the apical dendrite, click the left mouse button about half way to the right along the "fat" dendrite extending from the soma. If the "click_site1" box then shows something other than "/CA3/apical_14," click the middle button in the same place to remove the electrode and try again. With a little practice, you will find it easy to record from any desired part of the cell.

Next, click on "Iclamp" under the "ELECTROPHYSIOLOGY" heading. Underneath, a dialog box labeled "inject (nanoAmps)" should appear, with a value of "0" in the box. To change the field of a dialog box, use the mouse to move the cursor into the dialog box. Then, use the keyboard right and left arrow keys to move the "^" symbol to the right of the place where you wish to make the changes. Then use the "Backspace" key to back up over anything you wish to change, and type in the changes to set the injection current to 0.2 nA. To cause the changes to be entered, you must then hit "Enter" while the cursor is in the field of the dialog box (*see* **Note 2**).

At this point, the SIMULATION CONTROL PANEL will look like that in **Fig. 2**. Now, click on the soma to plant an injection electrode in the soma. Finally, click on "run" in the SIMULATION CONTROL PANEL to start the simulation. While the simulation is running, you will see the compartments in the cell diagram change color to represent the value of the membrane potential, V_m. During the simulation run, the Cell Window will resemble that in **Fig. 3**. The plots below the cell diagram show the membrane potential in the two compartments. The lower one shows the beginning of a burst of action potentials in the soma that terminates after about 50 ms. If we were to run the simulation longer, we would see another burst in about a second. The curve above, which has been displaced by 100 mV for clarity, shows the membrane potential in the middle of the apical dendrite.

The description of the model given by the "help" display explains the process by which action potentials in the soma propagate to the dendrites to open calcium channels, which then activate potassium channels to produce these results. It also suggests other experiments that may be performed with this simulation, as well as the procedure for plotting other quantities of interest, such as the intracellular calcium concentration. When you are finished with this simulation, click on "quit" in the menu bar. Although it may be tempting to terminate a GENESIS simulation by clicking on the "X" at the right end of a window's title bar, this is not recommended, as it will not allow GENESIS to terminate cleanly.

The Kinetikit simulation in *Scripts/kinetikit* is another "kit" for building simulations using a graphical interface. This program is used for modeling biochemical kinetic reactions *(8)* and provides a convenient interface for the use of the objects in the GENESIS kinetics library. The *README* file in this directory provides the details. Other demonstrations and tutorials in the *Scripts* directory that you might want to try are *squid*, *burster*, *purkinje*, and *piriform*. These have their own *README* files and help displays and are run in a similar manner to that described above (*see* **Note 3**).

3.2. Creating New Simulations with GENESIS

Graphical environments for building models and simulations, such as Neurokit and Kinetikit, and those planned for GENESIS 3.0 (http://www.genesis-sim.org/GENESIS/G3/), make it possible to create and run simulations with little or no programming. Nevertheless, they have their limitations, and at one time or another, most serious users of GENESIS need the flexibility provided by the GENESIS scripting language to construct their own simulation scripts. A GENESIS simulation script constructs simulations from building blocks taken from the GENESIS objects library and links them together with messages. Fortunately, the modular object-oriented nature of GENESIS makes it easy to modify existing scripts. Once you have learned a few basics of GENESIS scripting and have some good example simulation scripts to get you started, you can create most of your simulations by simple "hacking" of existing scripts. That is the approach taken in the procedure described in this chapter and by most GENESIS modelers. The following brief description of the process for constructing a compartmentalized single-cell model is covered in more detail in the GENESIS Modeling Tutorial section "Building a cell the easy way."

3.2.1. Creating a Model Neuron from a Script

The principal steps in creating a realistic model of a neuron (*see* **Note 4**) are as follows:

1. Determine the morphology of the dendritic structure to be implemented in the model cell, and construct a compartmentalized passive cell model.
2. Set the passive membrane parameters [membrane resistance, membrane capacitance, axial resistance, and membrane resting potential (R_m, C_m, R_a, and E_k, respectively)] for each of the compartments.
3. Link compartments for the soma and dendrites together with appropriate messages to make a cell.
4. Populate the compartments with voltage-activated and synaptically activated ionic conductances ("channels") or other related neural elements.
5. Provide external inputs (current injection, synaptic connections to spike trains, etc.) and a means of recording the membrane potential or other variables of interest.

This procedure is illustrated in the following analysis of a GENESIS script for simulating a simple "generic" fast spiking cortical interneuron. The simulation uses the GENESIS cell reader to perform **steps 2–4** above to create a simple two-compartment neuron with a dendrite compartment, a soma, and an axon. The dendrite contains synaptically activated excitatory and inhibitory channels, and the soma contains "squid-like" voltage-activated Hodgkin–Huxley ***(16)*** sodium and potassium channels, plus a spike generator element that acts like the intial part of an axon. This may be used to provide synaptic input to another cell in a network. Other statements in the simulation script implement **step 5**.

Locate the "GENESIS Tutorials directory" (*Tutorials*) that contains the GENESIS tutorials and "cd" to it in a terminal window. "ls" will show you that it contains a *cells* directory, which contains other directories having various GENESIS cell models. To begin, "cd" to the directory *FSecell* and run the *FScell* simulation by typing "genesis FScell." You should be rewarded with a display similar to that in **Fig. 4**.

The CONTROL PANEL of the GUI allows a steady injection current to be applied to the cell soma, and/or synaptic activation to be applied to the dendrite compartment. In the synaptic input form, "gmax" is used to set the maximal conductance of the excitatory synaptically activated channel (a GENESIS synchan). Poisson-distributed background activation can be applied to the synchan by setting "Frequency" (the average activation frequency) to

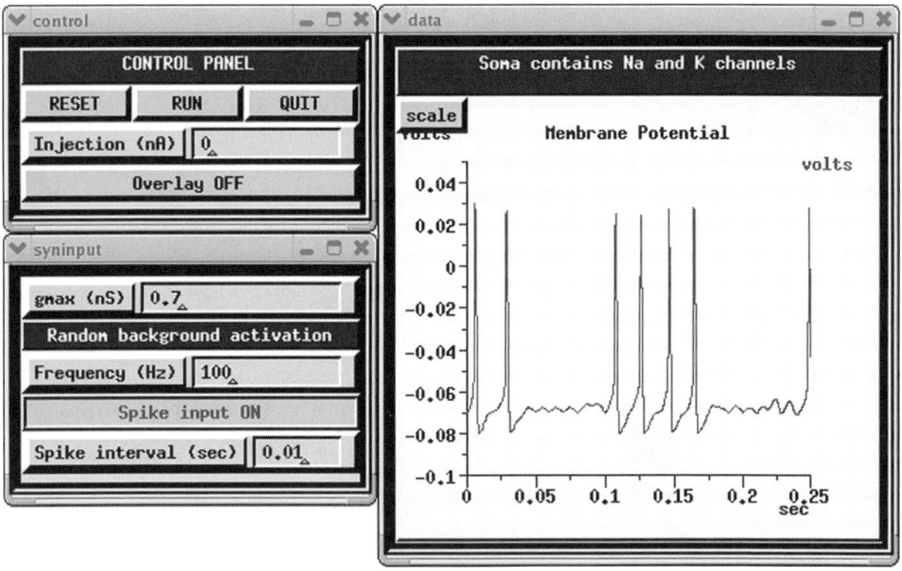

Fig. 4. The FScell simulation display. The membrane potential plot arises from a combination of a random background activation at an average frequency of 100 Hz and a continuous train of spikes at 0.01 s intervals applied to the excitatory channel in the dendrite compartment.

a non-zero value. This sets the frequency field of the synchan to that value. The "Spike input ON/OFF" toggle turns on and off the application of trains of spikes at the specified "Spike interval." You may vary the injection current from the default value of 0.3 nA by editing the value in the "Injection" dialog box. Experiment with the simulation by changing various parameters, then clicking "RESET," and then "RUN." Notice that the "scale" button on the graph lets you change the scales for the graph axes.

3.2.2. Understanding the Simulation Script

Now, it is time to understand the *FScell.g* script. In the listing below, line numbers have been added to facilitate the description. These do not appear in the actual *FScell.g* file.

```
1 //genesis - FScell.g
2 /*=================================================================
3 Creates a neuron from a specified cell parameter file, with channels
4 taken from the prototypes created by the file protodefs.g. When used
5 with the GUI functions defined in graphics.g, it allows an injection
6 current and a variety of synaptic inputs to be applied.
7 =================================================================*/
8
9 // Simple fast spiking interneuron model
10 // Customize these strings and parameters to simulate another cell.
11
12 str cellfile = "cell.p"
13 str cellpath = "/cell"
14 // Define the places for current injection and synaptic input
15 str injectpath = {cellpath} @ "/soma" // Use '@' to concatenate
     strings
16 str synpath = {cellpath} @ "/dend/Ex_channel"
17 // For no synaptic input, set
18 // str synpath = ""
19
20 float tmax = 0.25                     // simulation time in sec
21 float dt = 20e-6                     // simulation time step in sec
22 float injcurrent = 0.3e-9          // current injection in Amperes
23
24 // If there is synaptic input, define gmax for the
   target synchan
25 float gmax // value will be obtained from cell created
   by readcell
26 float spikeinterval = 0.01 // Default spike train interval
27
```

```
// Label for the graph
str graphlabel = "Soma contains Na and K channels"

// Create library of prototype elements used by the cell reader
include protodefs

//=================================
// Function Definitions
//=================================

function step_tmax
// Add other commands here that you want to execute for each run
        step {tmax} -time
end

function do_quit
// Add any other commands here to execute before quitting
        quit
end

function set_frequency(frequency) // random background activation
    float frequency
    setfield {synpath} frequency {frequency}
end

//=================================
// Main simulation section
//=================================

 setclock 0 {dt}                  // set the simulation clock

// read a cell parameter file and create cell in "cellpath"
readcell {cellfile} {cellpath}

// Set up circuitry for synaptic input if synapse exists
if ({synpath} != "")
    gmax = {getfield {synpath} gmax}
    create spikegen /spiketrain
    setfield /spiketrain thresh 0.5 abs_refract {spikeinterval}
    addmsg /spiketrain {synpath} SPIKE
end

// Include the functions used to build the GUI
include graphics

// make the control panel
```

```
74 make_control
75
76 // make the graph to display soma Vm and pass messages to it
77 make_Vmgraph
78 addmsg {cellpath}/soma /data/voltage PLOT Vm *volts *red
79
80 /* Set the initial injection current to the cell */
81 setfield {injectpath} inject {injcurrent}
82
83 colorize // function defined in graphics.g adds colors to widgets
84 check    // self-consistency check of the simulation elements
85 reset    // Initialize all simulation elements
```

This script illustrates some important features of the GENESIS scripting language that are explained in detail in the GENESIS Reference Manual:

- Comments are indicated either with a C-style comment block (e.g., lines 2–7) or with "//" for a comment that only extends to the end of the line (e.g., line 9 or the end of line 20).
- *include* statements are used to include the contents of other scripts (e.g., lines 32 and 71). Although these are GENESIS scripts that have the extension *.g*, it is not necessary to give the extension. Note the separation of the graphical and non-graphical code into separate scripts. This makes it possible to easily modify simulations to run in "batch mode" by including a script that writes the output to files, instead of to a graphical interface.
- As with most programming languages, typed variables are declared and assigned values (e.g., lines 12 and 20), and function definitions can be made (lines 38–41).
- Conditional execution is provided with the *if* statement (lines 63–76) and other flow control statements described in the Reference Manual.
- Line 65 illustrates the creation of a spike generator called "*/spiketrain*" from a spikegen object.
- Lines 66 and 81 illustrate the setting of object fields (internal variables).
- Lines 50, 67, and 78 illustrate the syntax for adding messages between objects. This is the way that they communicate the values of their fields with one another.
- Note the use of braces ("curly brackets") to force the evaluation of variables or functions. This is used by the GENESIS parser to distinguish between the string "frequency" and the value of the variable "frequency" in line 50.

The statement "include protodefs" merges in the contents of the *protodefs.g* file, which is also in the current (*Tutorials/cells/FScell*) directory. This file, which we will examine shortly, creates the prototype compartments and channels from which the cell reader will construct the cell. The included file, *graphics.g*, contains the definition of the functions *make_control* and *make_Vmgraph* that are used to make the control panel and the graph for plotting the soma membrane potential V_m.

The cell is constructed with the single command on line 62:

```
readcell {cellfile} {cellpath},
```

which creates a neutral placeholder object */cell* (the cellpath) and builds the cell under it, according to the specifications in the cell parameter file (the cellfile) *cell.p*:

```
// cell.p - Cell parameter file used in the FScell simulation
// Format of file :
// x,y,z,dia are in microns, all other units are SI (Meter Kg Sec Amp)
// In polar mode 'r' is in microns, theta and phi in degrees
// readcell options start with a '*'
// The format for each compartment parameter line is :
// name parent r     theta    phi     d     ch     dens
// in polar mode, and in cartesian mode :
// name parent x     y     z     d     ch     dens
// For channels, "dens" = maximum conductance per compartment area
// For spike elements, "dens" is the spike threshold Coordinate mode
*relative
*cartesian
*asymmetric

//                       Specifying constants
*set_compt_param          RM          0.33333
*set_compt_param          RA          0.3
*set_compt_param          CM          0.01
*set_compt_param      EREST_ACT     -0.07

// For the soma, use the leakage potential (-0.07 + 0.0106) for Em
*set_compt_param    ELEAK    -0.0594
soma none 30 0 0 30 Na_hh_tchan 1200 K_hh_tchan 360 spike 0.0
// The dendrite has no H-H channels, so ELEAK = EREST_ACT
*set_compt_param    ELEAK    -0.07
dend soma 100 0 0 2 Ex_channel 0.795775 Inh_channel 0.397888
```

The comments in the file give a brief description of the format of a cell parameter file, and further details are given in the documentation for *readcell*. The *readcell* options used here specify that relative Cartesian coordinates will be used with asymmetric compartments. If the option "*symmetric" had been used, then the cell would be constructed using symmetric compartment (symcompartment) objects. The choice of relative coordinates means that for lines such as

```
soma none 30 0 0 30
dend soma 100 0 0 2,
```

the soma, which is the start of the cell and has no parent, has its end at the (x, y, z) coordinates of (30, 0, 0) in micrometers. As the cell starts at (0, 0, 0), this means that the soma has a length of 30 μm and a diameter of 30 μm. The "dend" section will be connected to its parent, the soma, through its own axial resistance R_a, with suitable messages established between the two compartments. As relative coordinates are being used, the end of the dendrite compartment will lie 100 μm along the x-axis from the end of the soma compartment.

With the *FScell* simulation running, give these commands to the GENESIS prompt:

```
showfield /cell/soma -all
showfield /cell/dend -all
```

The "-all" option causes the compartment start (x0, y0, z0) and end (x, y, z) coordinates to be displayed, along with the length, diameter, and other compartment fields. Verify that these values are what you would expect.

Neural compartments have fields for the membrane resistance R_m, the membrane capacitance C_m, and the axial resistance R_a. These are calculated and set in each compartment by the cell reader, using the dimensions of the compartment and the parameters RM, CM, and RA. This is done using the "*set_compt_param" option of the cell parameter file. You can verify that this was correctly done by using the *showfield* commands above.

The global variable EREST_ACT is the nominal resting potential of the cell. Normally, we would expect to set the compartment E_m field, which represents the "battery" in series with the membrane resistance R_m, to the value of EREST_ACT. However, Hodgkin and Huxley found it necessary to set E_m to a leakage potential E_{leak} that compensates for current flow through other channels (such as chloride channels) that were not explicitly taken into account in their model ***(16)***. E_{leak} is set to a value that results in no net current flow when the cell is at EREST_ACT. The cell reader takes care of this by not only setting EREST_ACT to the value specified in the cell parameter file but allowing the use of another parameter ELEAK, which if specified, gives an alternate value E_m but allows V_m to be initialized to EREST_ACT on reset, instead of E_m. (For further details of the initializaton of V_m on reset, see the documentation for compartment.) Therefore, the cell parameter file above sets ELEAK to EREST_ACT + 0.0106 for the soma, but sets it to EREST_ACT for the dendrite, where there are only synaptically activated channels.

Finally, note the list of channels and their conductance densities that follow the compartment coordinates and diameter. For the soma, these are the values used by Hodgkin and Huxley for the squid giant axon, with the sodium channel (*Na_hh_tchan*) given a maximum conductance of 1200 siemens/m^2 of surface area and the potassium channel (*K_hh_tchan*) given a value of 300 siemens/m^2. The cell reader scales these by the surface area to set the "Gbar" fields of the channel elements and also sets up the necessary messages between the channels and the compartment that contains them.

To see the messages that have been set up by the cell reader (plus the PLOT message established by *FScell.g*), give the following commands:

```
showmsg /cell/soma
showmsg /cell/dend
```

3.2.3. Creating Channel Prototypes

The remaining important aspect of the use of GENESIS that needs to be understood is the way the cell reader is informed about the properties of the elements that the cell parameter file calls *Na_hh_tchan*, *K_hh_tchan*, *spike*, *Ex_channel*, and *Inh_channel*. This is done in the *protodefs.g* file.

The cell reader builds the cell by making copies of "prototypes" of the various elements that will be used, replacing the default values of parameter fields with values taken from the cell parameter file. For example, when constructing a soma with several attached dendrite compartments, it will make multiple copies of a generic compartment prototype and then set the data fields in each compartment to the appropriate values. Likewise, a cell having Hodgkin–Huxley Na channels in several compartments will get these channels from copies of the single Na-channel prototype, setting the value of the maximum channel conductance Gbar for each copy, making use of the specified conductance density and dimensions of the compartment that contains the channel.

The cell reader expects to find this library of prototype elements as a set of subelements of the neutral element */library*. Thus, we need a script that will create */library* and fill it with a prototype compartment, one copy of each of the different channel types we will use, and a spike generator. Although the statements that are needed to set up the prototype library could go into your main simulation script, it is customary to make a separate script for this and to then use *include* to bring it into the simulation. This script is often called *protodefs.g*, although you may give it any name you like in your own simulations. The listing for *protodefs.g* (with added line numbers) is shown below.

```
// protodefs.g - Definition of prototype elements for "FScell"
/* Included files are in genesis/Scripts/neurokit/prototypes */
// file for standard compartments
include compartments
// include definitions of functions to create H-H tabchannels
include hh_tchan
/* hh_tchan.g assigns values to the global variables EREST_ACT,
   ENA, EK, and SOMA_A. The first three will be superceded by
   values defined below. The value of SOMA_A set in hh_tchan.g is
   not relevant, as the cell reader calculates the compartment area.
*/
EREST_ACT = -0.07          // resting membrane potential (volts)
ENA   = 0.045               // sodium equilibrium potential
EK   = -0.082              // potassium equilibrium potential
/* file for synaptic channels */
include synchans
/* file which makes a spike generator */
include protospike
// Make a "library element" to hold the prototypes to be used
// by the cell reader to add compartments and channels to the cell.
if (!{exists /library}) // But, only if it doesn't already exist
   create neutral /library
end
// The prototype elements in the library are not needed to
// calculate anything, so we disable the library
disable /library
// To ensure that all subsequent elements are made in the library
pushe /library
/* Make a prototype compartment. The internal fields will be set
   by the cell reader, so they do not need to be set here. The
```

```
   make_cylind_compartment function is defined in compartments.g.
*/
make_cylind_compartment

/* Functions in hh_tchan.g are used to create prototype H-H
   tabchannels "Na_hh_tchan" and "K_hh_tchan"
*/
make_Na_hh_tchan
make_K_hh_tchan

// Make prototype excitatory channel, "Ex_channel" (in synchans.g)
make_Ex_channel /* synchan with Ek = 0.045, tau1 = tau2 = 3 msec */

// Make a prototype inhibitory channel, "Inh_channel"
make_Inh_channel // synchan with Ek = –0.082, tau1 = tau2 = 20 msec

// Make spike generator (spikegen) element "spike"
//(in protospike.g)
make_spike

pope // Return to the original place in the element tree
```

Note the use of several statements to include the files *compartments.g*, *hh_tchan.g*, *synchans.g*, and *protospike.g*. These files are not in the *Tutorials/cells/FScell* directory but are found in the *genesis/Scripts/neurokit/prototypes* directory. Normally, GENESIS sets its search path (SIMPATH) to include this directory, so that these and many other files with prototype definitions can be accessed from any directory.

These files contain definitions of some global variables for channel reversal potentials and the like, plus function definitions that create the prototype elements. For example, *hh_tchan.g* sets some default values for the cell resting potential EREST_ACT, the sodium reversal potential ENA, and the potassium reversal potential EK. It also declares the functions *make_Na_hh_tchan* and *make_K_hh_tchan* to make the Na and K channel elements, respectively. The *protodefs.g* file assigns different values to these variables **after** *hh_tchan.g* is included but **before** the functions that create the prototype channels are called. Usually, the file containing the functions to create prototype channels will have some means of shifting the voltage scale for activation and time constant curves by specifying a variable for the voltage offset in equations that depend on the membrane potential. Often, the variable EREST_ACT is used for this purpose, as it can also represent the nominal resting potential of the cell for which the active channel models were developed. By using this as a global variable and changing it before calling the channel creation functions that are defined in the channel files, you can shift these curves. This can be very useful when you use a channel developed for one cell model in a different cell (*see* **Note 5**).

3.2.4. Creating the Graphics

The file *graphics.g* can be used somewhat blindly, as long as you are happy with the default control panel and the graph that it creates. After including it, all you need to do is to invoke the *make_control* and *make_Vmgraph* functions and to send appropriate PLOT messages to the */data/voltage* xgraph object. The GENESIS Modeling Tutorial section "Building a cell the easy way" gives some further analysis of this script.

It is a good idea to keep the commands that involve graphics in a separate file, as we have done here. If you should later want to modify your simulation to run without graphics, as you might want to do when making long simulation runs on another networked computer, then you only need to make small changes to the main simulation script. This will also allow you to easily add alternate Java-based user interfaces when GENESIS 3.0 is available. If you need to modify this file for your own customized GUI, refer to the GENESIS Reference Manual section on the XODUS Graphical Interface and the links given there for documentation for the XODUS "widgets" xbutton, xtoggle, xlabel, and xgraph. Chapter 14 of *The Book of GENESIS* also explains some of the XODUS commands that were used to create a similar interface for *genesis/Scripts/tutorials/tutorial3.g*, and Chapter 22 gives a very detailed treatment of XODUS.

3.2.5. Modifying the Simulation Scripts

The *FScell* simulation is a good starting point for constructing your own single-cell simulations by modifying the main *FScell.g* script, providing an alternate cell parameter file, and making any changes needed to *protodefs.g* to create necessary prototype elements. The *graphics.g* file provided here will not need any modification. We describe here the procedure for modifying this simulation to reproduce the results for the Traub et al. ***(15)*** hippocampal CA3 region pyramidal cell model described above in **Subheading 3.1**.

The *Tutorials/cells/traubcell* directory contains a cell parameter file *CA3.p* that describes the compartmentalized cell model as specified in the original paper and lists the channels and their conductance densities in each compartment. It also contains a file *traub91proto.g* that defines the functions that create the various channels referenced in *CA3.p*. To "hack" the *FScell* simulation scripts into a simulation of the Traub model, it would be best to create your own directory to which you will copy the files that you will modify for your simulation. For example, assuming that you want to create a subdirectory in your home directory called *newcell* and it does not already exist, you might do the following from within the *Tutorials/cells/* directory:

```
mkdir ~/newcell
cp FScell/* ~/newcell
cp traubcell/CA3.p ~/newcell
cp traubcell/traub91proto.g ~/newcell

cd ~/newcell
cp FScell.g traub.g
```

At this point, you may use your favorite text editor to modify the copy of *FScell.g* called *traub.g*. These are the suggested changes:

1. To use the Traub model cell parameter file, change line 12 to

   ```
   str cellfile = "CA3.p"
   ```

2. As *CA3.p* does not specify a synaptically activated channel, change lines 16–18 to

   ```
   //str synpath = {cellpath} @ "/dend/Ex_channel"
   //For no synaptic input, set
   str synpath = ""
   ```

3. For a more appropriate label on the graph, change line 29 to

   ```
   str graphlabel = "Traub et~al. 1991 CA3 pyramidal cell"
   ```

The next step is to modify the *protodefs.g* file so that it will create prototype objects that are needed by *CA3.p*, instead of those used in the *FScell* model. Briefly examine the new cell parameter file to see what elements are used in the simulation. In this case, *CA3.p* uses the option "* symmetric," which means that we need to make a symcompartment prototype. The compartments use channels *Na*, *Ca*, *K_DR*, *K_AHP*, *K_C*, and *Ca_conc*. This calls for these changes in *protodefs.g*:

1. To load the channel definitions for the Traub model (*see* **Note 6**), instead of those for the Hodgkin–Huxley Na and K channels, change line 8 to

   ```
   include traub91proto
   ```

2. In this case, we are using a set of channels that were originally intended to be used with the cell model, so there is no need to modify the resting potential or ionic reversal potentials. Delete lines 10–18 that change EREST_ACT, ENA and EK, and the comment block before them.

3. To create a prototype symcompartment, change line 44 to

```
make_cylind_symcompartment
```

4. To create the prototype channels, change lines 49 and 50 to

```
make_Ca_conc
make_Na
make_Ca
make_K_DR
make_K_AHP
make_K_C
make_K_A
```

5. For appearances sake, remove the lines dealing with synchan and spike objects. There is no harm in creating these prototypes, but they would not be used in constructing the cell. These are lines 20–24 and 52–59. At this point, you should be able to run the *traub.g* script with GENESIS and reproduce the results plotted in **Fig. 3**.

4. Notes

1. The main index to the Tutorials package has a link to "Introduction to UNIX or Linux and the graphical desktop." This is a short guide to getting around in the UNIX command line environment for Windows users. It has the basic information needed to perform some of the steps that are carried out in a terminal window. Most graphical desktops for UNIX or Linux will have an icon for a terminal window on the taskbar. If your taskbar does not have an icon for a terminal window, look for it in the System Tools submenu of the Main Menu. If you are using Microsoft Windows with Cygwin-X installed, you start Cygwin-X by clicking on an icon on the desktop or from an entry in the Start menu. Then, locate the menu entry for a Cywin "xterm" to bring up a terminal window.
2. This feature of XODUS will be unfamiliar to many new users. If you ever suspect that the old values are being used in the running of a simulation, move the cursor back into the dialog box, hit "Enter," and repeat the run.
3. In some cases, you will want to copy a set of GENESIS simulation scripts into a directory of your own, so that you can modify the scripts or run simulations that create output files. (The permissions of the GENESIS directories are usually set so that you cannot write into them.) For this, you can create your own *Scripts* directory with "`mkdir ~ /Scripts`" and then copy a directory tree with a command such as

```
cp -r /usr/local/genesis/Scripts/purkinje ~/Scripts.
```

4. The GENESIS Modeling Tutorial sections "Making synaptic connections" and "Creating large networks with GENESIS" extend this procedure to the creation of networks of biologically realistic neurons.

5. Note that EREST_ACT is also used in the cell parameter file to give E_m and the starting V_m (initVm). And, if you include more than one file with functions to create channels, these files may set different values for EREST_ACT or the ionic reversal potentials. If this is the case, you should be careful to reset these variables to the desired values, after including the file and before invoking the channel creation functions defined in that file.
6. The *Scripts/neurokit/prototypes* has many other channel prototype definition files, including *traub91chan.g*, which is similar to *traub91proto.g*, but has longer (and more descriptive) channel names. The GENESIS Modeling Tutorial section "Creating your own channel models" describes the procedure for creating these files.

Acknowledgments

This work was supported by NIH grant R01 NS049288-01.

References

1. Bower, J.M. (1995) Reverse engineering the nervous system: an in vito, in vitro, and in computo approach to understanding the mammalian olfactory system,8 in *An Introduction to Neural and Electronic Networks*, 2nd Edition (Zornetzer, S., Davis, J., and Lau, C., eds.), Academic Press, San Diego, CA, pp. 3–28.
2. DeSchutter, E., and Bower, J.M. (1994) An active membrane model of the cerebellar purkinje cell: I simulation of current clamps in slice. *J. Neurophysiol.* **71**, 375–400.
3. Protopapas, A.D., and Bower, J.M. (1997) Pyramidal cell response to patterns of synaptic input believed to underlie gamma oscillations, in *Computational Neuroscience: Trends in Research 1997* (Bower, J., ed.), Plenum Press, New York, pp. 177–181.
4. Wilson, M., and Bower, J.M. (1992) Cortical oscillations and temporal interactions in a computer simulation of piriform cortex. *J. Neurophysiol.* **67**, 981–995.
5. Santamaria, F., Tripp, P.G., and Bower, J.M. (2007) Feedforward inhibition controls the spread of granule cell-induced Purkinje cell activity in the cerebellar cortex. *J. Neurophysiol.* **97**, 248–263.
6. De Schutter, E., and Schmolen, P. (1998) Calcium dynamics in large neuronal models, in *Methods in Neuronal Modeling: From Ions to Networks*, 2nd Edition (Koch, C., and Segev. I., eds.), MIT Press, Cambridge, MA, pp. 211–250.
7. Bhalla, U.S. (2004) Signaling in small subcellular volumes. II. Stochastic and diffusion effects on synaptic network properties. *Biophys. J.* **87**, 745–753.
8. Bhalla, U.S., and Iyengar, R. (1999) Emergent properties of networks of biological signaling pathways. *Science* **283**, 381–387.
9. Blackwell, K.T., and Hellgren-Kotaleski, J. (2002) Modeling the dynamics of second messenger pathways, in *Neuroscience Databases, A Practical Guide* (Kotter, R., ed.), Kluwer Academic Publishers, Boston, MA, pp. 63–80.

10. Jaeger, D. (2001) Accurate reconstruction of neuronal morphology, in *Computational Neuroscience: Realistic Modeling for Experimentalists* (De Schutter, E., ed.), CRC Press, Boca Raton, FL, pp. 159–178.
11. Nenadic, Z., Ghosh, B.K., and Ulinski., P. (2003) Propagating waves in visual cortex: a large scale model of turtle visual cortex. *J. Comput. Neurosci.* **14**, 161–184.
12. Stricanne, B., and Bower, J.M. (1998) A network model of the somatasensory system and cerebellum exploring recovery from peripheral lesions at various developmental stages in rats. *Soc. Neurosci. Abstr.* **24**, 669.
13. Bower, J.M., and Beeman, D. (1998) *The Book of GENESIS: Exploring Realistic Neural Models with the GEneral NEural SImulation System*, 2nd Edition, Springer-Verlag, New York.
14. Beeman, D. (2005) GENESIS Modeling Tutorial. *Brains, Minds, and Media* 1: bmm220(urn:nbn:de:0009-3-2206).
15. Traub, R.D., Wong, R.K., Miles, R., and Michelson, H. (1991) A model of a CA3 hippocampal pyramidal neuron incorporating voltage-clamp data on intrinsic conductances. *J. Neurophysiol.* **66**, 635–650.
16. Hodgkin, A., and Huxley, A. (1952) A quantitative description of membrane current and its application to conduction and excitation in nerve. *J. Physiol. (London)* **117**, 500–544.

8

Simulator for Neural Networks and Action Potentials

Description and Applications

Douglas A. Baxter and John H. Byrne

Summary

A key challenge for neuroinformatics is to devise methods for representing, accessing, and integrating vast amounts of diverse and complex data. A useful approach to represent and integrate complex data sets is to develop mathematical models [Arbib (*The Handbook of Brain Theory and Neural Networks*, pp. 741–745, 2003); Arbib and Grethe (*Computing the Brain: A Guide to Neuroinformatics*, 2001); Ascoli (*Computational Neuroanatomy: Principles and Methods*, 2002); Bower and Bolouri (*Computational Modeling of Genetic and Biochemical Networks*, 2001); Hines et al. (*J. Comput. Neurosci.* **17**, 7–11, 2004); Shepherd et al. (*Trends Neurosci.* **21**, 460–468, 1998); Sivakumaran et al. (*Bioinformatics* **19**, 408–415, 2003); Smolen et al. (*Neuron* **26**, 567–580, 2000); Vadigepalli et al. (*OMICS* **7**, 235–252, 2003)]. Models of neural systems provide quantitative and modifiable frameworks for representing data and analyzing neural function. These models can be developed and solved using neurosimulators. One such neurosimulator is simulator for neural networks and action potentials (SNNAP) [Ziv (*J. Neurophysiol.* **71**, 294–308, 1994)]. SNNAP is a versatile and user-friendly tool for developing and simulating models of neurons and neural networks. SNNAP simulates many features of neuronal function, including ionic currents and their modulation by intracellular ions and/or second messengers, and synaptic transmission and synaptic plasticity. SNNAP is written in Java and runs on most computers. Moreover, SNNAP provides a graphical user interface (GUI) and does not require programming skills. This chapter describes several capabilities of SNNAP and illustrates methods for simulating neurons and neural networks. SNNAP is available at http://snnap.uth.tmc.edu.

Key Words: Synaptic plasticity; modulation; bursting; second messengers; excitability; numerical methods; nonlinear dynamics.

From: *Methods in Molecular Biology, vol. 401: Neuroinformatics*
Edited by: Chiquito Joaquim Crasto © Humana Press Inc., Totowa, NJ

1. Introduction

Processing of information by neural networks emerges from interactions among (1) the intrinsic biophysical properties of neurons, (2) the pattern of synaptic connections within a network, and (3) the physiological properties of the synaptic connections. Moreover, the performance of neural networks is often modified over time by modulatory inputs. Thus, to better understand the nervous system, it is critical to understand the ways in which neural circuits are organized, what principles underlie their function, and the consequences of modulatory influences.

The nonlinearity, diversity, and dynamic nature of neural circuits provide formidable challenges to arriving at a synthetic understanding of how circuits operate and adapt. The specific role that any one process plays in the overall behavior of a network can be difficult to assess because of such factors as interacting nonlinear feedback loops and inaccessibility of the process for experimental manipulation. One approach to overcoming these problems is to mathematically model and simulate the biophysical and biochemical properties of neurons and synapses within networks. The output of the resynthesized network and its sensitivity to manipulation can be examined and thereby provide insights into neural function that may be difficult to achieve otherwise. The development and analysis of models can be enhanced by providing flexible and user-friendly neurosimulators that are capable of simulating key biochemical and biophysical aspects of neuronal and synaptic function and their modulation.

Several neurosimulators are available (*see* **Chapters 6** and **7**) *(1–5)* (for recent reviews of neurosimulators, see **refs** *6,7*). This chapter describes the neurosimulator "simulator for neural networks and action potentials" (SNNAP) *(8,9)*. With SNNAP, voltage-dependent conductances are described by Hodgkin–Huxley (HH)-type equations, and the contributions of time-dependent synaptic conductances are described by second-order differential equations. SNNAP incorporates equations for simulating neuronal modulation and synaptic plasticity. Parameters, initial conditions, and output options are passed to the simulator through modular ASCII files. These files can be modified by commonly available text editors or by a graphical user interface (GUI) that is part of SNNAP. The modular design facilitates the incorporation of existing files into new simulations and allows users to develop libraries of files describing distinctive cell types and neural networks. In this chapter, several commonly used capabilities of SNNAP are illustrated by simulating neurons with various biophysical properties and by simulating simple neural networks with various modes of synaptic transmission and plasticity. In addition, current flow in a branched neuron with multiple compartments is simulated. Finally, the features of a relatively complex neural network are simulated. This network summarizes empirical data that describe a central pattern generator (CPG), which mediates feeding in the marine mollusc *Aplysia*, and illustrates the ways in which computational models can contribute to neuroinformatics.

2. Operational Features of SNNAP

A full description of the operation, features, and capabilities of SNNAP is beyond the scope of this chapter. Rather, this chapter focuses on those aspects of SNNAP that are useful for developing realistic models of neurons, synapses, and neural networks. For example, SNNAP can simulate both integrate-and-fire cells (*.cell) and HH-type neurons (*.neu) (*see* **Note 1**), but only the HH-type neurons are discussed here. Although a few basic operations are described, users are referred to the *SNNAP Tutorial Manual* for step-by-step instructions. The tutorial is included in the download. In addition, the download includes a library of example simulations (/snnap8/examples). These examples provide practical illustrations of the capabilities and features of SNNAP and provide starting points for developing new models. The input files for all models in this chapter are in the example simulations. Additional models are available at the ModelDB Web site (*see* **Chapter 6**) *(10,11)*.

2.1. Hardware and Software

SNNAP is available at http://snnap.uth.tmc.edu/. SNNAP is written in the programming language Java and will run on any computer with a properly installed Java virtual machine (JVM) (*see* **Note 2**). Two pitfalls should be avoided when installing and running SNNAP. First, Java does not interpret blank characters in path names. Thus, folders such as /Program Files or /My Documents are not appropriate for installing SNNAP. Second, European conventions often replace the decimal point in real numbers with a comma. Java does not accept commas in real numbers, and thus, the default system settings of some computers must be modified. For additional details, see the SNNAP Web site.

2.2. Input/Output Controls

A key feature of SNNAP is the embedding of equations for simulating neural networks within the neurosimulator. To develop models, users select equations from lists (*see* **Fig. 1**), which are specified in an array of input files (*see* **Fig. 2B**), and enter values for parameters (*see* **Fig. 1D**). Thus, no programming skills are necessary to use SNNAP (*see* **Note 2**).

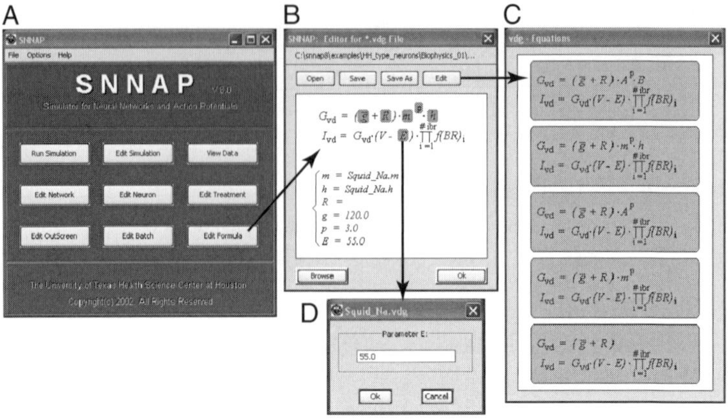

Fig. 1. Graphical user interface (GUI). All aspects of developing models and running simulations can be controlled through a GUI. (**A**) The main control panel provides access to the neurosimulator (Run Simulation) and to editors for each type of input file. For example, to edit a formula, users select Edit Formula, which evokes a file manager (not shown). (**B**) Once the appropriate file is selected (in this example Squid_Na.vdg), the equation and its parameters appear in a new window. (**C**) To change the type of equation, users select Edit, which evokes a new window with a list of equations. Users select a new equation by clicking on the formula. (**D**) Values for the parameters are entered by selecting the appropriate button (e.g., *E*) and typing the value into a pop-up box.

SNNAP is versatile and provides equations that are sufficient for simulating a broad spectrum of biophysical and biochemical processes. For example, SNNAP provides two methods for calculating the time- and voltage-dependent activation and inactivation of ionic conductances. One method is based on the rate-constant method used by Hodgkin and Huxley ***(12)*** (*.m and *.h files), whereas the second method is based on Boltzman functions used by Byrne ***(13,14)*** (*.A and *.B files). Each type of input file provides a list of several equations, which allows users to simulate various neuronal and synaptic properties.

Operational control parameters for a simulation and selected equations with their associated parameters are passed to the neurosimulator through an array of modular plain-text (ASCII) files (*see* **Fig. 2A**). These files are organized hierarchically (*see* **Fig. 2B**). The modular design allows users to incorporate existing files into new simulations and to develop libraries of input files that describe specific types of neurons, synapses, and neural networks (*see* **Note 3**).

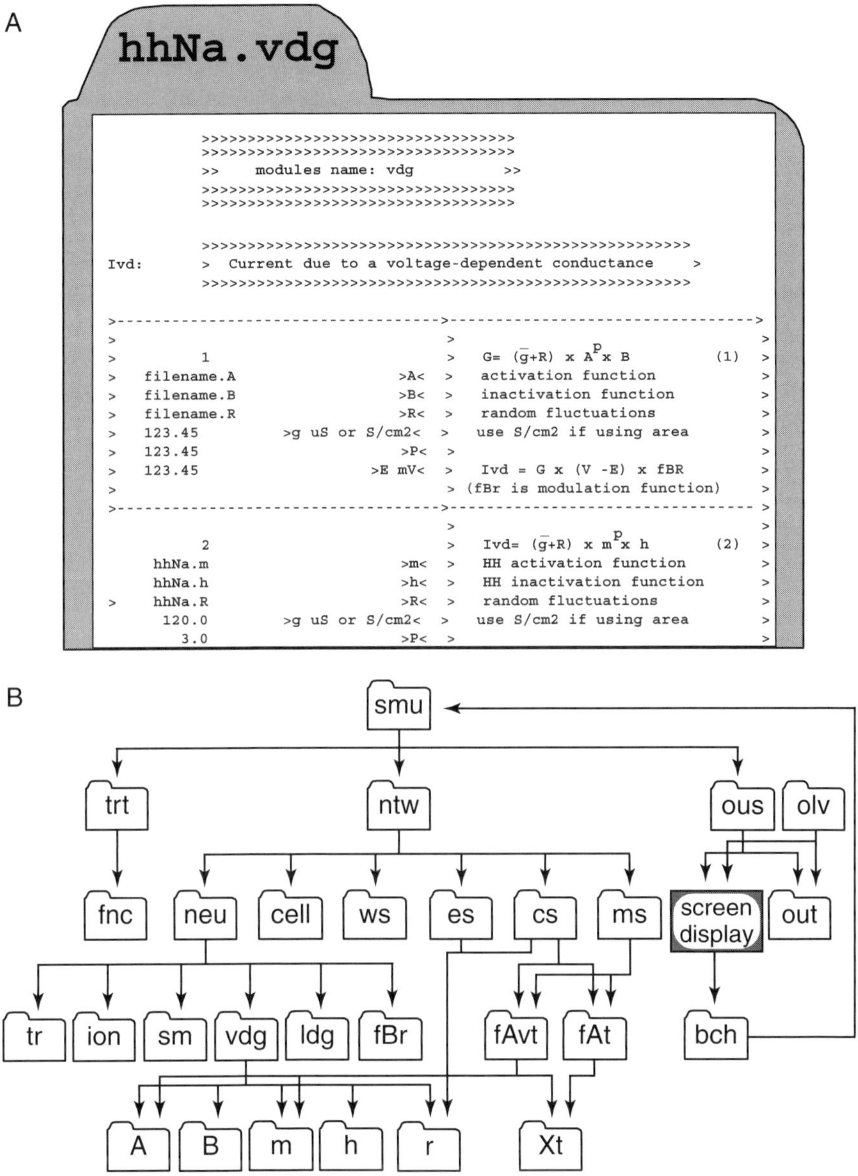

Fig. 2. Input/output (I/O) files. (**A**) Equations, parameters, initial conditions, and operational controls for a simulation are passed to the neurosimulator through a set of plain-text (ASCII) files. This example (hhNa.vdg) illustrates a portion of the Na^+-conductance input file for the Hodgkin–Huxley (HH) model. The right column of the input file provides informational comments. The left column provides information

Within the hierarchical organization (*see* **Fig. 2B**), the simulation file (*.smu) is at the top and directs all other relevant files. The other major files are (1) network (*.ntw), which specifies the neurons in a network and the types and pattern of their synaptic connections; (2) treatment (*.trt), which specifies the timing, targets and magnitudes of external currents, modulatory stimuli, and/or voltage-clamp protocols (*see* **Note 4**); and (3) output (*.ous), which specifies options for the online screen display and/or storing data to a file (*.out). A few additional input files include the following: (1) neuron (*.neu), which specifies the properties of HH-type neurons such as its membrane capacitance, morphological features, membrane conductances (*.vdg and *.ldg), intracellular pools of ions (*.ion), transmitter (*.tr) and second messengers (*.sm), and the modulatory relationships among elements within the neuron (*.fBr). (2) Voltage-dependent conductance (*.vdg), which specifies the maximal conductance and reversal potential of an ionic current. To fully describe

Fig. 2. necessary for the selected equation, such as values for parameters (e.g., maximum conductance, <g>) and names of related files (e.g., hhNa.m). (**B**) The input files are organized in a hierarchical structure. Arrows indicate the relationships among the files. For example, a simulation file (*.smu) indicates which treatment (*.trt), network (*.ntw), and output (*.ous) files are to be used. The other input files are as follows: *.ntw specifies which neurons (*.neu, *.cell) and synaptic connections (*.ws, *.es, *.cs, *.ms) are included in a network; *.ous file specifies which variables are displayed on the screen and/or saved to disk (*.out); *.trt file controls experimental manipulations; *.fnc specifies functions (e.g., sin wave) that control current- and voltage-clamp protocols; *.olv controls displaying data from previous simulations (View Data in **Fig. 1A**); *.neu defines properties of HH-type neurons, including voltage- (*.vdg) and light-dependent (*.ldg) conductances, intracellular pools of ions (*.ion), second messengers (*.sm), and transmitter (*.tr); *.cell defines properties of integrate-and-fire cells; *.ws defines synaptic connections among integrate-and-fire cells; *.es defines electrical coupling between compartments within a neuron and/or between neurons within a network; *.cs defines chemical synapses; *.ms defines modulatory synapses; *.fBr defines modulatory relationships among *.ion, *.sm, *.vdg, and *.tr; *.fAvt defines synapses that are both time- and voltage-dependent; *.fAt defines synapses that are only time-dependent; *.bch controls batch operations; *.A and *.B define Boltzman functions for describing voltage- and time-dependent activation and inactivation of conductances; *.m and *.h define HH-type rate-constant equations for voltage- and time-dependent activation and inactivation of conductances; *.r introduces noise into conductances; and *.Xt defines synaptic strength.

a voltage-dependent conductance, *.vdg also specifies several related files, including *.m, *.h, *.A, and *.B, which specify voltage- and time-dependent activation and inactivation, and *.r, which specifics random fluctuations in the conductance. (3) Files for synaptic connections include electrical synapse (*.es, with related file *.r), chemical synapse (*.cs, with related files *.fAt, *.fAvt, *.Xt, and *.r), and modulatory synapse (*.ms). Annotated examples of all input files are available in the /snnap8/examples, and these examples provide users information about features of SNNAP, which are not covered in this chapter (e.g., *.cell and *.ws), and provide users with a starting point for developing models of their own.

Two methods are available for editing the input files. First, the files can be edited with the GUI that is incorporated into SNNAP (*see* **Fig. 1**). The GUI is based on modern point-and-click operations with the mouse and pop-up windows for text entry. Second, the files can be edited with commonly available plain-text editors. Each method offers its own advantages. The GUI is simpler to use, provides graphical representations of the model, and is better suited for novice users or for educational environments. The plain-text editors are faster and more versatile. For example, plain-text editors offer features such as undo, copy, paste, and find/replace, which are not yet incorporated into the GUI. Moreover, plain-text editors allow advanced users to annotate input files. The two methods are interchangeable, and users can take advantage of both.

2.3. Getting Started

Once JVM, SNNAP, the tutorial, and the example simulations are installed, there are two methods for beginning a simulation (*see* **Note 2**). Users can double click on the snnap8.jar file to launch SNNAP, or users can type java–jar snnap8.jar using the command prompt with the active directory positioned at the /snnap8 folder. Launching SNNAP evokes the main-control window (*see* **Fig. 1A**). The main-control window provides access to the various editors that are used for developing models and controlling simulations and provides access to the neurosimulator. For example, to run a simulation, users select the Run Simulation button, which evokes the simulation window (*see* **Fig. 3B3**). In the simulation window, the File button allows users to select and load a simulation (*.smu). Once a simulation is loaded, users press the Start button to begin the numerical integration and to display the results on the screen. An extensive selection of examples and step-by-step instructions for how to construct and run simulations are provided in the *SNNAP Tutorial Manual* (*see* **Note 3**).

A1 $$\frac{dV_i}{dt} = \frac{I_{Stim} - \sum_{j=1}^{m} Ivd_{ij} - \sum_{k=1}^{n} Ies_{ik} - \sum_{k=1}^{n}\sum_{l=1}^{o} Ics_{ikl}}{Cm_i}$$

A2 $$Ivd_{ij} = \left(\bar{g}vd_{ij} + R_{ij}\right) A_{ij}^{p} B_{ij} \left(V_i - E_{ij}\right) \prod_{q=1}^{z} f[REG_q]$$

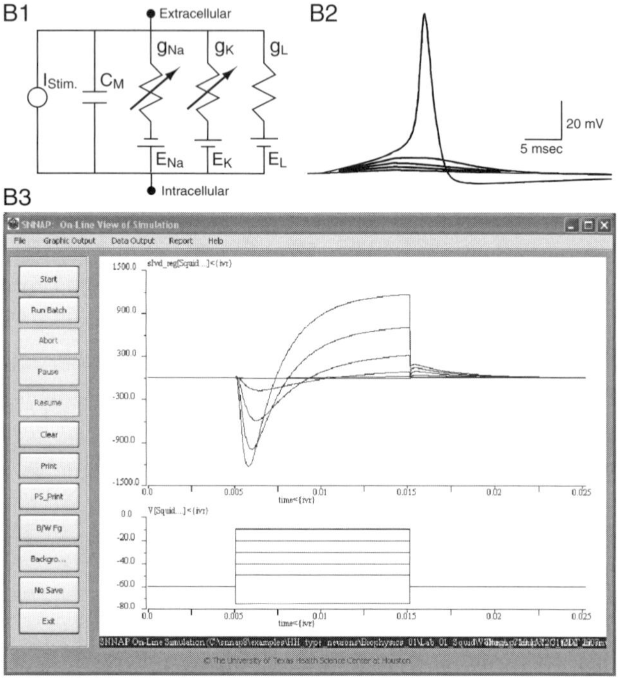

Fig. 3. Neuronal excitability. (**A1**) Ordinary differential equation (ODE) describing the membrane potential of a Hodgkin–Huxley (HH)-type neuron. C_M is the membrane capacitance. I_{Stim} is an extrinsic stimulus current. *Ivd*, *Ies*, and *Ics* are currents because of voltage-dependent conductances, electrical synapses, and chemical synapses, respectively. *i*, *j*, *k*, and *l* are indexes of the neuron, voltage-dependent conductance, presynaptic cell, and synapse, respectively. (**A2**) General equation describing an ionic current. The conductance (*gvd*) is modified by the addition of a random number (*R*), dimensionless voltage- and time-dependent activation and inactivation terms (*A* and *B*, respectively), a driving force (*V–E*), and the product of one or more functions (Π*f[REG]*) that link *gvd* to regulatory agents such as second messengers and/or ions. Many of these terms are optional and are included as needed to describe the unique features of a given ionic conductance. (**B1**) Equivalent electrical circuit of the HH model ***(12)***. The circuit diagram illustrates an external stimulus current (I_{Stim}), the membrane capacitance (C_M), time- and voltage-dependent conductances to Na^+(g_{Na}) and K^+(g_K), and leak conductance (g_L). Each conductance is associated with a battery

2.4. Integration Methods and Parameters

The forward Euler (FE) method is used to numerically integrate the set of differential equations. The accuracy of the FE method, as with any method for numerical integration, is dependent on the choice of the appropriate integration time step (for discussion of integration routines, see **ref. *15***). To avoid errors that arise if the integration time step is too large, users should first make a series of trial simulations beginning with a comparatively large time step and then rerun the simulation with progressively smaller time steps. Users should continue reducing the time step until further reductions do not change the results. The integration time step is specified in the *.sum file.

3. Capabilities of SNNAP

SNNAP was developed for researchers, educators, and students who wish to carry out simulations of neural systems without the need to learn a programing language. SNNAP is both powerful enough to simulate the complex biophysical properties of neurons, synapses, and neural networks and sufficiently user friendly that a minimal amount of time is spent in learning to operate the neurosimulator.

Fig. 3. (E), which represents the driving force. (**B2**) Simulating an action potential with the HH model. Results from five simulations are superimposed. With each simulation, the amplitude of I_{Stim} is increased. The largest stimulus evokes an action potential. SNNAP provides a method for making publication-quality figures. Once the results are displayed (Panel **B3**), users select the PS_Print button, which generates a postscript file of the displayed data, and this file can be modified by postscript compatible editors. (**B3**) Simulation of voltage-clamp experiment with the HH model. This panel illustrates the simulation window, which is evoked by the Run Simulation button in **Fig. 1A**. To run a simulation, first select File, which evokes a drop-down list of options (not shown). From this list, select Load Simulation, which evokes a file manager (not shown). The file manager is used to select the desired simulation. Once the simulation is loaded, users select the Start button to begin the numerical integration of the model and to display the results on the screen. In this example, the batch mode of operation was used (Run Batch) to run six simulations and superimpose the results. The batch mode (*.bch) allows users to automatically modify parameters or combinations of parameters and rerun a simulation. The upper traces illustrate the total membrane current of the HH model, and the lower traces illustrate the membrane potential during the voltage-clamp steps.

3.1. Intrinsic Properties of Neurons

The basic computational element of a SNNAP model is a neuron (*.neu) (*see* **Note 5**). Neurons include voltage-dependent conductances (*.vdg), transmitters (*.tr), intracellular pools of ions (*.ion), and second messengers (*.sm) (*see* **Note 6**). Neurons also make and receive synaptic connections, including chemical (*.cs) (*see* **Note 7**), modulatory (*.ms), and electrical (*.es) synapses. In addition, neurons are targets for treatments (*.trt) such as current injections, voltage-clamping protocols, and applications of modulators. A neuron can be conceptualized as a single, isopotential cell (*see* **Fig. 3**) or as a compartment within a larger, multicompartment model of a neuron (*see* **Fig. 4**).

The ordinary differential equation (ODE) that defines the membrane potential of a HH-type neuron is illustrated in **Fig. 3A1**. The membrane potential is determined by extrinsic stimuli (I_{Stim}), currents generated by voltage-dependent conductances ($\sum Ivd$), and electrical ($\sum Ies$) and chemical ($\sum Ics$) synaptic

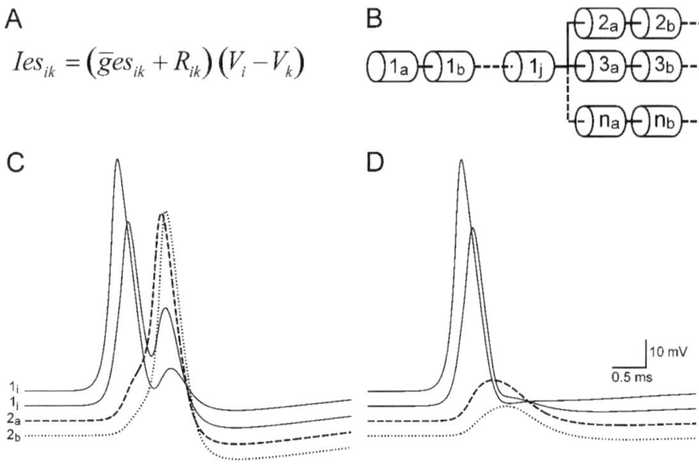

Fig. 4. Multicompartment model. (**A**) Equation defining the electrical coupling between neurons/compartments. Note that the values of ges_{ik} and ges_{ki} are specified independently. Thus, SNNAP can simulate asymmetrical coupling between neurons. (**B**) Schematic illustration of a model by Luscher and Shiner ***(16)***. A primary axon (1) branches into a variable number of daughter branches (2, 3, ... *n*). Each branch contains ten compartments (a–j). The compartments are identical and are represented by a Hodgkin–Huxley (HH) model. (**C**) Simulation of a branch point with seven daughter branches. The spike propagates through the branch point. (**D**) Simulation of a branch point with nine daughter branches. The spikes fail to propagate through the branch point.

inputs. A general form of the equation for a voltage-dependent current is illustrated in **Fig. 3A2**. The equation includes terms for the ionic conductance (*gvd*), a random number variable (*R*), voltage- and time-dependent activation and inactivation functions (*A* and *B*, respectively), an equilibrium potential (*E*), and modulation by one or more functions of intracellular regulatory agents ($\prod f[REG]$).

The electrical properties of a neuron can be summarized in an equivalent electrical circuit. For example, the equivalent electric circuit for the HH model is illustrated in **Fig. 3B1** (for recent reviews of the HH model, see **refs** ***17,18***). The circuit diagram illustrates an external stimulus current (I_{Stim}), the membrane capacitance (C_M), two voltage- and time-dependent ionic conductances (g_{Na} and g_K), and a leakage conductance (g_L). Each conductance is associated with a battery. Simulation of the HH model reproduces many of the biophysical properties of neurons, including a threshold for generating all-or-nothing action potentials (spikes) (*see* **Fig. 3B2**). The membrane currents of the HH model are revealed by simulating a voltage-clamp experiment (*see* **Fig. 3B3**) (*see* **Note 8**). SNNAP allows users to plot any variable or combination of variables, either as a time-series display (*see* **Fig. 3B3**) or as a phase-plane display.

The equations and parameters that define voltage-dependent currents are found in the *.vdg, *.r, *.A, *.B, *.m, *.h, and *.fBr files. In the *.vdg file, five equations are available for defining voltage-dependent currents (*see* **Fig. 1C**). The *.A and *.B files each offer nine equations, and *.m and *.h files each offer 21 equations. In addition, ionic conductances can be modulated by one or more intracellular regulatory agents (*see* **Fig. 5**), and the *.fBr file offers seven equations. With these extensive and diverse options, users can simulate the firing properties of a wide array of neurons (*see* **Figs 3B, 5B, 5C**, and **6**) ***(8,9,23–28)***. For a general discussion of neuronal firing properties, see McCormick ***(18)***.

3.2. Chemical Synaptic Transmission and Homosynaptic Plasticity

Figure 7A1 illustrates a general equation for currents generated by an increase-conductance synapse (*see* **Note 7**). The equation includes terms for the synaptic conductance (*gcs*), a random number variable (*R*), offset and scaling constants (k_1 and k_2, respectively), a time-dependent activation function (*A*), a voltage-dependent activation function (*W*), and an equilibrium potential (*E*). The equations and parameters that define a chemical synaptic connection are found in the *.cs, *.fAt, *.fvAt, *.Xt, *.tr, and *.fBr files. For a general discussion of synaptic potentials and integration, see Byrne ***(29)***.

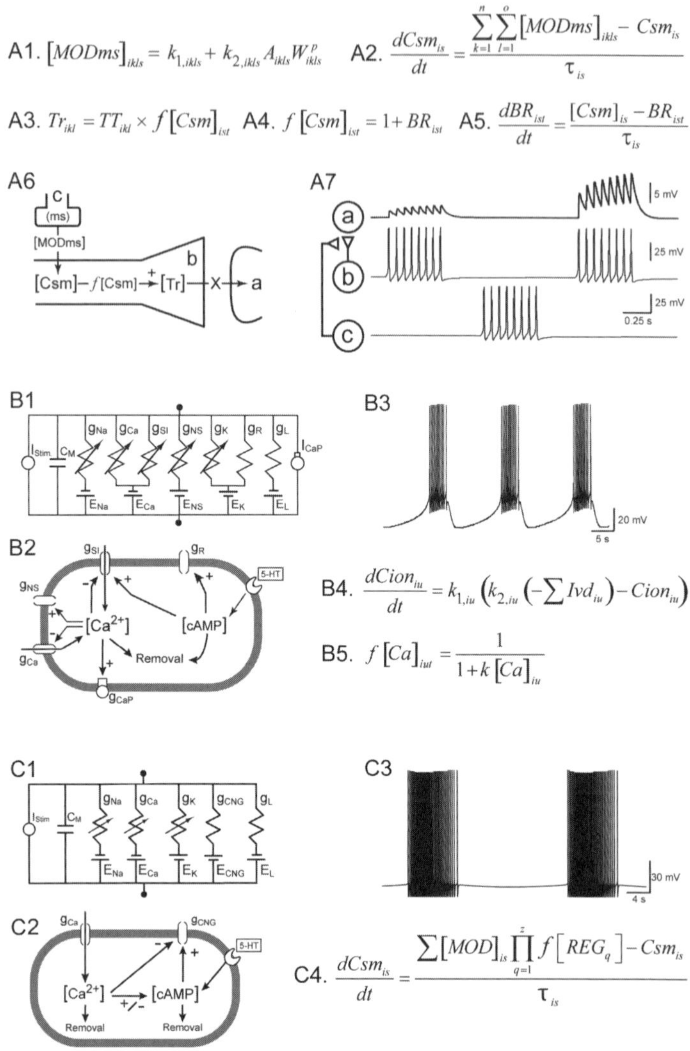

Fig. 5. Modulation. (**A1**) General equation for a modulatory synapse. Terms are defined in **Fig. 7A1**. *s* is an index for the pool(s) of second messenger in post-synaptic cell. Presynaptic activity at a modulatory synapse drives the accumulation of a modulatory agent [*MODms*]. (**A2**) An equation describing the concentration of a second messenger. The accumulation of a second messenger is driven by the modulatory agent(s) released by the synapse(s) linked to the second messenger and/or application of an extrinsic treatment (not shown). (**A3**) An equation describing the concentration of transmitter. Changes in transmitter release are produced through the term X in **Fig. 7A2**. One of the options is $X = Tr$, and one of the options for Tr is to make Tr

3.2.1. Excitatory and Inhibitory Synaptic Potentials and Homosynaptic Plasticity

The *.fAt (and *.fvAt) file offers several options for defining the time-dependent activation of synaptic conductances (the A term in **Fig. 7A1** and **7C1**). The most versatile option is a second-order differential equation (*see* **Fig. 7A2**) in which Y is the variable describing the second-order system, X, the forcing function, and τ, the time constant (for additional details, see **refs** ***30–32***).

Fig. 5. a function of the concentration of a second messenger (*f[Csm]*) (Panels **A4** and **A5**). (**A4** and **A5**) Equations describing a *positive* relationship between the concentration of a second messenger and its target (i.e., up-regulation). t is an index for the target(s) of the second messenger, and BR is the concentration- and time-dependent function that links synaptic strength to the second messenger. (**A6**) Schematic of a model for heterosynaptic plasticity. Activity in a modulatory synapse (*ms*) produces a modulatory agent [*MODms*], which in turn drives the accumulation of a second messenger [*Csm*] in the postsynaptic cell. The second messenger modulates the magnitude of *Tr*, which in turn alters the magnitude of the synaptic driving function (X) (*see* **Fig. 7A2**). (**A7**) Simulation of heterosynaptic plasticity. Two bursts of activity in cell b produce excitatory postsynaptic potentials (EPSPs) in cell a. Modulatory cell c makes a connection with cell b, and a burst of activity in cell c facilitates the b-to-a connection. (**B1**) Equivalent electrical circuit of the bursting neuron R15 in the *Aplysia* (for details, see **ref.** ***19***). (**B2**) Schematic illustration of the relationships among membrane conductances (g) and intracellular pools of ions [Ca^{2+}] and second messengers [cAMP]. Note that intracellular regulatory agents (e.g., [Ca^{2+}]) can affect multiple targets (e.g., g_{NS} and g_{Ca}) each with a different action (e.g., up-regulate, +, g_{NS} versus down-regulate, –, g_{Ca}), and multiple intracellular regulatory agents (e.g., [Ca^{2+}] and [cAMP]) can converge on a common target (e.g., g_{SI}). (**B3**) Simulation of bursting in R15. (**B4**) An equation describing the concentration of an intracellular pool of ions. k_1 and k_2 are constants, and u is the index for the ion pool. One or more ionic currents contributes to a pool (ΣIvd). (**B5**) An equation describing a *negative* relationship between the concentration of an intracellular ion and its target (i.e., down-regulation). (**C1**) Equivalent electrical circuit of a hypothetical neuron. (**C2**) Schematic of a hypothetical neuron in which bursting involves a cyclic-nucleotide-gated conductance (g_{CNG}). Note that intracellular regulators can influence each other; for example, the bimodal regulation of [cAMP] by [Ca^{2+}] (Panel **C4**) ***(20)***. (**C3**) Simulation of bursting in the presence of serotonin (5-HT). (**C4**) An equation describing the concentration of a second messenger that is regulated by both the presence of a modulator [*MOD*] and the product of one or more functions of intracellular regulators ($\Pi f[REG]$).

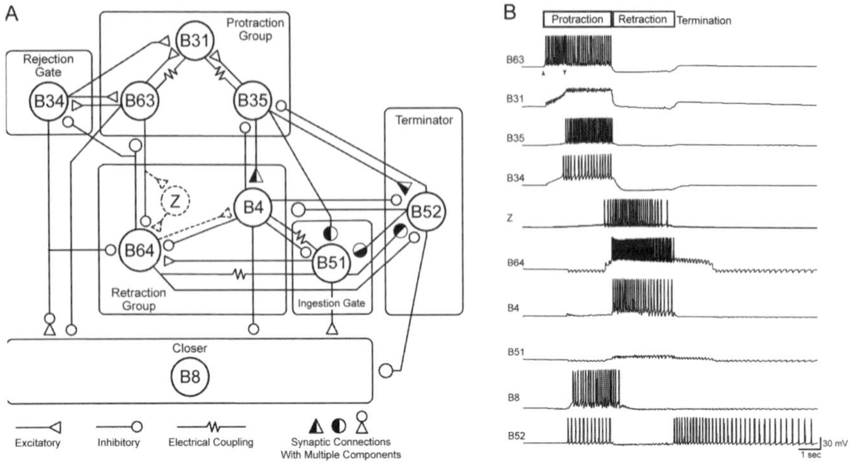

Fig. 6. Neural network. (**A**) Schematic illustrating several identified neurons and synaptic connections within a central pattern generator (CPG) that controls feeding in *Aplysia* (for description of the CPG, see **refs** ***21,22***). (**B**) Simulation of neural activity. A brief extrinsic depolarizing current pulse, which is applied to cell B63 (arrow heads), initiates a complex pattern of activity within the neural network. This pattern of activity is similar to activity that is observed empirically. This model and simulations summarize empirical data from at least 12 published studies.

In the simplest case, $X = 1$ during a presynaptic spike, otherwise $X = 0$. When X is an impulse, the explicit solution of this second-order system is the alpha function that is often used to describe the dynamics of synaptic conductance changes ***(33)***. However, X does not have to be simply 1 or 0, and the *.Xt file offers several options for defining X. For example, X can equal an activity-dependent function such as the plasticity function *PSM* in **Fig. 7A3**. As the amplitude of X increases, so will synaptic strength, and conversely, as the amplitude of X decreases, so will synaptic strength (*see* **Fig. 7B1**; see also **ref.** ***34***). Thus, by making X a variable and linking its amplitude to presynaptic activity, it is possible to simulate homosynaptic plasticity.

Several additional methods can be used to implement homosynaptic plasticity. First, if $X = 1$ for the duration of a presynaptic spike, then increases (or decreases) in the spike duration will increase (or decrease) synaptic strength. Second, the amplitude of X can be linked to a pool of transmitter (*Tr*), and *Tr* (*.tr) can be modulated by intracellular ions (*.ion and *.fBr). As ions accumulate during presynaptic activity, synaptic strength will increase (or decrease). Finally, the amplitude of X can equal *PSMxTr*, where *Tr* can be

A1 $$Ics_{ikl} = \left(\bar{g}cs_{ikl} + R_{ikl}\right)\left(k_{1,ikl} + k_{2,ikl}A_{ikl}W_{ikl}^{p}\right)\left(V_i - E_{ikl}\right)$$

A2 $$\frac{d^2Y_{ikl}}{dt^2} = \frac{X_{ikl} - 2\tau_{ikl}\frac{dY_{ikl}}{dt} - Y_{ikl}}{\tau_{ikl}^2}$$

A3 During a presynaptic spike $$\frac{dPSM_{ikl}}{dt} = -\frac{PSM_{ikl}}{\tau_d}$$ In the absence of a presynaptic spike $$\frac{dPSM_{ikl}}{dt} = \frac{\left(1 - PSM_{ikl}\right)}{\tau_r}$$

B1

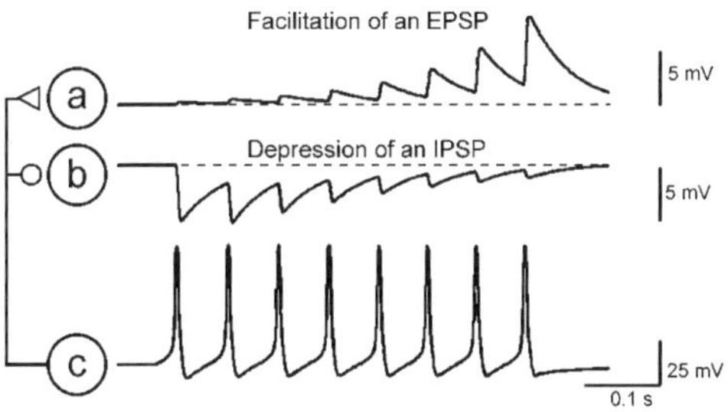

B2

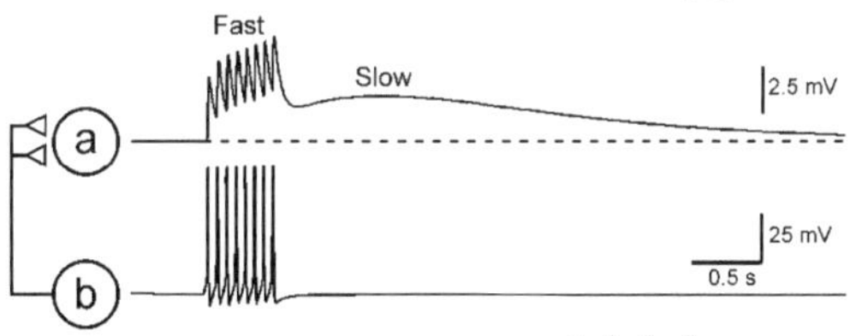

B3

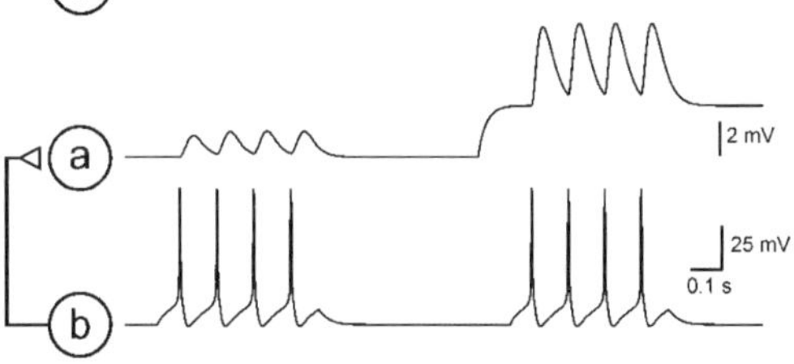

C1 $$Ics_{ikl} = \left(\bar{g}cs_{ikl} + R_{ikl}\right)\frac{1}{1 + k_{ikl}A_{ikl}W_{ikl}^{p}}\left(V_i - E_{ikl}\right)$$

C2

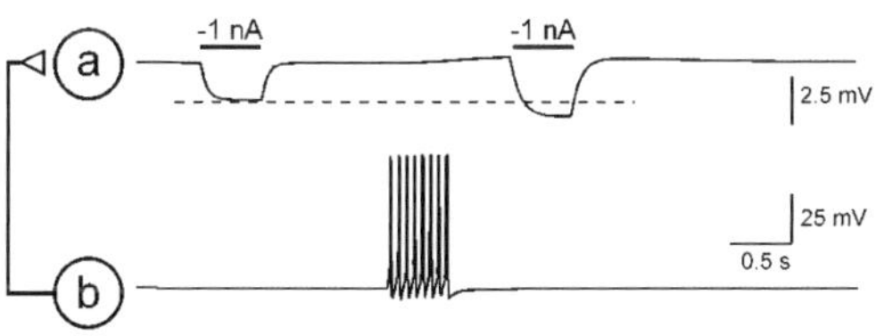

modulated by one or more intracellular agents or *Tr* can be equal to another *PSM* function. The interactions among *X*, *Tr*, *PSM* functions, intracellular regulatory agents, and spike duration allow users to simulate complex forms of homosynaptic plasticity.

3.2.2. Multicomponent Postsynaptic Potentials

Each postsynaptic cell *i* can receive up to *o* synaptic connections from each presynaptic cell *k* (*see* **Fig. 3A1**). The properties of each synaptic connection are defined independently. Thus, users can simulate synaptic connections with multiple, distinct components. For example, a synaptic connection can have both a fast and a slow component (*see* **Fig. 7B2**). The fast versus slow components are created by selecting different time constants for the time-dependent

Fig. 7. Chemical synaptic transmission and homosynaptic plasticity. (**A1**) General equation for an increase-conductance postsynaptic potential (PSP). Currents resulting from chemical synapses are calculated in a manner similar to voltage-dependent currents (**Fig. 3A2**). k_1 and k_2 are constants. *A* and *W* are time- and voltage-dependent functions, respectively. Several options are available for the term *A*, and they are defined in the *.fAt file. If the term *W* is included, users specify either *.A or *.m file in the *.fAvt file (**Fig. 2B**). (**A2**) Second-order differential equation for a time-dependent change in synaptic conductance. See text for details. (**A3**) ODE describing a method for homosynaptic plasticity. Several options are available for defining the term *X* in Panel **A2**. One option is *X* = *PSM*, which provides a mechanism for simulating homosynaptic plasticity. Plasticity develops during a presynaptic spike with the time constant τ_d, and in the absence of a spike, plasticity recovers with the time constant τ_r. (**B1**) Homosynaptic plasticity of EPSPs and IPSPs (inhibitory postsynaptic potentials). Cell c makes an inhibitory connection with cell b and an excitatory connection with cell a. If τ_d (Panel **A3**) is negative, then PSPs facilitates during presynaptic activity, and conversely, positive values produce depression. (**B2**) PSPs with multiple components. Cell b makes two excitatory connections with cell a. One of the connections has fast kinetics, whereas the other has slow kinetics. The kinetics of a PSP are determined by τ in Panel **A2**. (**B3**) Time- and voltage-dependent PSP. The postsynaptic cell (a) is depolarized during the second burst of presynaptic activity, and this depolarization increases the amplitude of the EPSPs. (**C1**) General equation for a decrease-conductance PSP. Terms are defined in Panel **A1**. (**C2**) Simulation of a decrease-conductance synaptic connection. The input conductance of the postsynaptic cell (a) is monitored by injecting –1 nA current pulses. A burst of spikes in the presynaptic cell (b) produces a slight depolarization of cell a and reduces its resting conductance.

activation functions for the two distinct components (e.g., τ in **Fig. 7A2**). Many other combinations are possible. For example, postsynaptic potentials (PSPs) can have both excitatory and inhibitory components, or they may have both increase- and decrease-conductance components *(35)*.

3.2.3. Time- and Voltage-Dependent PSPs

Some synaptic conductances are both time and voltage dependent. The term W (*see* **Fig. 7A1** and **7C1**) defines the voltage-dependent properties of a synaptic current. To simulate a synaptic conductance with voltage-dependent properties, users specify a *.fvAt file (rather than a *.fAt file). In the *.fvAt, users specify both a time-dependent function and a *.A (or *.m) file. The *.A (or *.m) file provides a list of equations that can be used to define the voltage-dependent properties of synaptic conductances. For example, the amplitude of the EPSP in **Fig. 7B3** is increased when the postsynaptic cell (a) is depolarized. In this example, the voltage-dependent component of the synaptic activation was defined by the function $1/(1+\exp^{V+58})$. With the *.A and *.m files, users have many options for defining the voltage-dependent properties of a synapse. In addition, voltage-dependent synaptic conductances can incorporate all of the features of homosynaptic plasticity that are described above.

3.2.4. Decrease-Conductance PSPs

In addition to increasing the conductance of a postsynaptic cell (*see* **Fig. 7A1**), synaptic inputs also can decrease the conductance of a postsynaptic cell. Decrease-conductance PSPs are often mediated by intracellular second-messenger systems that regulate one or more membrane conductance. SNNAP provides two methods for simulating decrease-conductance PSPs. First, users can simulate modulatory synapses (*.ms) that drive the accumulation of a second messenger that in turn can be linked to one or more membrane conductance. Second, the *.fAt (and *.fvAt) file offers users a simpler method. Users can select the equation that is illustrated in **Fig. 7C1**. The equation includes terms for the synaptic conductance (gcs), a random number variable (R), a scaling constant (k), a time-dependent activation function (A), a voltage-dependent activation function (W), and an equilibrium potential (E). In this function, the synaptic conductance is inversely related to the activation function(s). Thus, as the activation of the synapse increases (A and/or W), the synaptic conductance decreases (*see* **Fig. 7C2**). The extent to which the decrease conductance causes a change in the membrane potential will depend on the equilibrium potential of the synapse. Control of the kinetics and plasticity at decrease-conductance synapses are identical to those for increase-conductance synapses.

In addition to chemical synaptic connections, SNNAP can simulate modulatory synapses with similar biophysical properties (*see* **Fig. 5A**). Because any number of the biophysical properties can be combined (e.g., homosynaptic plasticity, increase- and/or decrease-conductance, and chemical and/or modulatory), SNNAP provides users with the versatility to simulate synaptic contacts with a very wide array of distinctive properties.

3.3. Modulation

Modulation is mediated through intracellular pools of ions and/or second messengers (*see* **Note 6**). A neuron can have multiple pools of ions and/or second messengers. The dynamics of an ion pool are defined in the *.ion file. The *.ion file provides two options. One of the options was adapted from Epstein and Marder *(36)* and is illustrated in **Fig. 5B4**. The accumulation of an ion is driven by one or more voltage-dependent currents (the term ΣIvd in **Fig. 5B4**), which users specify in the *.neu file, and the removal of an ion is defined to be a first-order process. The second option (not shown) was adapted from Gingrich and Byrne *(37)* and includes additional expressions for removal of an ion through active uptake and diffusion.

The dynamics of a second messenger pool are defined in the *.sm file, which also provides two options (*see* **Fig. 5A2** and **5C4**). The accumulation of a second messenger is driven by treatments and/or by modulatory synapses (*ms*) (*see* **Fig. 5A**). For example, in **Fig. 5A6**, activity in the cell c releases a modulatory transmitter [*MODms*] that drives the production of a second messenger [*Csm*] in cell b. Alternatively, in **Fig. 5B2** and **5C2**, the accumulation of a second messenger [cAMP] is driven by the exogenous application(s) of serotonin (5-HT). The accumulation of a second messenger also can be regulated by other ion and/or second messenger pools. For example, in **Fig. 5C2**, accumulation of cAMP is regulated by both the intracellular levels of Ca^{2+} and the exogenous application of 5-HT. Moreover, the Ca^{2+}-mediated regulation of cAMP is bimodal (+/−). Low concentrations of Ca^{2+} enhance (+) the accumulation of cAMP, whereas higher concentrations a Ca^{2+} reduce (−) the accumulation of cAMP. Thus, intracellular regulatory agents can have complex interactions.

The modulatory relationship between intracellular pools and their targets is defined by functions in the *.fBr file, which provides several options for up- and down-regulation (*see* **Fig. 5A4, 5A5**, and **5B5**). The targets for modulation are membrane conductances (*see* **Fig. 3A2**), transmitters (*see* **Fig. 5A6**), and second messengers (*see* **Fig. 5C2** and **5C4**).

The interactions among ion and/or second messenger pools and their targets can manifest diverse dynamics and can endow neurons with complex biochemical and electrical properties. For example, the accumulation of an ion during spike activity can feedback and down (or up-)-regulate membrane conductances (*see* **Fig. 5B2**). Such feedback loops between spike activity and ion pools often underlie bursting activity in neurons. A model of a bursting neuron is illustrated in **Fig. 5B**, which simulates bursting in the R15 neuron of *Aplysia* ***(19,20)***. A more complex mechanism for bursting is illustrated **Fig. 5C**, which simulates bursting by combining a metabotropic receptor, cyclic-nucleotide-gated conductance (g_{CNG}), and intracellular Ca^{2+} ***(38)***.

3.3.1. Heterosynaptic Plasticity

The properties of a modulatory synapse (e.g., time-dependent activation, voltage-dependent activation, and homosynaptic plasticity) are identical to those of a chemical synapse (*see* **Fig. 7**) and are defined in the *.fAt, *.fvAt, *.Xt, and *.tr files. However, modulatory synapses do not directly alter the conductance of their postsynaptic targets. Rather, presynaptic activity at a modulatory synapse drives the accumulation of a modulatory agent [*MODms*] according to the general equation in **Fig. 5A1** (*.ms). The modulatory agent can be conceptualized as a modulatory transmitter (e.g., 5-HT), and the concentration of the modulatory agent, in turn, drives the accumulation of a second messenger in the postsynaptic cell (*see* **Fig. 5A2**). This second messenger can be linked to membrane conductances (*see* **Fig. 5B2**), to other second messengers, and/or to a pool of transmitters (*see* **Fig. 5A3** and **5A6**). The functions that define the link between the concentration of the second messenger (*f[Csm]*) and the pool of transmitter (*Tr*) are in the *.fBr file. An example of a modulatory function that up-regulates the pool of transmitter is illustrated in **Fig. 5A4** and **5A5**. Such a modulatory function can be used, for example, to simulate heterosynaptic facilitation (*see* **Fig. 5A6** and **5A7**). Activity in the modulator cell c drives the accumulation of a second messenger [*Csm*] in cell b, which in turn up-regulates the amplitude of *Tr* and increases synaptic strength. Thus, a burst of activity in cell c facilitates the synaptic connection from cell b to its target (cell a) (*see* **Fig. 5A7**). The magnitude and kinetics of the heterosynaptic facilitation are determined by functions and parameters in the *.ms, *.sm, and *.fBr files (*see* **Fig. 5A1–5A5**) and the level of spike activity in the modulatory cell (cell c in **Fig. 5A6 and 5A7**).

3.3.2. Intracellular Ion Pools

Equations for intracellular pools of ions allow users to simulate activity-dependent modulation. For example, an ion may accumulate during spike activity, and the ion pool, in turn, can modulate (up- and/or down-regulate) membrane conductances (*see* **Fig. 5B**), the size of a transmitter pool, and/or the accumulation of another second messenger (*see* **Fig. 5C**). The relationships between ion pools and their targets can be relatively complex (*see* **Fig. 5B2**). More than one voltage-dependent current can contribute to an ion pool (termed Σ *Ivd* in **Fig. 5B4**), and an ion pool can modulate more than one conductance. The actions of the ion pool can up- and/or down-regulate a conductance. In addition, ion pools can regulate the accumulation of second messengers (*see* **Fig. 5C2** and **5C4**).

The positive and/or negative feedback that can be simulated using ion pools is critical for modeling bursting activity in neurons. For example, in a model of the bursting neuron R15 in *Aplysia* (*see* **Fig. 5B1** and **5B3**), Ca^{2+} accumulation during a burst of spikes down-regulates the slow inward conductance (g_{SI}), which underlies the slow depolarization that drives spike activity. The negative feedback loop between g_{SI} and [Ca^{2+}] endows R15 with an intrinsic ability to spontaneously burst (i.e., an endogenous burster). Ion pools also can be used to simulate ion-regulated conductances such as the Ca^{2+}-dependent K^+ current ***(23,39,40)***.

3.3.3. Interactions Between Intracellular Ions and Second Messengers

The actions of a single intracellular pool can diverge and modulate multiple targets, and the actions of multiple intracellular pools can converge and modulate a common target (*see* **Fig. 5B2**). In addition, intracellular pools can interact and modulate the accumulation of second messengers (*see* **Fig. 5C**). For example, the accumulation of a second messenger (e.g., [cAMP]) can be a function of both a modulatory agent (e.g., serotonin, 5-HT) and the intracellular concentration of Ca^{2+} (*see* **Fig. 5C2** and **5C4**). Moreover, the modulatory relationships among intracellular pools can be complex, such that the target is up-regulated over a select range of concentrations and down-regulated at other concentrations.

The rich dynamical relationship between intracellular pools and their targets allows users to simulate relatively complex biochemical processes. For example, the bimodal modulation of a second messenger in combination with a cyclic-nucleotide-gated conductance (g_{CNG}) can provide a method for simulating a conditional burster, that is, a cell that only bursts in the presence of a modulatory agent (*see* **Fig. 5C1** and **5C3**). In the presence of 5-HT, the model of **Fig. 5C1** and **5C2** produces bursting activity (*see* **Fig. 5C3**). The presence of 5-HT drives the accumulation of cAMP, which activates g_{CNG}. The current that flows through the g_{CNG} depolarizes the cell and initiates spiking. The accumulation of Ca^{2+} during the spike activity has several effects. The initial low levels of Ca^{2+} enhance the accumulation of cAMP and thereby enhance spiking. As the spiking continues and Ca^{2+} levels increase, however, the higher levels of Ca^{2+} inhibit both g_{CNG} and the accumulation of cAMP. This negative feedback terminates spiking, and the cell remains silent until the levels of Ca^{2+} have been reduced sufficiently to allow g_{CNG} to be activated once again. Thus, the interactions among membrane conductances and intracellular pools of ions and second messengers lead to oscillations in the membrane potential (*see* **Fig. 5C3**).

3.4. Multicompartment Models

The current generated by electrical coupling between neurons is calculated by the equation illustrated in **Fig. 4A**. Conceptually, the electrical coupling can be between neurons in a neural network (*see* **Fig. 6**) or between compartments within a single cell (*see* **Fig. 4B**). Compartmental modeling is used to simulate the flow of currents in time and space within the complex morphology that characterizes neurons ***(33,41–45)***. With this method, a neuron is modeled as a system of electrically coupled compartments, and the biophysical properties of each compartment are defined to match the unique morphology and/or channel distribution of a segment of the cell. The electrical coupling features of SNNAP can be used to simulate the properties of branched neurons ***(8,39,40,46–49)***. **Figure 4B** illustrates an SNNAP implementation of a model originally developed by Luscher and Shiner ***(16)***. The model originally was used to examine the ways in which increasing the number of branches influences spike propagation. All compartments have the same dimensions and electrical properties. In the simulations, a spike propagates through a branch point with seven branches (*see* **Fig. 4C**) but not through a branch point with nine branches (*see* **Fig. 4D**).

3.5. Neural Networks

Bringing together many capabilities of SNNAP, it is possible to simulate relatively complex neural networks. For example, the neural network that is illustrated in **Fig. 6A** represents elements of a CPG that mediates feeding in the marine mollusk *Aplysia* (for reviews, see **refs** ***21,22,50***). Many of the cells in the CPG have complex biophysical properties. For example, cells B31, B34, B64, and B51 generate plateau potentials, and cell B52 has a strong tendency for postinhibitory rebound excitation. Many of the cells are electrically coupled and function as modules or subnetworks. Moreover, the chemical synaptic connections within the network are diverse and complex. Some synaptic connections undergo activity-dependent facilitation, whereas others express depression, and many of the synapses have multiple components. The versatility of SNNAP makes it possible to develop models of cells and synapses that match the empirical data describing this CPG, and simulation of the model (*see* **Fig. 6B**) closely matches empirical observations. The model can be used to test whether the present understanding of the circuit is sufficient to account for empirical observations and to evaluate the contributions of component processes to the overall activity of the neural network. In addition, the model provides a quantitative summary of data that were published in a dozen or more papers over the span of several decades.

4. Conclusions

Computational models offer a useful tool to neuroinformatics. For example, the HH model consists of only 4 ODEs, 6 algebraic expressions, and 28 parameters, yet this simple model summarizes data that were published in four papers. Moreover, subsequent simulations of the HH model revealed aspects of neuronal function that were not originally considered ***(51)***. Thus, models provide neuroscientists with a method for gaining new insights into neural function and provide neuroinformatics with a quantitative, concise, and dynamic method for representing large and complex data sets ***(52–54)***. Many such models are available at the ModelDB Web site (*see* Chapter 6) ***(10,11)***, and similar databases that provide public access to models are an increasingly important aspect of neuroinformatics ***(55)***. As neurosimulators continue to develop and to play an important role in neuroinformatics, several issues should be addressed. For example, current neurosimulators do not offer adequate methods for annotating models and linking models to data bases ***(52,54–56)***. In addition, current neurosimulators (e.g., GENESIS, NEURON, and SNNAP) do not offer methods for interoperability. The input/output (I/O) files for one neurosimulator are not compatible with another. Because each neurosimulator embodies a unique set of advantages, a greater benefit could be realized with a degree of interoperability (*see* Chapter 5) ***(41,52,57–60)***.

Notes

1. SNNAP can simulate networks that contain both HH-type neurons (*.neu) and integrate-and-fire-type cells (*.cell). Users are provided with several rules that govern plasticity in the synapses of integrate-and-fire-type cells (*.ws).
2. SNNAP is written in Java and runs on most computers. No programming skills are necessary to use SNNAP, and novice users can learn to use the program quickly. The software, example files, and tutorial are available at http://snnap.uth.tmc.edu.
3. SNNAP includes a suite of over 100 example simulations that illustrate the capabilities of SNNAP and that can be used as a tutorial for learning how to use SNNAP, as an aid for teaching neuroscience and as a starting point for developing new models.
4. SNNAP simulates a number of experimental manipulations, such as injecting current into neurons, voltage-clamping neurons, and applying modulators to neurons. In addition, SNNAP can simulate noise applied to any conductance (i.e., membrane, synaptic, or coupling conductances).
5. SNNAP simulates levels for biological organization, which range from second messengers within a cell to multicompartment models of a single cell to large-scale networks. Within a neural network, the number of neurons and synaptic connections (electrical, chemical, and modulatory) is limited only by the memory available in the user's computer.
6. SNNAP simulates intracellular pools of ions and/or second messengers that can modulate neuronal processes such as membrane conductances and transmitter release. Moreover, the descriptions of the ion pools and second-messenger pools can include serial interactions as well as converging and diverging interactions.
7. Chemical synaptic connections can include a description of a pool of transmitter that is regulated by depletion and/or mobilization and that can be modulated by intracellular concentrations of ions and second messengers. Thus, users can simulate homosynaptic and heterosynaptic plasticity.
8. SNNAP includes a Batch Mode of operation, which allows users to assign any series or range of values to any given parameter or combination of parameters. The Batch Mode automatically reruns the simulation with each new value and displays, prints, and/or saves the results.
9. Morphological parameters, such as the length and diameter of neuronal compartments, can be included in descriptions of neurons, and SNNAP will calculate the membrane area and scale membrane conductances accordingly.

Acknowledgments

We thank Drs. E. Av-Ron and G. Phares for helpful comments on an earlier draft of this manuscript. This work was supported by NIH grants R01RR11626 and P01NS38310.

References

1. Bower, J. M. and Beeman, D. (eds) (1998) *The Book of Genesis: Exploring Realistic Neural Models with the GEneral NEural SImulation Sytem*, Second Edition. Springer-Verlag, New York, NY.
2. Bower, J. M., Beemand, D., and Hucka, M. (2003) GENESIS simulation system, in *The Handbook of Brain Theory and Neural Networks* (Arbib, M. A., ed.). The MIT Press, Cambridge, MA, pp. 475–478.
3. Hines, M. L. and Carnevale, N. T. (1997) The NEURON simulation environment. *Neural Comput.* **15**, 1179–1209.
4. Hines, M. L. and Carnevale, N. T. (2001) NEURON: a tool for neuroscientists. *Neuroscientist* **7**, 123–135.
5. Carnevale, N. T. and Hines, M. L. (2005) *The NEURON Book.* Cambridge University Press, Cambridge, UK.
6. Hayes, R. D., Byrne, J. H., and Baxter, D. A. (2003) Neurosimulation: tools and resources, in *The Handbook of Brain Theory and Neural Networks*, Second Edition (Arbib, M. A., ed.). MIT Press, Cambridge, MA, pp. 776–780.
7. Kroupina, O. and Rojas, R. (2004) *A Survey of Compartmental Modeling Packages.* Free University of Berlin, Institute of Computer Science, Technical Report B-04-08. Available at http://www.inf.fu-berlin.de/inst/ag-ki/ger/b-04-08.pdf.
8. Ziv, I, Baxter, D. A., and Byrne, J. H. (1994) Simulator for neural networks and action potentials: description and application. *J. Neurophysiol.* **71**, 294–308.
9. Hayes, R. D., Byrne, J. H., Cox, S. J., and Baxter D. A. (2005) Estimation of single-neuron model parameters from spike train data. *Neurocomputing* **65-66C**, 517–529.
10. Hines, M. L., Morse, T., Migliore, M., Carnevale, N. T., and Shepherd, G. M. (2004) ModelDB: a database to support computational neuroscience. *J. Comput. Neurosci.* **17**, 7–11.
11. Migliore, M., Morse, T. M., Davison, A. P., Marenco, L., Shepherd, G. M., and Hines, M. L. (2003) ModelDB: making models publicly accessible to support computational neuroscience. *Neuroinformatics* **1**, 135–139.
12. Hodgkin, A. L. and Huxley, A. F. (1952) A quantitative description of membrane current and its application to conduction and excitation in nerve. *J. Physiol. (Lond.)* **117**, 500–544.
13. Byrne, J. H. (1980) Analysis of ionic conductance mechanisms in motor cells mediating inking behavior in *Aplysia californica. J. Neurophysiol.* **43**, 630–650.
14. Byrne, J. H. (1980) Quantitative aspects of ionic conductance mechanisms contributing to firing pattern of motor cells mediating inking behavior in *Aplysia californica. J. Neurophysiol.* **43**, 651–668.
15. Mascagni, M. V. (1989) Numerical methods for neuronal modeling, in *Methods in Neuronal Modeling* (Koch, C. and Segev, I., eds). MIT Press, Cambridge, MA, pp. 439–486.

16. Luscher, H.-R. and Shiner, J. S. (1990) Computation of action potential propagation and presynaptic bouton activation in terminal arborizations of different geometries. *Biophys. J.* **58**, 1377–1388.
17. Baxter, D. A., Canavier, C. C., and Byrne, J. H. (2004) Dynamical properties of excitable membranes, in *From Molecules to Networks: An Introduction to Cellular and Molecular Neuroscience* (Byrne, J. H. and Roberts, J., eds). Academic Press, San Diego, CA, pp. 161–196.
18. McCormick, D. A. (2004) Membrane potential and action potential, in *From Molecules to Networks: An Introduction to Cellular and Molecular Neuroscience* (Byrne, J. H. and Roberts, J., eds). Academic Press, San Diego, CA, pp. 115–140.
19. Butera, R. J., Clark, J. W., Canavier, C. C., Baxter, D. A., and Byrne, J. H. (1995) Analysis of the effects of modulatory agents on a modeled bursting neuron: dynamic interactions between voltage and calcium dependent systems. *J. Comput. Neurosci.* **2**, 19–44.
20. Yu, X., Byrne, J. H., and Baxter, D. A. (2005) Modeling interactions between electrical activity and second-messenger cascades in *Aplysia* neuron R15. *J. Neurophysiol.* **91**, 2297–2311.
21. Cropper, E. C., Evans, C. G., Hurwitz, I., Jing, J., Proekt, A, Romero, A., and Rosen, S. C. (2004) Feeding neural networks in the mollusc *Aplysia*. *Neurosignals* **13**, 70–86.
22. Elliott, C. J. and Susswein, A. J. (2002) Comparative neuroethology of feeding control in molluscs. *J. Exp. Biol.* **205**, 877–896.
23. Baxter, D. A., Canavier, C. C., Clark, J. W., and Byrne, J. H. (1999) Computational model of the serotonergic modulation of sensory neurons in *Aplysia*. *J. Neurophysiol.* **82**, 2914–2935.
24. Komendantov, A. O. and Kononenki, N. I. (2000) Caffeine-induced oscillations of the membrane potential in *Aplysia* neurons. *Neurophysiology* **32**, 77–84.
25. Pelz, C., Jander, J., Rosenboom, H., Hammer, M., and Menzel, R. (1999) IA in Kenyon cells of the mushroom body of honeybees resembles shaker currents: kinetics, modulation by K^+, and simulation. *J. Neurophysiol.* **81**, 1749–1759.
26. Steffen, M. A., Seay, C. A., Amini, B., Cai, Y., Feigenspan, A., Baxter, D. A., and Marshak, D. W. (2003) Spontaneous activity of dopaminergic retinal neurons. *Biophys. J.* **85**, 2158–2169.
27. Wustenberg, D. G., Boytcheva, M., Grunewald, B., Byrne, J. H., Menzel, R., and Baxter, D. A. (2004) Current- and voltage-clamp recordings and computer simulations of Kenyon cells in the honeybee. *J. Neurophysiol.* **92**, 2589–2603.
28. Xiao, J., Cai, Y., Yen, J. Steffen, M., Baxter, D. A., Feigenspan, A., and Marshak, D. (2004) Voltage clamp analysis and computational model of dopaminergic neurons from mouse retina. *Vis. Neurosci.* **21**, 835–849.
29. Byrne, J. H. (2004) Postsynaptic potentials and synaptic integration, in *From Molecules to Networks: An Introduction to Cellular and Molecular Neuroscience* (Byrne, J. H. and Roberts, J., eds). Academic Press, San Diego, CA, pp. 459–478.

30. Jack, J. J. B. and Redman, S. J. (1971) The propagation of transient potentials in some linear cable structures. *J. Physiol. (Lond.)* **215**, 283–320.
31. Rall, W. (1967) Distinguishing theoretical synaptic potentials computed for different soma-dendritic distributions of synaptic inputs. *J. Neurophysiol.* **30**, 1138–1168.
32. Wilson, M. A. and Bower, J. M. (1989) The simulation of large-scale neural networks, in *Methods in Neuronal Modeling* (Koch, C. and Segev, I., eds). The MIT Press, Cambridge, MA, pp. 291–334.
33. Rall, W. and Agmon-Snir, H. (1998) Cable theory for dendritic neurons, in *Methods in Neuronal Modeling*, Second Edition (Koch, C. and Segev, I., eds). MIT Press, Cambridge, MA, pp. 27–92.
34. Phares, G. A., Antzoulatos, E. G., Baxter, D. A., and Byrne, J. H. (2003) Burst-induced synaptic depression and its modulation contribute to information transfer at *Aplysia* sensorimotor synapses: empirical and computational analyses. *J. Neurosci.* **23**, 8392–8401.
35. White, J. A., Ziv, I., Cleary, L. J., Baxter, D. A., and Byrne, J. H. (1993) The role of interneurons in controlling the tail-withdrawal reflex in *Aplysia*: a network model. *J. Neurophysiol.* **70**, 1777–1786.
36. Epstein, I. R. and Marder, E. (1990) Multiple modes of a conditional neural oscillator. *Biol. Cybern.* **63**, 25–34.
37. Gingrich, K. J. and Byrne, J. H. (1985) Simulation of synaptic depression, posttetanic potentiation and presynaptic facilitation of synaptic potentials from sensory neurons mediating gill-withdrawal reflexes in *Aplysia*. *J. Neurophysiol.* **53**, 652–669.
38. Huang, R. C. and Gillette, R. (1993) Co-regulation of cAMP-activated Na^+ current by Ca^{2+} in neurones of the mollusc *Pleurobranchaea*. *J. Physiol. (Lond.)* **462**, 307–320.
39. Cataldo, E., Brunelli, M., Byrne, J. H., Av-Ron, E., Cai, Y., and Baxter, D. A. (2005) Computational model of touch mechanoafferent (T cell) of the leech: role of afterhyperpolarization (AHP) in activity-dependent conduction failure. *J. Comput. Neurosci.* **18**, 5–24.
40. Lombardo, P., Scuri, R., Cataldo, E., Calvani, M., Nicolai, R., Mosconi, L., and Brunelli, M. (2004) Acetyl-L-carnitine induces a sustained potentiation of the afterhyperpolarization. *Neuroscience* **128**, 293–303.
41. Backwell, K. T. (2005) A new era in computational neuroscience. *Neuroinformatics* **3**, 163–166.
42. Segev, I. and Rall, W. (1998) Excitable dendrites and spines: earlier theoretical insights elucidate recent direct observations. *Trends Neurosci.* **21**, 453–460.
43. Segev, I. and Schneidman, E. (1999) Axons as computing devices: basic insights gained from models. *J. Physiol. (Paris)* **93**, 263–270.

44. Shepherd, G. M. (2004) Electrotonic properties of axons and dendrites, in *From Molecules to Networks: An Introduction to Cellular and Molecular Neuroscience* (Byrne, J. H. and Roberts, J., eds). Academic Press, San Diego, CA, pp. 91–113.
45. Shepherd, G. M. (2004) Information processing in complex dendrites, in *From Molecules to Networks: An Introduction to Cellular and Molecular Neuroscience* (Byrne, J. H. and Roberts, J., eds). Academic Press, San Diego, CA, pp. 479–497.
46. Cai, Y, Baxter, D. A., and Crow, T. (2003) Computational study of enhanced excitability in *Hermissenda*: membrane conductances modulated by 5-HT. *J. Comput. Neurosci.* **15**, 105–121.
47. Flynn, M., Cai, Y., Baxter, D. A., and Crow, T. (2003) A computational study of the role of spike broadening in synaptic facilitation of *Hermissenda. J. Comput. Neurosci.* **15**, 29–41.
48. Moss, B. L., Fuller, A. D., Sahley, C. L., and Burrell, B. D. (2005) Serotonin modulates axo-axonal coupling between neurons critical for learning in the leech. *J. Neurophysiol.* **94**, 2575–2589.
49. Susswein, A. J., Hurwitz, I., Thorne, R., Byrne, J. H., and Baxter, D. A. (2002) Mechanisms underlying fictive feeding in *Aplysia*: coupling between a large neuron with plateau potentials and a spiking neuron. *J. Neurophysiol.* **87**, 2307–2323.
50. Kabotyanski, E. A., Ziv, I., Baxter, D. A., and Byrne, J. H. (1994) Experimental and computational analyses of a central pattern generator underlying aspects of feeding behavior in *Aplysia. Neth. J. Zool.* **44**, 357–373.
51. Guttman, R., Lewis, S., and Rinzel, J. (1980) Control of repetitive firing in squid membrane as a model for a neuroneoscillator. *J. Physiol. (Lond.)* **305**, 377–395.
52. Arbib, M. A. (2003) Neuroinformatics, in *The Handbook of Brain Theory and Neural Networks* (Arbib, M. A., ed.). The MIT Press, Cambridge, MA, pp. 741–745.
53. Arbib, M. A. and Grethe, J. S. (eds) (2001) *Computing the Brain: A Guide to Neuroinformatics*. Academic Press, San Diego, CA.
54. Shepherd, G. M., Mirsky, J. S., Healy, M. D., Singer, M. S., Skoufos, E., Hines, M. S., Nadkarni, P. M., and Miller, P. L. (1998) The human brain project: neuroinformatics tools for integrating, searching, and modeling multidisciplinary neuroscience data. *Trends Neurosci.* **21**, 460–468.
55. Kotter, R. (2001) Neuroscience databases: tools for exploring brain structure-function relationships. *Philos. Trans. R. Soc. Lond. B.* **356**, 1111–1120.
56. Gardner, D., Toga, A. W., Ascoli, G. A., Beatty, J. T., Brinkley, J. F., Dale, A. M., Fox, P. T., Gardner, E. R., George, J. S., Goddard, N., Harris, K. M., Herskovits, E. H., Hines, M. L., Jacobs, G. A., Jacobs, R. E., Jones, E. G., Kennedy, D. N., Kimberg, D. Y., Mazziotta, J. C., Miller, P. L., Mori, S., Mountain, D. C., Reiss, A. L., Rosen, G. D., Rottenberg, D. A., Shepherd, G. M., Smalheiser, N. R., Smith, K. P., Strachan, T., Van Essen, D. C., Williams, R. W., and Wong, S. T. (2003) Towards effective and rewarding data sharing. *Neuroinformatics* **1**, 289–295.

57. Finney, A. and Hucka, M. (2003) Systems biology markup language: level 2 and beyond. *Biochem. Soc. Trans.* **31**, 1472–1473.
58. Shapiro, B. E., Hucka, M., Finney, A., and Doyle, J. (2004) MathSBML: a package for manipulating SBML-based biological models. *Bioinformatics* **20**, 2829–2831.
59. Webb, K. and White, T. (2005) UML as a cell and biochemistry modeling language. *Biosystems* **80**, 283–302.
60. Weitzenfeld, A. (2003) NSL neural simulation language, in *The Handbook of Brain Theory and Neural Networks* (Arbib, M. A., ed.). The MIT Press, Cambridge, MA, pp. 784–788.

9

Data Mining Through Simulation

Introduction to the Neural Query System

William W. Lytton and Mark Stewart

Summary

Data integration is particularly difficult in neuroscience; we must organize vast amounts of data around only a few fragmentary functional hypotheses. It has often been noted that computer simulation, by providing explicit hypotheses for a particular system and bridging across different levels of organization, can provide an organizational focus, which can be leveraged to form substantive hypotheses. Simulations lend meaning to data and can be updated and adapted as further data come in. The use of simulation in this context suggests the need for simulator adjuncts to manage and evaluate data. We have developed a neural query system (NQS) within the NEURON simulator, providing a relational database system, a query function, and basic data-mining tools. NQS is used within the simulation context to manage, verify, and evaluate model parameterizations. More importantly, it is used for data mining of simulation data and comparison with neurophysiology.

Key Words: Simulation; computer modeling; neural networks; neuronal networks; databasing; query systems; data mining; knowledge discovery; inductive database.

1. Introduction

Knowledge discovery and data mining (KDD) is a process of seeking patterns among the masses of data that can be stored and organized in modern computer infrastructure. KDD arose in the commercial sector, where information was amassed for accounting and legal reasons over decades before being recognized as a valuable resource, figuratively a gold mine of information. Since the advent of sequence searching in the early 1980s *(1)*, similar techniques have

From: *Methods in Molecular Biology, vol. 401: Neuroinformatics*
Edited by: Chiquito Joaquim Crasto © Humana Press Inc., Totowa, NJ

been developed and adapted over the past decade as massive information veins associated with the genomes of human, fly, mouse, and others have come online.

As suggested by its commercial history, data mining grew out of databasing: data had been collected, and now something has to be done with it. This history resulted in data-mining techniques arising apart from the underlying databasing approach. In this paradigm, the database provides data for data mining but is not itself altered by the data-mining endeavor (*see* **Fig. 1 A**). This limitation has been recently addressed by proposing the concept of inductive databases, which utilizes a two-way flow of information between database and data-mining tools *(2)*. The inductive database will include metadata calculated from the base data, which is then used as base data for further exploration.

Scientific data mining differs in several ways from the more highly developed commercial applications *(3)*. Differences include a higher reliance on complex numerical manipulations and less clear-cut goals, requiring that data be analyzed and reanalyzed from different perspectives. In these respects, the scientific KDD process is more free-form than the commercial variety.

Neurobiologists are faced with the intellectual aim of understanding nervous systems, perhaps as complex a task as science faces. At the same time, modern techniques, including the genome projects, are making more and more information available, raising the question of what can best be done with it. In addition to the sociological perils common to cooperation in any scientific field, the emerging field of neuroinformatics must also confront unusual interoperability challenges because of the diverse types of data generated. These arise from different techniques as well as from the many orders of magnitude in time and space covered by different investigations of even a single subsystem *(4)*.

Data-mining tools are typically chosen on an ad hoc basis according to the task. These tools include various algorithmic constructions as well as traditional

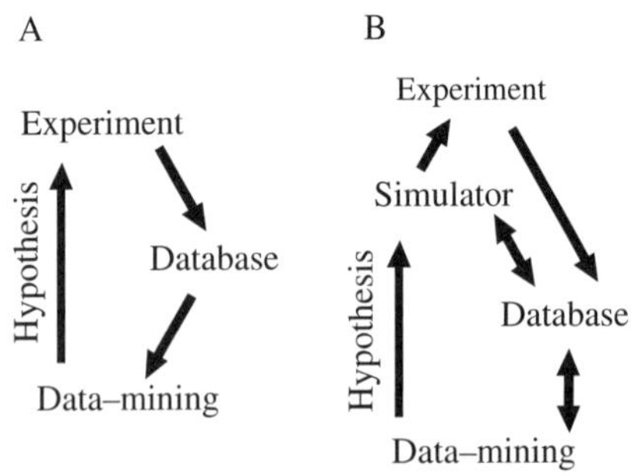

Fig. 1. Views of data mining. (**A**) Database centered and (**B**) simulator centered.

statistical techniques. Although some statistical procedures are used, the data-mining enterprise differs from statistics in that the data are multi-valued and do not generally fit a recognizable statistical distribution *(5)*. Thus, statistical procedures that depend on the distribution, such as the Student's *t*-test, often cannot be used. Those that are used are non-parametric. For example, mean and standard deviation lose meaning when working with a bimodal distribution.

Many data-mining tools in the commercial world are text oriented or symbolic, providing clustering or classifications by meaning or symbols *(6)*. This symbol-oriented approach extends into science in the realm of genomics and proteomics, whose objects of interest are described by a circumscribed list of symbols representing nucleotides and amino acids *(7,8)*. In neuroscience, spatially oriented tools, some developed from geography, can be utilized in areas such as development *(9)*, neuroanatomy *(10–12)*, and imaging *(13,14)*. Taxonomic thinking, involving ontologies and part–whole relations, requires semantically oriented tools *(15)*. Other areas of neuroscience, such as electrophysiology, are primarily numerical. Numerical data-mining tools include numerical integration and differentiation *(16,17)*, wavelet and spectroscopic analyses *(18)*, numerical classification methods such as principal component and independent component analysis *(19)*, and standard statistical methods. Spike trains, being a time series of discrete events, require yet other approaches for analysis *(20)*.

On the algorithmic side, iterative or recursive search programs are used to cluster or associate data or to make decision trees. Another major class of algorithm involves machine learning algorithms such as simulated annealing, genetic algorithms, and artificial neural networks (ANNs). The category of ANNs has potential for confusion in the present context. Realistic neural simulation includes realistic neural networks that seek to replicate the actual physiology of a set of connected neurons. By contrast, ANNs are typically used as tools for finding regularities in data. Although the original inspiration for ANNs came in part from neuroscience, these networks are not primarily used as direct models of nervous systems.

We have developed database and data-mining facilities in a neural query system (NQS) implemented within the NEURON simulation program. We suggest that realistic simulation of neural models will be the ultimate excavator in the data-mining endeavor, providing causality in addition to correlation *(21)*. One use of NQS is to manage models: to analyze existing simulations and to assist in development of new models. However, the main use for this package will be analysis of large-volume experimental data with both retrospective and online analysis and correlation of simulations related to such data.

2. Materials

NQS compiles into the NEURON simulator as a NMOD module containing C-code for rapid processing. A hoc module is then loaded that makes use of these low-level procedures. The NQS package is available at http://senselab.med.yale.edu/senselab/SimToolDB/default.asp. A user manual is included with the package.

3. Methods

As noted above, traditional data flow is one-way from databases to data-mining tools (*see* **Fig. 1 A**). The loop closes as data-mining insights are used to develop hypotheses that suggest new experiments. In this traditional view, simulation would be considered as just another data-mining tool, to be called from the data-mining suite to assess correlations. However, realistic simulation differs from other tools in that it provides causal explanations rather than simple correlations. For this reason, the simulator is intercalated in the hypothesis return loop in **Fig. 1 B**. In addition to testing hypotheses generated by mining experimental data, simulation will also generate new hypotheses, either directly or through mining of simulation output.

NQS is a simulator-based and simulator-centric databasing system with added data-mining tools. It is not a full database management system, because it is meant primarily as single-user system that does not have to handle problems of data-access control and data security. It also does not presently have the indexing and hashing capabilities of a full database management system although these may be added in the future. NQS provides some spreadsheet functionality. However, several features characterize it as a database rather than a spreadsheet system: the presence of a query language, the ability to handle many thousands of records, data structuring, handling of non-numeric data, and capacity for relational organization across database tables.

Using the simulator to control embedded databasing and data-mining software facilitates the use of simulation as a focusing tool for knowledge discovery (*see* **Fig. 1 B**). Several considerations suggest such a central role for neural simulation. First, there is the practical consideration of data quantity. Neural simulation, as it becomes more and more realistic, must become a massive consumer of experimental data, requiring a reliable pipeline. In addition to being dependent on experiment, simulation is itself an experimental pursuit, differing in this respect from the closed-form analytic models of traditional physics *(22)*. As we experiment on these models, the simulator becomes a producer of vast quantities of simulation data that must also be managed and organized.

In **Fig. 1 B**, the two-headed arrow between "Simulation" and "Database" denotes not only the flow of experimental and simulation data but also the management of simulation parameters. Simulation parameters can be stored in NQS databases, separated from experimental and simulation data. Neural simulations, particularly network simulations, are highly complex and can be difficult to organize. Once set up, it can be hard to visualize the resulting system to verify that the system has been organized as planned. Storing parameters in a database is a valuable adjunct for checking and visualizing parameter sets and for storing and restoring them between simulations. Data mining can then be used both for comparing parameters among sets of simulations and for relating changes in parameter sets to changes in model dynamics.

As compared to many data-mining tools, neural simulation tends to be very computationally intensive, particularly when large parameter searches are undertaken. (Machine learning and ANN algorithms are also computationally intensive.) Providing databasing and data-mining tools within the simulator allows partial data analysis to be done at runtime. This permits simulation results to be immediately compared with salient experimental features discovered through data mining. Simulations that are a poor match can then be aborted prematurely. Similarly, in a parameter learning context, using a terrain search or evolutionary algorithm, fitness can be determined on the fly by using a variety of measures calculated with a set of data-mining tools.

Many neuroscience applications generate spatial data that do not map directly onto the rectangular array used in relational database tables. Some data relations will be lost or obscured in remapping. NQS allows object storage in cells, permitting pointers and other indicators of non-rectangular relations. Completely non-rectangular data storage formats can be implemented as an adjunct to the database tables, which then use pointers to indicate locations in this supplementary data geometry. For example, the dendritic tree is stored in NEURON as a tree structure with parent and children pointers. In the course of reducing the tree to rectangular form, a variety of rectangular representations can be used, which then use pointers to maintain correspondence with the original tree format. These object adjuncts are useful for drilling down into raw data when doing data mining. However, it remains unclear how basic database functions such as sort and select can be extended to make use of such non-rectangular objects in a useful way.

3.1. NQS Functionality

NQS is implemented as a series of low-level routines that operates on vectors that are maintained as the parallel columns of a table. NQS thereby provides access to a variety of vector (array) manipulation tools built into NEURON. These vector functions permit convolution, numerical differentiation and integration, and basic statistics. Additional data-mining tools have been added on by compiling C-language code that can be directly applied to the numerical vectors used as columns for a table. Vector-oriented C-language code is readily available from a variety of sources ***(23,24)***. Such code can be compiled into NEURON after adding brief linking headers.

NQS handles basic databasing functionality including (1) creating tables; (2) inserting, deleting, and altering tuples; and (3) data queries. More sophisticated databasing functionality such as indexing and transaction protection are not yet implemented. Databasing commands in NQS provide (1) selection of specified data with both numerical and limited string criteria; (2) numerical sorting; (3) printing of data slices by column and row designators; (4) line, bar, and scatter graphs; (5) import and export of data in columnar format; (6) symbolic spreadsheet functionality; (7) iterators over data subsets or over an entire table; (8) relational selections using criteria across related tables; and (9) mapping of user-specified functions onto particular columns.

Among basic databasing functions, querying is the most complex. A query language, although often regarded as a database component and thereby denigrated as a data-mining tool, is a critical aspect of data mining. Structured query language (SQL), consistent with its commercial antecedants, is focused on returning a limited number of instances rather than encouraging serial transformations of data. The NQS select command is designed to focus on numerical comparisons. Owing to the importance of geometric information in neuroscience, inclusion of geometric criteria will be an additional feature that would be desirable in further development of NQS.

The NQS select()command is similar to the commands related to the WHERE and HAVING subfunctions of SQL's SELECT. NQS syntax naturally differs from that of SQL as it must follow the syntax of NEURON's object-oriented hoc language. An NQS database table is a template in hoc.

The NQS select() command takes any number of arguments in sets. Each set consists of a column name, a comparative operator such as "<" or "==," and one or two arguments depending on the operator. Multiple criteria in a single select() statement are handled with an implicit AND. A flag can be set to use OR on the clauses. A command can also begin with "&&" or "||" to return the union or intersection of the selected rows with previously selected rows (cf. SQL UNION and INTERSECT subcommands). Although the NQS select() does not replicate the agglutinative syntax of SQL SELECT, this functionality can be effected by serial application of NQS's select(), sort(), and stat() functions. Inner join between tables is implemented to permit formation of relational databases consisting of multiple tables.

3.2. Usage Examples

We have been using NQS to explore system parameters, simulation results, and the relation between parameters and simulation output. One seemingly trivial but important usage is to design, implement, and verify the connectivity of a neuronal network. Although this task is fairly simple in a basic network that typically has only two classes of cells, one inhibitory and the other excitatory, the complexity increases enormously as multiple cell types are included, each with its own connectivity patterns with each other type. Individual tables can be developed for each postsynaptic cell type or region of network, thus allowing networks to be built up incrementally. This has proved important for rapidly loading large models with tens of thousands of cells and tens of millions of synapses *(25)*. In this context, the table columns identify presynaptic cell type and index as well as postsynaptic receptor type and dendritic location. Other implementation-specific information, such as pointer to the actual connection object, can be included as additional columns *(26)*.

The primary use of the NQS package is analysis of large data sets, derived from either simulation or physiology. Although NQS makes it easy to manage the large data sets, there remains much work to be done in providing adequate techniques for displaying and assessing of these high-dimensional systems. In **Fig. 2**, we are evaluating simulated field potential data from a population of 132,000 1200-cell network simulations generated by 13 varying parameters over 2–3 values apiece *(27)*. Although the simulated field was originally generated to permit ready comparison to physiology, it became apparent that the field provided a useful abstract of the data otherwise contained in a raster plot or in multiple-simulated intracellular records, being far easier to compare between simulations than either of the alternatives.

Given the simulated field potentials, we extract population spikes and classify by height, width, and occurrence time to fill out NQS tables that allow summarized classification of each simulation result. In **Fig. 2**, we compare the attributes of two selected subpopulations representing pair-wise matched simulations differing by two alternative values (alleles) in one parameter (single neuron excitability). In **Fig. 2** (bottom), we graph a point for each simulation pair indicating the difference in duration of population spiking activity (x-axis) and the total number of population spikes (y-axis). In this case, the upper right quadrant represents the expected result: increased single-neuron excitability results in a greater number of spikes (positive y-axis) and a longer duration of activity (positive x-axis). The interesting observation here is that there are

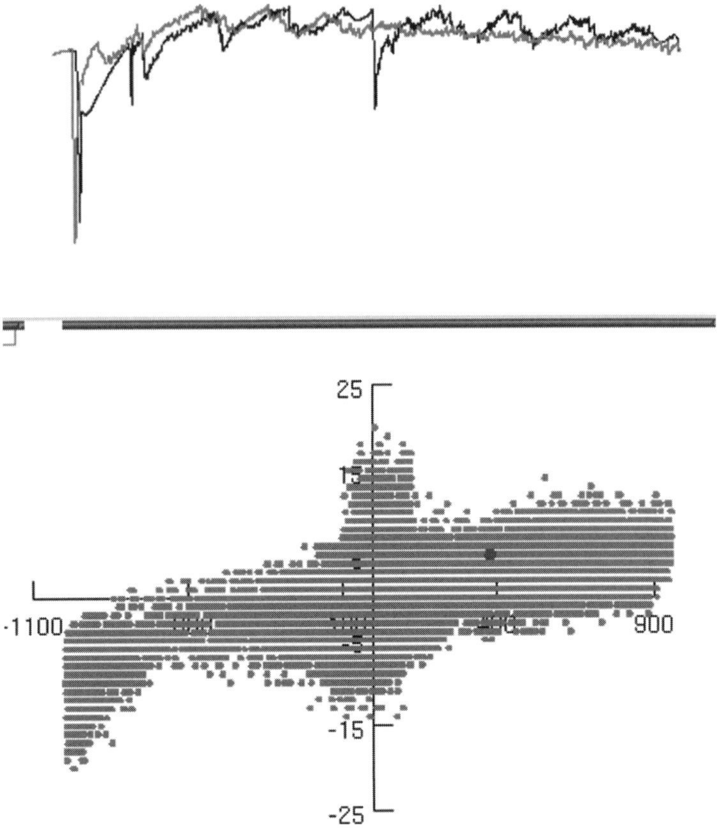

Fig. 2. Graphical use of neural query system (NQS). See text for details.

many simulation pairs in other quadrants as well, representing reduction in duration or spike number with increased neuron excitability.

In addition to a textual select() command, NQS permits graphical selection by selecting along the borders of the rectangle or by clicking at an individual point with mouse or cursor. In **Fig. 2**, we have clicked on a single pointer in the right upper quadrant that displays the associated simulation pair above (higher excitability is the teace with late spikes). This allows a rapid graphical search through hundreds of simulations selected according to their relationships across chosen attributes. Alternatively, we can map individual or paired simulations according to parameters rather than according to activity and thereby search in parameter space rather than in collapsed state space. Given the vast number of simulations in this case, it is also desirable to develop higher-order automated

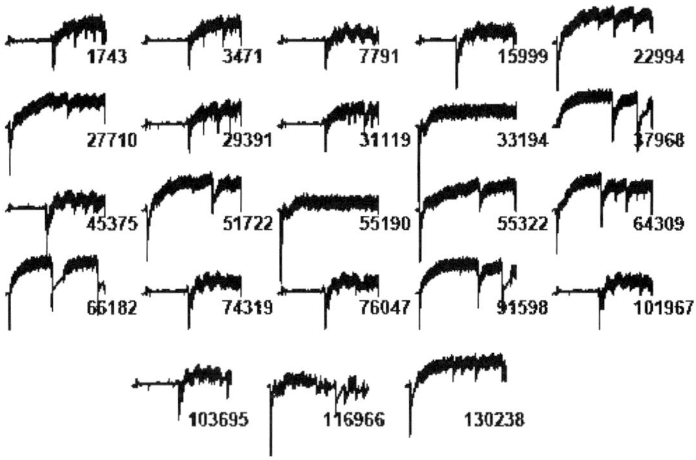

Fig. 3. Simulations selected according to complex criteria: here the presence of late population spikes.

selection protocols to find simulations with particular attributes. For example, in **Fig. 3**, we select for simulations with late population spike activity. Here, each trace is graphed with its unique ascension number.

4. Notes

Computer ubiquity is driving biology, together with most other fields, into the realm of big science. Unlike the big science of the past where large groups had to be organized and mobilized at a particular site, modern computer communication will allow groups to form and re-form ad hoc around particular problems and particular data sets. This informatics revolution is particularly welcome in neurobiology, where the stretch between levels of organization is so great and the amount of potential data so overwhelming in depth, breadth, and variety.

We argue here that simulation may offer a good focal point for working with data and have illustrated a simulator-embedded system for database manipulation. However, it is clear that the huge databases that will be required in neurobiology will need dedicated full-bandwidth database software. It is envisioned that NQS will play an intermediate role, together with standard database software. NQS might be used independently for the type of parameter definition and verification illustrated in the examples above. When saving results from large numbers of simulations or comparing with results from many experiments, NQS would read and write to disk through other database software but would still be able to rapidly manipulate and assess data online to compare simulations and experimental results.

NQS stores data in a standard relational database format. As noted above, this rectangular data structure has limitations compared with the dedicated internal data structures of a simulator such as NEURON. For example, NEURON stores a dendritic tree using pointers to create a tree-like format involving parent and child sections. However, major database systems, whether commercial or open source, require data in a rectangular, column/row format. Therefore, it is important to translate such data into tables in order to communicate between a simulator and other databases. One advantage of finding a simple, easily stored representation for models would be the ability to readily move a given model from one simulator to another. As compared with the use of a model meta-language, a database would require a more rigid format but would provide more rapid access and exploration *(28)*.

Acknowledgments

The author thanks Mike Hines and Ted Carnevale for continuing assistance with NEURON. This research was sponsored by NIH (NS045612 and NS032187).

References

1. Wilbur, W. J. and Lipman, D. J. (1983) Rapid similarity searches of nucleic acid and protein data banks. *PNAs* **80**, 726–730.
2. Imielinski, T. and Mannila, H. (1996) A database perspective on knowledge discovery. *Commun ACM* **39**, 58–64.
3. Han, J., Altman, R., Kumar, V., Mannila, H., and Pregibon, D. (2002) Emerging scientific applications in data mining. *Commun ACM* **45**, 54–58.
4. Churchland, P. and Sejnowski, T. (1994) *The Computational Brain*. MIT Press Cambridge, MA.
5. Hand, D. (1999) Statistics and data mining: intersecting disciplines. *ACM SIGKDD* **1**, 16–19.
6. Hirji, K. (2001) Exploring data mining implementation. *Commun ACM* **44**, 87–93.
7. Wei, G., Liu, D., and Liang, C. (2004) Charting gene regulatory networks: strategies, challenges and perspectives. *Biochem J* **381**, 1–12.
8. Winslow, R. and Boguski, M. (2003) Genome informatics: current status and future prospects. *Circ Res* **92**, 953–961.
9. Concha, M. and Adams, R. (1998) Oriented cell divisions and cellular morphogenesis in the zebrafish gastrula and neurula: a time-lapse analysis. *Development* **125**, 983–994.

10. Martone, M., Zhang, S., Gupta, A., Qian, X., He, H., Price, D., Wong, M., Santini, S., and Ellisman, M. (2003) The cell-centered database: a database for multi-scale structural and protein localization data from light and electron microscopy. *Neuroinformatics* **1**, 379–395.
11. Stephan, K., Kamper, L., Bozkurt, A., Burns, G., Young, M., and Kotter, R. (2001) Advanced database methodology for the collation of connectivity data on the macaque brain (cocomac). *Philos Trans R Soc Lond B* **356**, 1159–1186.
12. Senft, S. and Ascoli, G. (1999) Reconstruction of brain networks by algorithmic amplification of morphometry data. *Lect Notes Comput Sci* **1606**, 25–33.
13. Langer, S. (2002) Openrims: an open architecture radiology informatics management system. *J Digit Imaging* **15**, 91–97.
14. Megalooikonomou, V., Ford, J., Shen, L., Makedon, F., and Saykin, A. (2000) Data mining in brain imaging. *Stat Methods Med Res* **9**, 359–394.
15. Neill, M. and Hilgetag, C. (2001) The portable UNIX programming system (PUPS) and CANTOR: a computational environment for dynamical representation and analysis of complex neurobiological data. *Philos Trans R Soc Lond B* **356**, 1259–1276.
16. Zhu, J., Lytton, W., and Uhlrich, D. (1999) An intrinsic oscillation in interneurons of the rat lateral geniculate nucleus. *J Neurophysiol* **81**, 702–711.
17. Sekerli, M., Negro, C., Lee, R., and Butera, R. (2004) Estimating action potential thresholds from neuronal time-series: new metrics and evaluation of methodologies. *IEEE Trans Biomed Eng* **51**, 1665–1672.
18. Hazarika, N., Chen, J., Tsoi, A., and Sergejew, A. (1997) Classification of EEG signals using the wavelet transform. *Signal Processing* **59**, 61–72.
19. Delorme, A. and Makeig, S. (2004) EEGLAB: an open source toolbox for analysis of single-trial EEG dynamics including independent component analysis. *J Neurosci Methods* **134**, 9–21.
20. Victor, J. and Purpura, K. (1997) Metric-space analysis of spike trains - theory, algorithms and application. *Netw Comput Neural Syst* **8**, 127–164.
21. Gallagher, R. (2004) The binding agent for the life sciences soufflé. *Scientist* **18**, 4.
22. Wolfram, S. (1984) Computer software in science and mathematics. *Sci Am* **251**, 188–204.
23. Galassi, M., Davies, J., Theiler, J., Gough, B., Jungman, G., Booth, M., and Rossi, F. (2003) *GNU Scientific Library: Reference Manual*, 2 edn. Network Theory Limited, Bristol.
24. Press, W., Flannery, B., Teukolsky, S., and Vetterling, W. (1992) *Numerical Recipes In C: The Art of Scientific Programming*, 2 edn. Cambridge University Press, Cambridge.
25. Migliore, M., Cannia, C., Lytton, W., and Hines, M. (2006) Parallel network simulations with NEURON. *J Comput Neurosci* **21**, 119–129.

26. Lytton, W. (2006) Neural query system: data-mining from within the NEURON simulator. *Neuroinformatics* **4**, 163–176.
27. Lytton, W. and Stewart, M. (2005) A rule-based firing model for neural networks. *Int J Bioelectromagnetism* **7**, 47–50.
28. Goddard, N., Hucka, M., Howell, F., Cornelis, H., Shankar, K., and Beeman, D. (2001) Towards NeuroML: model description methods for collaborative modelling in neuroscience. *Philos Trans R Soc Lond B* **356**, 1209–1228.

10

Computational Exploration of Neuron and Neural Network Models in Neurobiology

Astrid A. Prinz

Summary

The electrical activity of individual neurons and neuronal networks is shaped by the complex interplay of a large number of non-linear processes, including the voltage-dependent gating of ion channels and the activation of synaptic receptors. These complex dynamics make it difficult to understand how individual neuron or network parameters—such as the number of ion channels of a given type in a neuron's membrane or the strength of a particular synapse—influence neural system function. Systematic exploration of cellular or network model parameter spaces by computational brute force can overcome this difficulty and generate comprehensive data sets that contain information about neuron or network behavior for many different combinations of parameters. Searching such data sets for parameter combinations that produce functional neuron or network output provides insights into how narrowly different neural system parameters have to be tuned to produce a desired behavior. This chapter describes the construction and analysis of databases of neuron or neuronal network models and describes some of the advantages and downsides of such exploration methods.

Key Words: Computational modeling; simulation; model database; neuronal network; membrane conductance; synapse strength; network behavior; dimension stacking; visualization.

1. Introduction

Neuronal networks—and even individual neurons—are frequently so complex that it is difficult to predict how changing one or several cellular or synaptic parameters will affect the electrical activity of the cell or circuit. This complexity is reflected in the way neurons and neuronal networks are described mathematically, with systems of coupled, non-linear differential equations that

From: *Methods in Molecular Biology, vol. 401: Neuroinformatics*
Edited by: Chiquito Joaquim Crasto © Humana Press Inc., Totowa, NJ

are usually well beyond analytical solution. At the level of single neurons, non-linear systems theory can provide some insights into how neuron activity is shaped by individual neuronal parameters, such as the maximal conductance of a particular membrane current *(1–4)*. However, non-linear systems theory requires significant simplification of the neuronal model in question, usually reducing the number of dynamic variables that describe the system from ten or more to just two, and thus does not adequately represent the complexity and dynamical richness of biological neurons. Furthermore, non-linear systems theory becomes less and less adequate the more the complexity of a system increases, such that its application cannot easily be generalized from the neuron to the network level.

The computational exploration method described here takes a different approach by leaving the full complexity of a neuron or network model intact and by systematically exploring the dependence of the system's behavior on its parameters through a large number—often many millions—of computational simulations (*see* **Fig. 1**). Each of these simulations assumes a different set of neuron or network parameters and determines how the neural system behaves—and responds to inputs—for that particular combination of parameters. The simulation results are then saved in a database that is searchable by neuron or network parameter and by salient features of the circuit behavior, allowing mining of the data set under different aspects *(5,6)*.

Beyond showing how changes in an individual parameter affect neuron or network behavior, model databases constructed in this manner have various useful applications. Perhaps the most obvious example is the identification of particular parameter combinations that reproduce a desired behavior of a neural system *(5)*. Such functional parameter combinations have traditionally been identified by hand-tuning neuron or network models one parameter at a time *(7–9)*, a process that requires significant experience and time and is not always successful. By searching a data set created through systematic exploration of a model's parameter space, one can quickly identify parameter combinations that generate a desired behavior without the need for extensive hand-tuning.

Another advantage of model neuron or model network databases is that—rather than providing just one parameter set that results in a given behavior—they can be used to delineate entire regions of parameter space, which generate neural system output in a desired range, for example, bursting neuron activity with a burst period and burst duration within experimentally observed bounds *(5,6,10,11)*. Analysis of the shape and extent of such functional subspaces in different directions of parameter space can provide

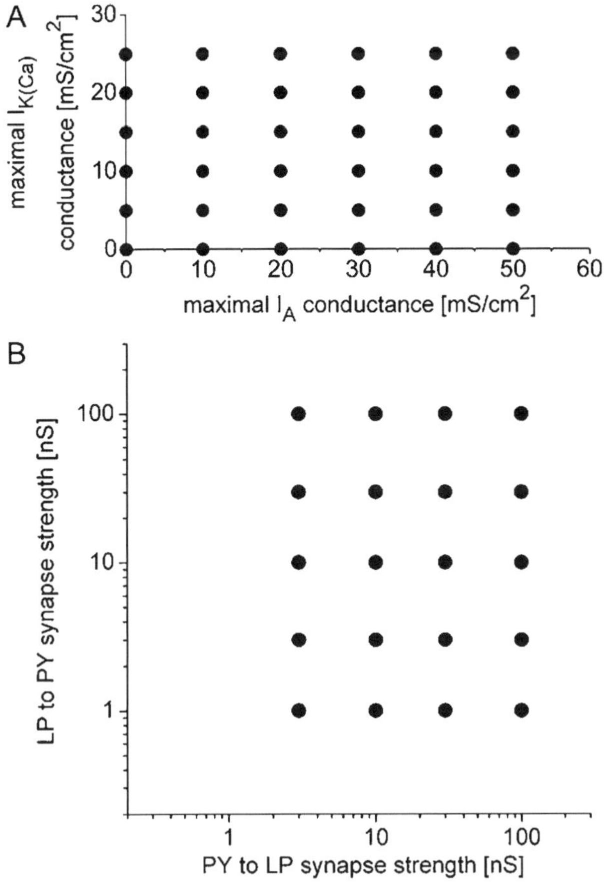

Fig. 1. Examples of parameter space exploration schemes. (**A**) A regular grid of simulations covering the maximal conductance space of a model neuron. Only two of eight dimensions are shown. Every dot symbolizes a simulation with a different combination of maximal conductances. (**B**) A quasi-logarithmic grid of simulations covering the synaptic conductance space of a network model. Note the logarithmic scale. Only two of ten dimensions are shown. LP and PY designate two neurons in the network.

valuable information about the sensitivity of a neural system to changes in individual parameters or combinations of parameters. In addition, such analysis tells us what properties of a neural system need to be relatively tightly tuned to produce functional behavior and what can vary more widely *(6)*.

2. Materials

2.1. Simulation Software

Model neuron or network databases can be constructed with any simulation software that is capable of stepping through a large number of different neuron or network parameter combinations in an automated fashion, of outputting and saving the desired activity measures for each parameter set and of running on the available simulation hardware platform. Databases have in the past been constructed with custom-designed C^{++} programs ***(5,6)***, with NEURON ***(12)*** (*see* Chapter 6) and with the GENESIS (*see* Chapter 7) software packages ***(11, 13)***, although using GENESIS under some circumstances requires re-starting the simulation program before each individual simulation because of program initialization issues.

Custom-made software, while requiring a larger initial programming investment, has the advantage of larger flexibility, which allows it to be optimized for simulation speed, a consideration that can be of crucial importance when large parameter spaces are to be explored. Furthermore, custom-made software can be programmed to include adaptive algorithms that minimize the simulation time per parameter set by monitoring the activity pattern generated by the simulated system during runtime and terminating each individual simulation as soon as initialization artifacts have decayed and a stable limit cycle is reached (*see* **Note 1**) ***(5)***.

2.2. Simulation Hardware

Database simulations of the scale described here—covering many millions of different parameter combinations—are usually executed on high-performance computer clusters. The inherently serial nature of the simulation task, however, does not require truly parallel computation, and smaller databases on the order of a few million different neurons or networks can be run with low priority in the background on the computers available in most laboratories. In fact, an earlier version of one of the first model neuron databases ***(5)*** was constructed in this manner, with the simulation of 1.7 million model neurons requiring a few weeks of background computation on 8–10 standard laboratory computers. A user-friendly client–server application that can be installed to run large numbers of NEURON simulations on any number of available laboratory computers as a screen saver is available at http://neuronpm.homeip.net. As computer power increases over the years, simulation of larger and larger databases will become more and more feasible and eventually routine.

2.3. Analysis Software

Analysis software for typical model neuron and model network databases needs to be able to detect events such as spikes and bursts and to automatically extract salient activity features such as spike or burst period, burst duration, number of spikes per burst, and relationships between the activity of different neurons in a network, such as delays between spikes in one cell relative to spikes in the other or phase relationships of synchronous oscillations in multiple neurons. Custom-made analysis software is again at an advantage, because it allows the analysis to be incorporated into the simulation process and thus does not require the saving of large amounts of voltage trace information for offline analysis. As previously mentioned, online analysis of simulation results can lead to significant reductions in the simulation time per parameter set, because it can lead to early detection of stable limit cycle activity (*see* **Note 1**).

2.4. Database Administration Software

An important part of the database approach to complex dynamical systems is that a lot of information about the activity of the model neuron or network in question get preserved for later analysis under various aspects, including questions not directly related to the initial purpose of the database. The types of information preserved in a database can include raw simulation results such as voltage traces or voltage maxima and minima (*see* **Fig. 2**) or snapshots of the model system's dynamic variables for a point on the limit cycle (*see* **Note 2**).

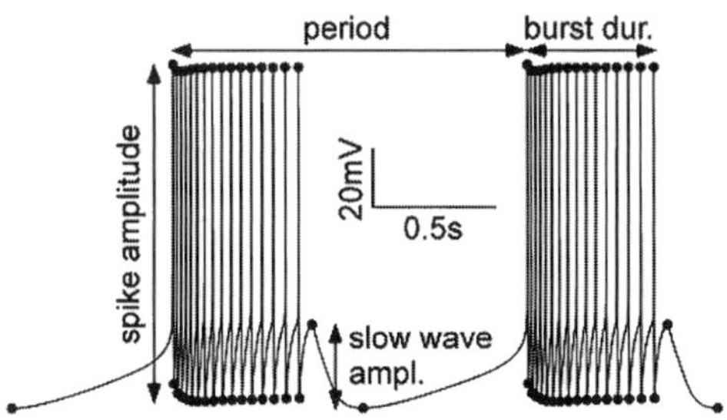

Fig. 2. Examples of the information saved for every parameter combination in a model neuron database. The dots indicate local voltage maxima and minima whose time and voltage values are saved. Additional information saved in the database includes duty cycle (burst duration divided by period), number of spikes per burst, area under the voltage trace per burst, resting potential (for silent models), spike frequency (for tonically spiking models), response to current injection, dynamic variable snapshot, and others.

In addition, characteristics of the system activity that were extracted from the raw data, such as spike or burst frequencies, can also be stored in a database.

Ideally, the software used for administering and searching a model database should allow the user to extract correlations between any combination of neuron or network parameter and activity feature. In practice, this is accomplished with custom-made C^{++}- or MATLAB (The MathWorks, Natick, MA)-based program packages or with commercially available database programs such as Access (Microsoft, Redmond, WA).

2.5. Visualization Tools

Information contained in a model neuron or network database can be difficult to access and analyze because of the high dimensionality of the parameter space. This can be overcome with a recently developed visualization tool that makes use of a technique called "dimensional stacking" to display regularly spaced data points from a high-dimensional space in the form of a lower dimensional—typically 2D—projection that makes previously hidden structures emerge *(14)*. Every entry in the database corresponds to one pixel in the 2D image, and the pixel color can be assigned to indicate any model characteristic of interest, such as activity type (spiking, bursting, and/or silent), discharge frequency, or spike amplitude, and so on. The example in **Fig. 3** shows the models in a single neuron database *(5)*—or pixels—arranged in a 2D image in $6 \times 6 = 36$ major blocks, with each block corresponding to a different combination of the maximal conductances of the membrane currents I_{CaT} (vertical axis) and $I_{K(Ca)}$ (horizontal axis), which each can take six different values. Within each of these blocks, pixels are arranged in 36 smaller blocks according to their values of the I_{CaS} and the I_{Kd} conductances and so on through two additional levels of organization. As **Fig. 3** indicates, 2D stacking displays can reveal how different types of neuronal or network activity are distributed in parameter space.

3. Methods

This section will describe the preparation steps necessary before a model database can be constructed and the steps involved in constructing and analyzing a database. It is assumed that a fully defined model neuron or network (collectively referred to as "neural model system") is used as a starting point.

3.1. Preparations

1. Decide which model parameters can vary in the database and over what ranges and which parameters to keep fixed. Ideally, one would want to vary all parameters of a neural system, but because parameters are by definition varied independently

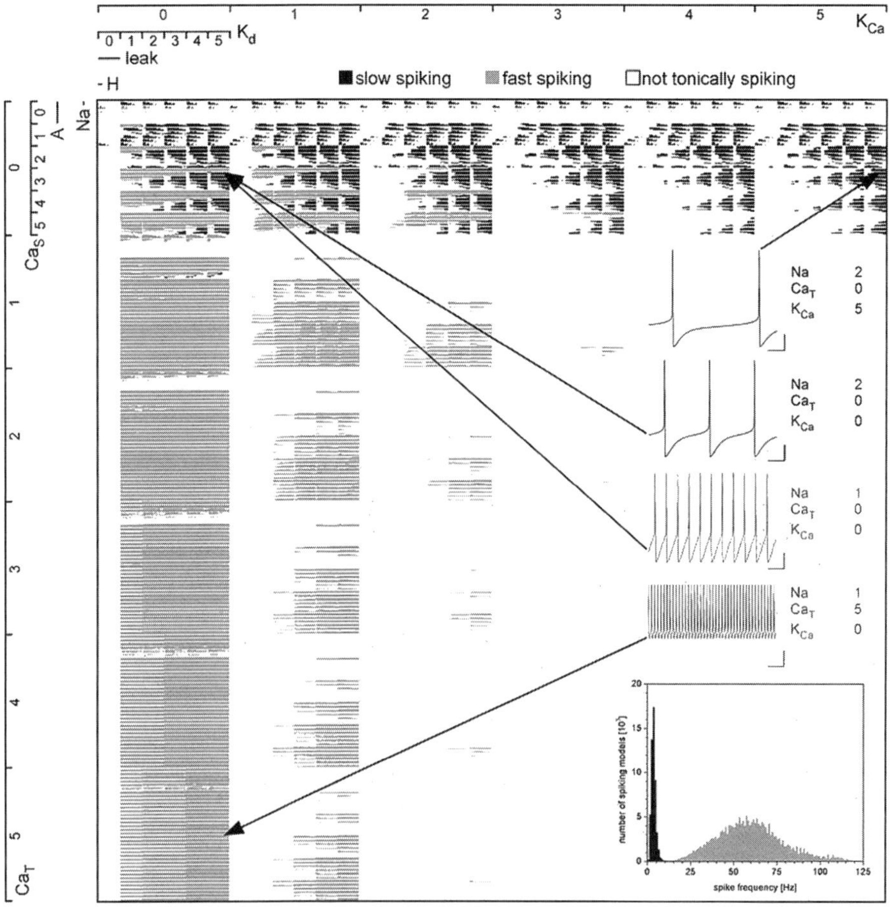

Fig. 3. Dimensional stacking visualization of a model neuron database that reveals the dependence of spike frequency on the underlying conductances. Scales at the top and left indicate the stacking order. The insets show voltage traces for selected conductance combinations and a spike frequency histogram of all tonic spikers in the database.

during database construction, large numbers of varied parameters quickly lead to combinatorial explosion (*see* **Note 3**). The decision which subset of parameters can vary will obviously be influenced by the type of scientific questions the database is supposed to address, but physiological constraints can also provide guidance. For example, in a mature biological neuron, the amounts of different ion channels in the neuronal membrane are likely to vary over larger ranges—because of channel turn over—than the total amount of membrane area, suggesting that varying the maximal conductances of different ion currents in a model may be more meaningful than varying the membrane capacitance. Another consideration is the amount of information available about different parameters. Some parameters may be known to be relatively tightly constrained from experimental observations, whereas others may be less well defined and may thus warrant exploration in the database. Known experimental ranges for neuron or network parameters can also be helpful in constraining the range over which a given parameter is varied.

2. Decide how many different values to explore for each varied parameter and how to distribute the values over the explored range. This decision calls for a compromise between computational feasibility and sufficient resolution in parameter space (*see* **Note 4**). Most existing databases have, however, been able to capture salient features of parameter space with relatively few explored values per varied parameter, typically on the order of three to six ***(5,6,11)***. Whether these values are best distributed linearly or logarithmically (or otherwise) over the covered parameter range depends on the type of parameter that is being varied (*see* **Fig. 1**). Although maximal membrane conductances are typically distributed linearly during database construction, a logarithmic distribution may make more sense for synaptic strengths, because small changes in synapse strength can have major impact when a synapse is weak, whereas even doubling the strength of an already strong synapse usually has little functional impact ***(15)***.
3. Generate a database simulation program that steps through a list of parameter combinations, simulates the spontaneous activity for each combination and/or activity under current injection or other perturbations (*see* **Note 5**), detects salient features of that activity (such as spikes, burst, voltage maxima and minima, etc.), classifies the activity based on these features (e.g., as silent, tonically spiking, bursting, or irregularly spiking for individual neurons or as rhythmic or non-rhythmic for a network), and saves the features, the classification result, and a snapshot of all dynamic variables in the model at the end of each simulation (*see* **Note 2**). An important consideration that can have major influence on the amount of simulation time needed per parameter combination is the choice of initial values for the dynamic variables of the system (*see* **Note 1**). A frequently used approach is to select a relatively hyperpolarized starting voltage, for example −65 mV, for each neuron in the model and to initialize all activation and inactivation variables to the steady-state values consistent with that voltage (*see* **Note 6**).
4. Test the simulation program on a random subset of parameter combinations from the entire parameter range to be covered with the database. Determine how long it typically takes for a simulation to reach a stable limit cycle (*see* **Fig. 4**), whether the salient activity features are detected correctly, whether the activity is classified correctly, and whether there are unexpected types of activity whose proper detection and classification requires modifications to the simulation program.
5. Once the simulation program is finished, run it on a random subset of parameter combinations on the same type of processor that will run the entire database

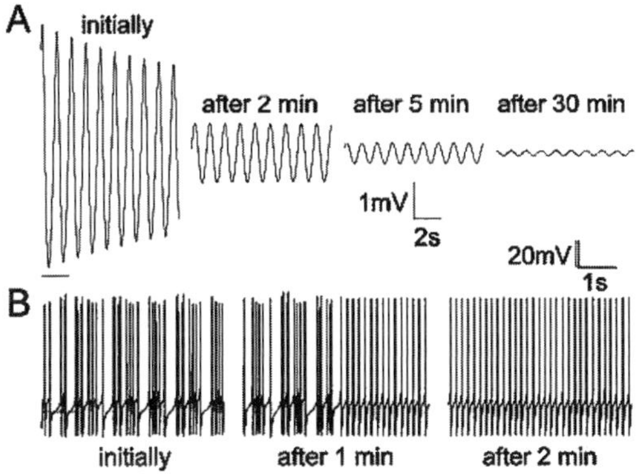

Fig. 4. Models that take a long time to reach steady state after initialization. (**A**) A silent model neuron that takes over 30 min to reach its steady state. (**B**) An initially irregularly firing model neuron that suddenly enters a tonically spiking limit cycle 1 min after initialization. Horizontal bars indicate −49 mV in (**A**) and −50 mV in (**B**). Adapted from **ref. 5**.

simulation to estimate how long simulation of all parameter combinations will take and how much data it will generate. If necessary, repeat **steps 1** and **2** above to reduce the number of parameter combinations and thus the total simulation time, or repeat **step 3** to reduce the amount of saved data per simulation.

3.2. Database Construction

1. Run the simulation program on all available processors until all parameter combinations have been simulated. Because database construction inherently does not require communication between different processors, all parameter combinations can be covered by dividing the entire list of simulations among the processors at the start of the simulation and having each processor work independently. If varying numbers of processors (on a shared cluster) or processors of different speeds are used, efficient distribution of the simulations among the different processes can be achieved with a master list of simulations that is accessed by one process at a time to acquire a short list of scheduled simulations that are then executed before that processor accesses the master list again.
2. Upon termination of all simulations, check whether the output files contain all combinations of parameters. Database simulations frequently run for days or weeks, and small subsets of simulations and results can get lost during simulation. It is therefore necessary to crosscheck the simulation results with the initial list of parameter combinations to determine which simulations need to be executed again.
3. Combine the simulation results from different processors in easily searchable files. Once all simulations for a database have been completed and missing simulations

have been re-run, database entries are often distributed over a large number of files, such that finding individual entries can require long search times. Re-organizing all results in a small number of database files with a fixed order of entries can facilitate subsequent database analysis.

4. Verify the classification results for a random subset of database models. Neural system models, especially for exotic parameter combinations, can sometimes show behavior that was not anticipated during the initial design of the classification algorithm. For example, the silent neuron in **Fig. 4A** can easily be mistaken for a tonically spiking neuron if classification is based on the intervals between voltage maxima alone *(5)*. If a significant number of models were misclassified, the classification algorithm may have to be modified and run again on saved raw activity data, such as local voltage maxima and minima, to reclassify misclassified models.
5. Perform offline data extraction steps. Database construction and analysis are exploratory methods, and it is not always a priori clear what types of activity feature will be worth extracting and analyzing further. By design, the general nature of the raw data typically saved during database construction, such as voltage maxima and minima, allows for later extraction of many higher-level activity features (*see* **Fig. 2**). Examples are the extraction of burst durations, number of spikes per burst, inter-spike interval profile during a burst, phase relationships between neurons in a rhythmic network, and so on, which can all be determined from the saved voltage maxima and minima offline and stored in secondary database files.

3.3. Database Analysis

The analysis steps necessary for extracting higher-level information from the saved data largely depend on the types of scientific questions addressed with the database, such that there is no general set of steps to be followed. A few possible types of analysis are described below.

1. Select parameter combinations that produce a desired range of behaviors. On the basis of the classification results and particular activity features, one can select a subset of models that exhibit a certain type of behavior, for example, bursting neurons with a burst period and duration within experimentally observed bounds *(5)*. Such a subset of neuron or network models can then be analyzed for their distribution in parameter space, for example, by determining their 1D distributions over the values of one of the system parameters. If models with similar behavior show a wide distribution over the values of a given parameter, that parameter is unlikely to be a major determinant of that type of behavior, whereas a narrow distribution indicates that the behavior sensitively depends on the parameter in question *(6,10)*.
2. Locate a model neuron or network that best matches a given experimentally observed behavior *(5)*. This alternative to time-consuming hand-tuning of a model requires the definition of a goodness-of-fit measure, which can be anything from a spike-count deviation per time window to the mean-square-root deviation in

membrane potential to a complex combination of many different measures. By comparing the desired activity profile to the stored information for each parameter combination, one can then determine the goodness-of-fit for each parameter combination and locate the particular point in parameter space that best reproduces the desired behavior (*see* **Note 4**).

3. Visualize database information through dimensional stacking. **Figure 3** shows an example of dimension stacking for a single neuron database *(5)* and illustrates the type of information that can be gained from this analysis. For example, the figure shows that the two distinct populations of fast and slow spikers are located in separate regions of conductance space, that almost all slow spikers lack I_{CaT} and that fast spiking—as opposed to other types of activity—is favored by small amounts of I_{KCa} in the neuronal membrane. An important issue with dimension stacking is the stacking order, that is, which model parameters determine the highest level of image organization and which parameters determine order at lower levels. Although **Fig. 3** clearly reveals the distributions of fast and slow spikers in conductance space, many other stacking orders for the same database would lead to an almost homogeneous distribution of fast and slow spikers and non-spikers in the 2D image. A stacking order optimization algorithm that overcomes that problem has recently been devised *(14)*.

4. Notes

1. Different model neurons or model networks may take different amounts of time to reach a stable steady state after initialization of the dynamic variables (*see* **Fig. 4**). A simulation strategy that allows enough simulation time for each model to reach steady state while minimizing overall simulation time uses an adaptive algorithm to terminate simulation in each model as soon as a steady state is reached. This can be achieved by running simulations in brief epochs of, for example, 1-s duration and by searching the accumulated simulation results for periodic spike or burst patterns or for constant membrane potential after each epoch *(5)*.
2. Database analysis can lead to the finding that additional simulations are necessary for a subset or all the parameter combinations tested. In this event, the simulation time needed for the initial state of the system to decay away can be saved if simulation can resume for each parameter combination where it left off during the initial database construction, namely on the limit cycle. It is advisable therefore to save a snapshot of each model, consisting of the values of all dynamic variables, at the end of each simulation *(5,6)*.
3. The number of simulations necessary for the exploration of a complex neural system can often be dramatically reduced by breaking up the system into known subsystems. For example, for a small neuronal network consisting of identified neurons whose behavior in synaptic isolation has been experimentally characterized, the model parameter spaces of the individual neurons can be explored first to locate parameter combinations that match the experimentally observed

neuron behavior *(6)*. The full network simulation then only has to be performed for combinations of these viable subsystems, not for the full network parameter space.

4. High-dimensional model neuron or network databases are often constructed with fairly low parameter value resolution. However, once a region of interest in parameter space has been identified, that area can be re-sampled with higher resolution. For example, this can be done to investigate an area that produces a desired activity type in more detail or to search the parameter space vicinity of a model that optimally reproduces experimental data for parameter combinations that do even better.
5. As for all numerical simulations, a decision has to made as to the integration method to be used and the integration time step, but because of the large number of simulations involved in database construction, a good balance between computational speed (calling for a large time step) and accuracy (calling for a small time step) is more crucial than usual. Regardless of the integration method, a series of test simulations with a random subset of models from the database grid and various time steps should be run. On the basis of these simulations, the most efficient time step to choose is the largest one that for the vast majority of models leads to negligible deviations of the activity pattern from the "true" pattern obtained with a very small time step. The percentage of models affected by deviations and the extent of these deviations should be quantified to assess the validity of the database simulation results *(5)*.
6. Other initialization schemes include initializing all activation variables to 0 and all inactivation variables to 1 (corresponding to the steady state for a very hyperpolarized membrane potential *(5)*) or using the final dynamic variable values from the previous simulation to initialize the next simulation during database construction. This latter approach works best for simulation strategies that step through all database entries in a way that minimizes parameter changes from one simulation to the next, because similar model parameters are likely to lead to similar limit cycles.

References

1. Guckenheimer, J., Gueron, S., and Harris-Warrick, R. M. (1993) Mapping the dynamics of a bursting neuron. *Philos. Trans. R. Soc. Lond. B.* **341,** 345–359.
2. Guckenheimer, J., HarrisWarrick, R., Peck, J., and Willms, A. (1997) Bifurcation, bursting, and spike frequency adaptation. *J. Comput. Neurosci.* **4,** 257–277.
3. Izhikevich, E. M. (2000) Neural excitability, spiking and bursting. *Int. J. Bifurcat. Chaos* **10,** 1171–1266.
4. Rush, M. E., and Rinzel, J. (1994) Analysis of bursting in a thalamic neuron model. *Biol. Cybern.* **71,** 281–291.
5. Prinz, A. A., Billimoria, C. P., and Marder, E. (2003) Alternative to hand-tuning conductance-based models: construction and analysis of databases of model neurons. *J. Neurophysiol.* **90,** 3998–4015.

6. Prinz, A. A., Bucher, D., and Marder, E. (2004) Similar network activity from disparate circuit parameters. *Nat. Neurosci.* **7,** 1345–1352.
7. De Schutter, E., and Bower, J. M. (1994) An active membrane model of the cerebellar Purkinje-cell. I. Simulation of current clamps in slice. *J. Neurophysiol.* **71,** 375–400.
8. Nadim, F., Olsen, Ø. H., De Schutter, E., and Calabrese, R. L. (1995) Modeling the leech heartbeat elemental oscillator. I. Interactions of intrinsic and synaptic currents. *J. Comput. Neurosci.* **2,** 215–235.
9. Traub, R. D., Wong, R. K. S., Miles, R., and Michelson, H. (1991) A model of a CA3 hippocampal pyramidal neuron incorporating voltage-clamp data on intrinsic conductances. *J. Neurophysiol.* **66,** 635–650.
10. Goldman, M. S., Golowasch, J., Marder, E., and Abbott, L. F. (2001) Global structure, robustness, and modulation of neuronal models. *J. Neurosci.* **21,** 5229–5238.
11. Gunay, C., Edgerton, J. R., and Jaeger, D. (2004) *2004 Abstract Viewer and Itinerary Planner*, Society for Neuroscience, Washington, DC.
12. Hines, M. L., and Carnevale, M. T. (1997) The NEURON simulation environment. *Neural Comput.* **9,** 1179–1209.
13. Bower, J. M., and Beeman, D. (1998) *The Book of GENESIS: Exploring Realistic Neural Models with the GEneral NEural SImulation System*, Springer, New York.
14. Taylor, A. L., Hickey, T. J., Prinz, A. A., Marder, E. (2006) Structure and visualization of high-dimensional conductance spaces. *J. Neurophysiol.* **96**, 891–905.
15. Prinz, A. A., Thirumalai, V., and Marder, E. (2003) The functional consequences of changes in the strength and duration of synaptic inputs to oscillatory neurons. *J. Neurosci.* **23,** 943–954.

III

Imaging

11

Brain Atlases and Neuroanatomic Imaging

Allan MacKenzie-Graham, Jyl Boline, and Arthur W. Toga

Summary

Quantifying the effect of a genetic manipulation or disease is a complicated process in a population of animals. Probabilistic brain atlases can capture population variability and be used to quantify those variations in anatomy as measured by structural imaging. Minimum deformation atlases (MDAs), a subclass of probabilistic atlases, are intensity-based averages of a collection of scans in a common space unbiased by selection of a single target image. Here, we describe a method for generating an MDA from a set of magnetic resonance microscopy images. First, the images are segmented to remove any non-brain tissue and bias field corrected to remove field inhomogeneities. The corrected images are then linearly aligned to a representative scan, the geometric mean of all the transformations is calculated, and a minimum deformation target (MDT) is produced by averaging the volumes in this new space. The brains are then non-linearly aligned to the MDT to produce the MDA. Finally, the images are linearly aligned to the MDA using a full-affine transformation to spatially and intensity normalize them, removing global differences in size, shape, and position but retaining anatomically significant differences.

Key Words: Mouse; MRI; probabilistic; variability; spatial normalization.

1. Introduction

The relatively close evolutionary relationship between humans and mice has prompted the use of mice as a model system for investigation of disease. The common house mouse, through manipulation, breeding, or the disruption of normal genetic pathways, has been widely used to model human diseases. The creation of more mutant mice each year comes with a concomitant increase in the information used to describe them. The challenge is to analyze and compare

From: *Methods in Molecular Biology, vol. 401: Neuroinformatics*
Edited by: Chiquito Joaquim Crasto © Humana Press Inc., Totowa, NJ

the enormous amounts of data in an efficient and reliable manner. Digital atlases can provide the framework for the organization, analysis, and publication of large collections of data and a framework for researchers studying normal, mutant, and transgenic animals.

Traditional atlas construction has typically involved sectioning, staining, and recording photomicrographs. Franklin and Paxinos *(1,2)*, Hof et al. *(3)*, and Valverde *(4)* have published this type of mouse brain atlas; however, recent advances have expanded the atlas concept *(5)* into dynamic organizational and analytical tools including digital and probabilistic atlases.

1.1. Spatial Normalization

Probabilistic brain atlasing is a research strategy for generating anatomical templates that retain information on the variation in anatomy between subjects *(6–9)*. However, variations in the size, shape, and position of anatomical structures are significantly confounding in brain-mapping studies. Spatial normalization is the process used to reduce anatomic variance across specimens by registering the spatial features of a source brain image to those of a reference brain image. Spatial normalization methods correct for global size, shape, and positional differences not smaller localized, anatomically significant differences.

There is no single representative brain nor a simple way of creating an "average" anatomy or 3D anatomic variations, let alone variations across strains, genetically manipulated animals, and disease states. The choice of an individual brain as a template for the normalization leads to biasing because of large registration errors in brains with significantly different morphology, such as developmental or disease-induced changes. Plainly stated, just selecting a representative normal brain as the reference brain will create an atlas that is biased toward normal brains, underrepresenting brains with anatomical abnormalities.

1.2. Minimum Deformation Atlases

Minimum deformation atlases (MDAs), a subclass of probabilistic brain atlases, can be used to create "average" representations of anatomy by intensity averaging multiple MRI scans (***10–14***). An MDA is a representation of a set of brains where the transformational distance between all the elements of that set have been minimized, removing the bias toward any single scan.

Linear spatial normalization methods have been developed *(11)* to create average templates. Methods based on affine registrations give rise to an unbiased common average space. However, the resulting average template is blurry because the constituent images are corrected for global size and shape only, leaving residual individual differences behind.

Alternate average creation methods, based on non-linear deformations, generally depend on the choice of a single group member as the target brain, again biasing the result. Recently, non-linear normalization strategies ***(12,13)*** have been used to generate averages relatively free of the bias from a single individual.

1.3. Protocol

The protocol described here uses an algorithm to produce an intensity-based population MDA that takes advantage of the unbiased nature of the linear spatial normalization combined with a simple non-linear normalization to produce both a high-dimensional and an unbiased average representation ***(14)***. In this process, image volumes are segmented to remove non-brain tissue ***(15)***, and field inhomogeneity is corrected ***(16)***. The result is MDA, an intensity-based probabilistic atlas for spatial and intensity normalization ***(17)***. For a schematic representation of the entire process, *see* **Fig. 1**.

2. Materials

2.1. Segmentation

1. Magnetic resonance imaging files in Analyze Image 7.5 file format (http://www.mayo.edu/bir/PDF/ANALYZE75.pdf) (*see* **Note 1**). Analyze Image format files come in header and image data pairs. The image files should be in either 8 or 16 bits per pixel for use with BrainSuite 2. All the image files need to have the same orientation (i.e., share the same anterior–posterior axis, etc.)
2. BrainSuite 2 magnetic resonance image analysis tool (http://www.loni.ucla.edu/Software/Software_Detail.jsp?software_id=19) ***(15)***: BrainSuite 2 was developed in Microsoft Visual C++ and makes use of OpenGL for displaying surface models. It provides high-performance simultaneous viewing of multiple data volumes and surfaces even when run on modest PC computer hardware. BrainSuite 2 runs under the Windows 2000 and Windows XP Professional operating systems.

2.2. Field Inhomogeneity Correction

1. Segmented image files in Analyze Image 7.5 file format: Accurate segmentation is important for this and subsequent steps.
2. LONI Pipeline Processing Environment (http://pipeline.loni.ucla.edu) ***(14)*** (*see* **Note 2**): The LONI Pipeline Processing Environment provides a unified framework for the organization, dissemination, and use of neuroimaging analysis procedures formed through the linking of these algorithms into a cohesive methodology.

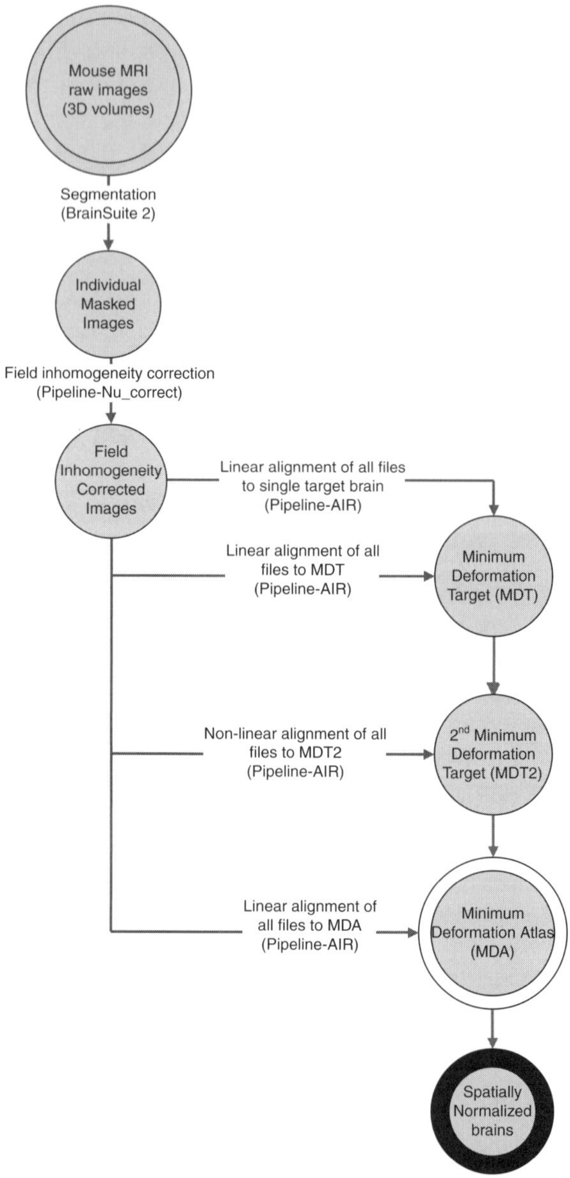

Fig. 1. **Schematic of the process of creating a minimum deformation atlas (MDA)**. An MDA was constructed from 32 postmortem high-resolution T2-weighted MRM images of mouse brain.

The environment deals with the diverse input and output requirements of various software packages, accommodates different file formats, and executes procedures across hardware environments as diverse as desktop PCs and supercomputers.

3. LONI Debabler (http://www.loni.ucla.edu/Software/Software_Detail.jsp?software_id=11) ***(18)***: The LONI Debabler contains file converters from Analyze Image to Medical Imaging Net CDF (MINC) and back. The N3 field inhomogeneity correction software requires the images to be in MINC format. The Pipeline contains file converters, but the Debabler will be required if the Pipeline is not used.
4. Nu_correct N3 field inhomogeneity correction software (http://www.bic.mni.mcgill.ca/software/) ***(16)***: The N3 field inhomogeneity correction algorithm does not make assumptions about the type of scan nor the relative ratios of white matter/gray matter/cerebrospinal fluid, making it ideal for performing bias field correction on non-T1-weighted, non-human images.

2.3. MDA

1. Bias field-corrected images in Analyze Image 7.5 file format.
2. LONI Pipeline Processing Environment.
3. Automated Image Registration 5.2.5 (http://air.bmapucla.edu/AIR5/) ***(19,20)***: The Automated Image Registration package is a robust and flexible image registration toolkit. The package allows for the manipulation of the transformations as well as the images themselves, permitting the user to concatenate them and resample their data only once, avoiding the loss of information that resampling entails (*see* **Note 3**).

3. Methods

The first step in producing an MDA is the segmentation of the constituent images. Accurate segmentation is crucial for a good atlas. We begin with an automated process and finish the segmentation manually to correct for any inaccuracies in the automated step.

Removing field inhomogeneities from the MRM images is the next major step. Differences in the intensity of the images should come from biologically significant differences in the tissue, not inhomogeneities in the magnetic field of the instrument. Field inhomogeneity correction removes these artifacts.

Lastly, the construction of the MDA is a multistep process involving the generation of a minimum deformation target (MDT) followed by the non-linear alignment of the constituent images to the MDT to produce the MDA (*see* **Fig. 2**). The MDA is then used as a target for the linear alignment of the images to correct for global differences in size, shape, position, and intensity, a process known as spatial and intensity normalization.

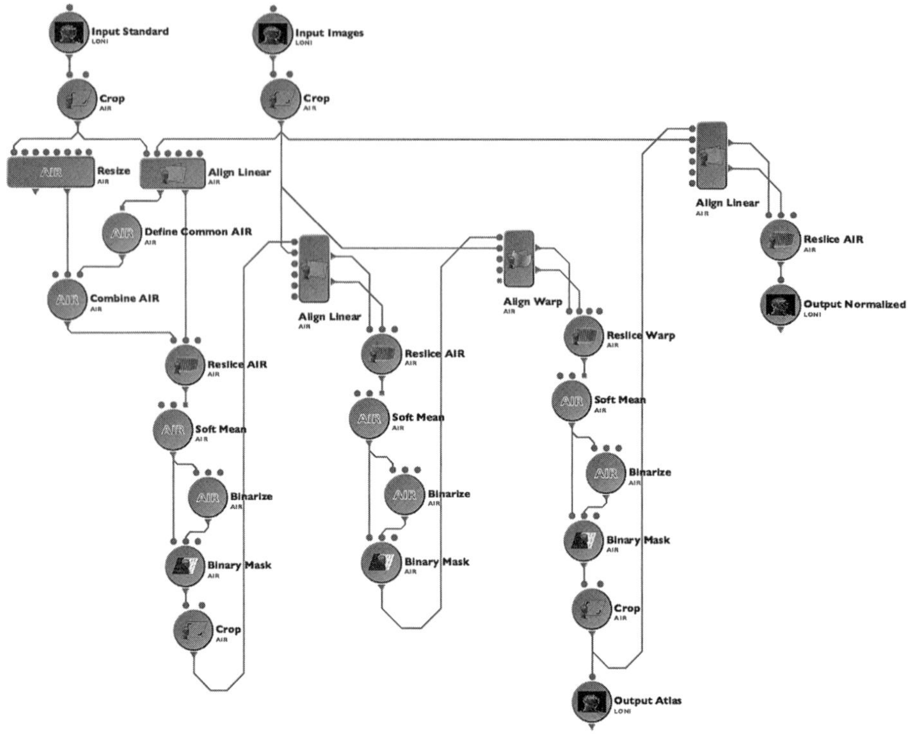

Fig. 2. **Creation of a minimum deformation atlas (MDA) in the LONI Pipeline Processing Environment**. An MDA was constructed from 32 postmortem high-resolution T2-weighted MRM images of mouse brain.

3.1. Segmentation

1. Open MRI image in BrainSuite 2.
2. From the BrainSuite menu, select Skull-stripping (BSE).
3. Press the "Do All" button in the BSE window and wait until the process is complete (*see* **Note 4**). A mask will appear on the image as a green outline.
4. From the Tools menu, select Mask Tool.
5. Edit the mask using the "Dilate" and "Erode" buttons on the Mask Tool. Manually edit the mask by checking the "Apply edits to mask" check box on the Mask Tool. Holding down the control key while pressing the left mouse button adds to the mask. Holding down the control key while pressing the right mouse button subtracts from the mask.
6. Apply the mask to the MRI by pressing the "Apply to MRI" button on the Mask Tool.

7. Save the now segmented file with a new name by selecting "Volume..." from the Save submenu in the File menu.

3.2. Field Inhomogeneity Correction

1. Create two text files, one that holds a list of your segmented image files and another that holds a list of file names for the field inhomogeneity corrected images. Make sure to include the full paths (*see* **Note 5**).
2. Open the LONI Pipeline application and select Connections... from the Window menu. Connect to the pipelinev4.loni.ucla.edu server by clicking on the "Connect" button.
3. Open the Field Inhomogeneity Correction pipeline from the LONI folder in the Library tab on the left by double-clicking on it.
4. Double click on the small blue circle above the module named "Input Image". Click the "Browse..." button and select the list of segmented image files to be the input. Click OK to close the Edit Value window.
5. Double click on the small blue arrow below the module named "Output Corrected". Click the "Browse..." button and select the list of field inhomogeneity corrected image file names to be the output (*see* **Note 6**). Click OK to close the Edit Value window.
6. Select Save as...from the File menu and save the modified pipeline.
7. Click the Run button (green triangle) on the bottom. The Pipeline will run, and the interface will show the state of the pipeline as it progresses. A grey outline indicates that a module is waiting to execute, a moving blue outline indicates that a module is running, and a green outline indicates successful completion. A red outline indicates a problem with the module. Right click on it to display the error. The pipeline may be aborted by clicking on the "stop" button (red square) on the upper right.
8. The input to the Analyze to MINC module will be the "input" images in the list. The images will be converted from Analyze Image 7.5 file format to Medical Imaging NetCDF.
9. The image files will then be field inhomogeneity corrected by the Nu_correct module.
10. The corrected files will be converted from MINC files to Analyze Image 7.5 files and output to the field in homogeneity corrected output list detailed in step 1.
11. The outlines of all of the modules will turn green when the pipeline has been completed.

3.3. MDA

1. Create two text files, one that holds a list of your field inhomogeneity corrected image files and another that holds a list of file names for the normalized images. Make sure to include the full paths (*see* **Note 5**).

2. Open the LONI Pipeline application and select Connections from the Window menu. Connect to the pipelinev4.loni.ucla.edu server by clicking on the "Connect" button.
3. Open the Minimum Deformation Atlas pipeline from the LONI folder in the Library tab on the left by double-clicking on it.
4. Double click on the small blue circle above the module named "Input Standard". Click the "Browse..." button and select a representative bias field corrected image file to be the standard (*see* **Note 7**). Click OK to close the Edit Value window.
5. Double click on the small blue circle above the module named "Input Image". Click the "Browse..." button and select the list of field inhomogeneity corrected image files to be the input. Click OK to close the Edit Value window.
6. Double click on the small blue arrow below the module named "Output Normalized". Click the "Browse..." button and select the list of normalized image file names to be the output (*see* **Note 6**). Click OK to close the Edit Value window.
7. Double click on the small blue arrow below the module named "Output Atlas". Type a path and name for the Atlas file (*see* **Note 6**) in the field provided. Click OK to close the Edit Value window.
8. Select "Save as..." from the File menu and save the modified pipeline.
9. Click the "Run" button (green triangle) on the bottom. The Pipeline will run and the interface will show the state of the pipeline as it progresses. The pipeline may be aborted by clicking on the "Stop" button (red square) on the bottom.
10. The input to the Crop module on the left will be the "standard" image selected, and the input to the Crop module on the right will be the list of the rest of the files. Cropping the images increases the robustness and decreases the duration of the registration by decreasing the search space of the algorithm. Accurate segmentation is crucial for this step.
11. The input files will be aligned to the representative image by an Alignlinear module with a full-affine transformation and a least squares with intensity-rescaling cost function. The full-affine transformation corrects for global differences in size, shape, and position. The intensity-rescaling cost function corrects for gross differences in intensity, preventing a higher intensity image from over-contributing to the MDT.
12. The geometric mean of all the transformations will be calculated by the Define Common AIR module, which will generate new linear transformation files for each of the input files into this common space, also know as the MDT.
13. The input files will be resampled to the MDT in the Reslice AIR module. Up until this point, the original data have not been modified. Reslice AIR will resample the data using the transformation calculated by Define Common AIR.
14. The Soft Mean module will create a soft mean of the resliced files that is the minimum deformation target (MDT) (*see* **Note 8**).
15. The MDT will then be used as the standard for the input images in a second linear alignment by an Alignlinear module using a full-affine transformation. Once

we have created an unbiased MDT, we begin the process of creating a high-dimensional average, the MDA. First is a linear alignment to the MDT followed by a non-linear alignment to the average resulting from the linear alignment.

16. The input files will be resampled and averaged to produce a second MDT2. The original input images are resampled using a new transformation aligning them to the MDT. They are not resampled twice but successively.
17. The MDT2 will then be used as the standard to which the input files are non-linearly aligned in the Align Warp module using a fifth-order polynomial warp.
18. The files will be resampled and averaged again to create a final MDA which is saved to the file name defined in step 7. Once more the original input images will be resampled using a non-linear transformation that aligns them to the MDT2. An example of the results is shown in **Fig. 3**.
19. The input files are then linearly aligned to the MDA to spatially and intensity normalize them by another Align linear module using a full-affine transformation. The spatial and intensity normalization step corrects global differences in size, shape, position, and intensity, retaining anatomically significant differences.
20. The input files will be resampled and output using the file names defined in the normalized output list detailed in step 1, spatially and intensity normalized.

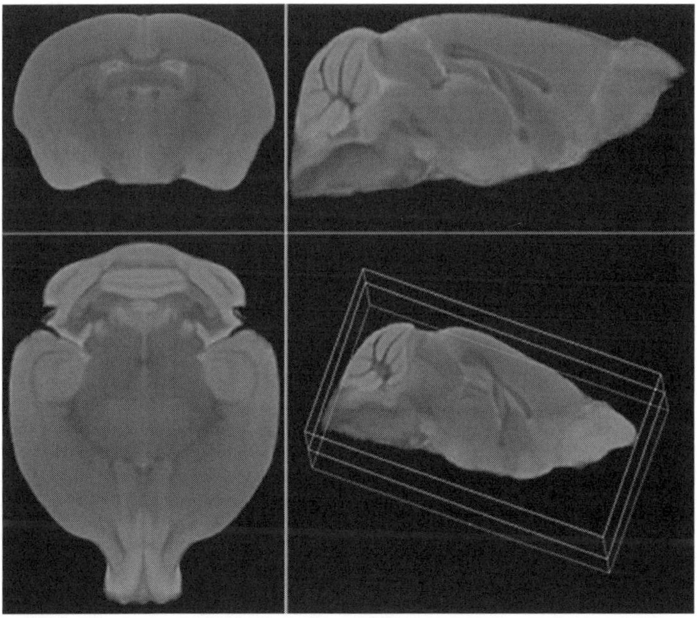

Fig. 3. **Minimum deformation atlas (MDA).** An MDA was constructed from 32 postmortem high-resolution T2-weighted MRM images of mouse brain.

21. The outlines of all of the modules will turn green when the pipeline has been completed.

4. Notes

1. There are many file converters available to and from the Analyze Image 7.5 file format. Makeaheader from AIR can produce headers for raw grayscale images with a data size of 8 or 16 bits per pixel.
2. Although the methods described here can be implemented in the absence of the LONI Pipeline Processing Environment, it greatly simplifies the process of creating and executing image processing algorithms. AIR, the BIC toolkit, and file converters are included in the Pipeline library of tools, allowing users to easily use the methods described here. In addition, the Pipeline contains image viewers to assess the data at any point during its processing. If desired, the same algorithm may be implemented in almost any scripting environment with access to AIR, the BIC toolkit, and file converters.
3. Although we use AIR 5.2.5, previous versions of AIR 5 should work. We suggest using the most recent version available.
4. Modifying parameters in the Brain Surface Extractor can lead to widely different results. Generally speaking, increasing the Diffusion Constant will increase the size of the mask, increasing the size of the Edge Constant will decrease the size of the mask, and increasing the Erosion Size will decrease the size of the mask. Unless you are confident with changing the parameters, run it with the defaults.
5. One can use a list of files instead of single file names to run the same process on each file in the list in the Pipeline. Create a text file (name it input list) that holds a list of the .img files to be processed. An example for the Microsoft Windows operating system is shown below; however, the same concept holds for Linux, UNIX, or Mac operating systems. For the Windows environment, prefix each item on the list with the drive letter (i.e., C:).

   C:\data\MRI_001.img
   C:\data\MRI_002.img
   C:\data\MRI_003.img
   C:\data\MRI_004.img

6. Select a directory different from the input directory for the output files. Output files will often have the same names as the input files and will overwrite them.
7. A representative image is one that is not markedly different from the group of images that make up the input images.
8. Errors in registration can bias the atlas in several different ways. In the worst-case scenario, the MDT can contain multiple brain images, each image acting as a target for a subset of the constituent images in subsequent registrations. This is often caused by decoys in the input images, unmasked tissue, or image artifacts. This result can also occur if the number of samples is small. A single misregistration

can bias the MDT with the results previously described. The larger the number of samples, the smaller the contribution of a single misregistration to the mean transformation. Occasionally, selection of a different standard image can lead to the creation of an unbiased atlas.

Acknowledgments

This work was generously supported by a research grant from the National Center for Research Resources U24 RR021760 and P41 RR13642 and the National Institute of Health U54 RR 021813. The authors acknowledge their deep appreciation to the members of the LONI and the Mouse Biomedical Informatics Research Network (Mouse BIRN), without whom this work would not have been possible.

References

1. Franklin, K.B.J. and G. Paxinos, *The Mouse Brain in Stereotaxic Coordinates*. 1997, San Diego, CA: Academic Press. p. xxii, (186) of plates.
2. Paxinos, G. and K.B.J. Franklin, *The Mouse Brain in Stereotaxic Coordinates*, 2nd ed. 2001, San Diego, CA: Academic Press. p. xxv, (264) of plates.
3. Hof, P.R., et al., *Comparative Cytoarchitectonic Atlas of the C57BL 6 and 129 Sv Mouse Brains*. 2000, Amsterdam and New York: Elsevier.
4. Valverde, F., *Golgi Atlas of the Postnatal Mouse Brain*. 1998, Vienna, Austria: Springer-Verlag.
5. Toga, A.W. and P.M. Thompson, Multimodal brain atlases, in *Advances in Biomedical Image Databases*, S. Wong, Editor. 1998, Dordrecht, the Netherlands: Kluwer Academic Press. p. 53–88.
6. Mazziotta, J.C., et al., A probabilistic atlas of the human brain: theory and rationale for its development. The International Consortium for Brain Mapping (ICBM). *Neuroimage*, 1995. **2**(2): 89–101.
7. Thompson, P.M., et al., Detection and mapping of abnormal brain structure with a probabilistic atlas of cortical surfaces. *J Comput Assist Tomogr*, 1997. **21**(4): 567–581.
8. Thompson, P.M., C. Schwartz, and A.W. Toga, High-resolution random mesh algorithms for creating a probabilistic 3D surface atlas of the human brain. *Neuroimage*, 1996. **3**(1): 19–34.
9. Thompson, P.M. and A.W. Toga, Detection, visualization and animation of abnormal anatomic structure with a deformable probabilistic brain atlas based on random vector field transformations. *Med Image Anal*, 1997. **1**(4): 271–294.
10. Ashburner, J. and K. Friston, Multimodal image coregistration and partitioning–a unified framework. *Neuroimage*, 1997. **6**(3): 209–217.
11. Kochunov, P., et al., Regional spatial normalization: toward an optimal target. *J Comput Assist Tomogr*, 2001. **25**(5): 805–816.

12. Kovacevic, N., et al., A three-dimensional MRI atlas of the mouse brain with estimates of the average and variability. *Cereb Cortex*, 2005. **15**(5): 639–645.
13. Ma, Y., et al., A three-dimensional digital atlas database of the adult C57BL/6J mouse brain by magnetic resonance microscopy. *Neuroscience*, 2005. **135**(4): 1203–1215.
14. Rex, D.E., J.Q. Ma, and A.W. Toga, The LONI pipeline processing environment. *Neuroimage*, 2003. **19**(3): 1033–1048.
15. Shattuck, D.W. and R.M. Leahy, BrainSuite: an automated cortical surface identification tool. *Med Image Anal*, 2002. **6**(2): 129–142.
16. Sled, J.G., A.P. Zijdenbos, and A.C. Evans, A nonparametric method for automatic correction of intensity nonuniformity in MRI data. *IEEE Trans Med Imaging*, 1998. **17**(1): 87–97.
17. Mackenzie-Graham, A., et al., *Cerebellar Cortical Atrophy in Experimental Autoimmune Encephalomyelitis*, 2006.
18. Neu, S.C., D.J. Valentino, and A.W. Toga, The LONI Debabeler: a mediator for neuroimaging software. *Neuroimage*, 2005. **24**(4): 1170–1179.
19. Woods, R.P., et al., Automated image registration: I. General methods and intrasubject, intramodality validation. *J Comput Assist Tomogr*, 1998. **22**(1): 139–152.
20. Woods, R.P., et al., Automated image registration: II. Intersubject validation of linear and nonlinear models. *J Comput Assist Tomogr*, 1998. **22**(1): 153–165.

12

Brain Mapping with High-Resolution fMRI Technology

Nian Liu

Summary

This chapter describes the use of high-resolution functional magnetic resonance imaging (fMRI) technology in brain odor mapping and a suite of informatics tools for building, databasing, and analyzing fMRI odor maps. OdorMapBuilder is a software program that extracts the olfactory signals that occurred in a particular layer of the olfactory bulb (OB), that is, the glomerular layer, from the 3D imaging data and generates 2D flat odor maps. Odor maps describe the odor-induced spatial activity patterns in the entire glomerular layer in the OB. OdorMapDB is a Web-based database system that serves as a centralized repository for the fMRI odor maps. OdorMapComparer is a software program that allows users to visually evaluate and statistically determine the similarity or difference between two fMRI odor maps being compared. Taken together, the fMRI technique and the related informatics tools play an important role in the study of the signal-encoding mechanisms in the olfactory system, which is essential to our understanding of the perception of smell.

Key Words: Odor map; olfactory bulb; spatial activity pattern; statistical analysis; functional imaging; neuroinformatics.

1. Introduction

The mammalian brain has distinctive anatomical structures and signaling projection pathways. These are responsible for varied brain functions, including the control of body movement and perception of environmental stimuli *(1)*. Alterations in these structures or functions may lead to neurological diseases, for example, movement disorders or abnormal sensory perceptions *(2)*. In sensory systems, external stimuli are transformed into neuronal signals that

From: *Methods in Molecular Biology, vol. 401: Neuroinformatics*
Edited by: Chiquito Joaquim Crasto © Humana Press Inc., Totowa, NJ

project from one brain region to the next through neural circuits and synaptic transmissions. A given type of perception involves similar sensory receptors in the peripheral nervous system and grossly the same regions in the brain. Some coding modalities, such as those in the somatic sensory and the visual systems, contain spatial information. Such information is preserved by spatially fixed projection pathways and targets in the brain. An increasing volume of evidence suggests that the perception of smell, one of the non-spatial sensory modalities, involves spatial coding mechanisms. The identities of odor stimuli are believed to be encoded as unique neural activity patterns distributed in different areas of the glomerular layer in the olfactory bulb (OB), which is the first relay station for the olfactory signals between the peripheral olfactory receptor neurons and the brain cortex. In this chapter, the olfactory system is used to demonstrate how high-resolution brain mapping may help investigators understand the signal-encoding mechanisms underlying the sensory perceptions.

Odor stimuli-elicited neuronal activities include a variety of transient and late-response signals in the OB. Each of these signal types may be detected with an appropriate experimental technique. For example, the membrane potential of neurons, which is a transient signal, can be recorded with impaled recording electrodes *(3)*. The neural circuits conveying the propagating membrane potentials may indeed play a direct role in our perceptions of the external stimuli. Intracellular cascades coupled with changes in the membrane potentials may involve a redistribution of inorganic ions such as calcium, a transient signal that can be detected with the calcium imaging technology *(4)*. The late-response signals that occurred in the OB include changes in gene expression and in energy consumption in the neurons. Over-expression of immediate early gene *c-fos* in the OB has been detected by in situ hybridization following odor exposure to the animals *(5,6)*. Using radiolabeled non-metabolizable 2-deoxyglucose (2DG) administered into the animals prior to the experiments, accumulation of 2DG in the neurons has been detected with autoradiography following 15–45 min of odor exposures *(7–9)*. Most recently, odor-induced changes in the metabolic state of neural tissues in the OB have been detected by measuring blood flow with the optical imaging and functional magnetic resonance imaging (fMRI) techniques *(10–13)*. These imaging techniques have a temporal resolution from half to several minutes. Investigations using these different techniques all support the notion that unique spatial patterns of neural activities occur in the OB of animals exposed to odor stimuli.

Global brain mapping is the key to systematically study the spatial activity patterns in the OB. Electrophysiology and optical imaging techniques examine neural activities occurring in limited areas and thus fail to provide a global view of the odor-specific activity patterns in the brain. 2DG autoradiography, *c-fos* in situ hybridization, and fMRI techniques allow global mapping. The first two methods require that the animals be killed after the trial to detect the signals, thus resulting in only one trial for an animal ***(6,7)***. The fMRI method is non-invasive, allowing the animals to be repeatedly tested with different odor types or concentrations. This is an advantage of the method, because intrasubject comparisons eliminate variables introduced by the subjects. Also, the fMRI method provides a better temporal resolution compared with 2DG and *c-fos*. Whereas 2DG and *c-fos* methods require that the animals be exposed to odor stimuli for several to 45 min, fMRI is able to detect the olfactory signals as short as 0.5 min after the onset of the odor stimuli ***(14)***. Thus, the fMRI approach provides a unique brain-mapping tool to reveal the nature of smell closely related to our experience.

Described here is the use of high-resolution fMRI technology in brain odor mapping. From an informatics perspective, this chapter details how fMRI signals are obtained from the OB, how the 3D imaging data are converted onto flat odor maps, how the maps are archived and presented as a resource to the public, and how they are quantitatively analyzed and compared. Altogether, the informatics tools facilitate experimental research on the olfactory-coding mechanisms and help explain the nature of smell perception.

2. BOLD-based fMRI Studies of the Olfactory Bulb

fMRI for brain odor mapping is based on the blood oxygen level-dependent (BOLD) contrast at a magnetic field strength of 7 Tesla ***(11)***. The experiments are performed on a horizontal-bore Bruker spectrometer (Bruker, Billerica, MA) with fast, low-angle, single-shot gradient-echo imaging sequence. The tested animals are anesthetized and placed in the magnetic field. An odor-delivery system generates a clean or odorized airflow that is administered to the animals. In each test, a series of coronal images are obtained, covering the entire OB in an anterior–posterior fashion.

The detected MRI signals contain two components: T1 and T2. The T1 signal is primarily affected by the tissue properties, that is, the regional homogeneity of the brain tissues. The T1-weighted images are referred to as the anatomical MRI images, in which the laminar structures are clearly identified ***(14)***. The anatomical images have an in-plane resolution of $100 \times 100\,\mu m$. The T2 signal is affected by the local changes of the metabolic-vascular state in response to neuronal activities. The T2-weighted images are referred to as the functional MRI or fMRI. High-resolution fMRI for odor mapping generates T2-weighted OB images, each with an in-plane resolution of $200 \times 200\,\mu m$ and slice thickness of 200 μm *(12)*. In the multi-slice mapping experiments, 15 consecutive images may be obtained from an adult mouse and 20 from an adult rat. fMRI images are traditionally overlain to the anatomical MRI images so that the odor-induced fMRI signals can be localized to specific brain regions. Data from an odor-mapping experiment are traditionally presented as a set of images showing both the anatomical structures and the fMRI signals.

3. Informatics Tools

3.1. OdorMapBuilder—Building fMRI Brain Odor Maps

Although the 3D fMRI data show the odor stimuli-invoked spatial activity patterns in the entire OB, it is difficult to intuitively describe the findings from large amounts of data sets. In addition, fMRI signals may occur not only in the glomerular layer but also in some other OB layers, for example, the nerve layer and the external plexiform layer, in which the incoming axons from the olfactory receptors and the dendrites of the mitral cells are located, respectively. This inevitably hinders the interpretation of results and comparative analyses of the experimental data. A brain-mapping program called Caret has been developed to localize the neural activities that occurred in the primary visual cortex *(15)*. The convoluted cerebral cortex is reconstructed as an inflated flat sheet, in which the cortical activities, regardless the layer, are "mapped" to the surface of the brain *(16)*. Brain odor mapping in the OB, however, differs from those in the cerebral cortex. The odor-induced neural activities that involve encoding of the olfactory signals occur in the glomerular layer of the OB. Thus, any tools for odor mapping should be able to extract information from a particular layer of the brain (*see* **Fig. 1**).

The strategy for brain odor mapping is to identify the glomerular layer in the anatomical MRI images and apply the geometric coordinates of the layer to the corresponding fMRI images (*see* **Fig. 1**). The mapping program should allow investigators to use the computer mouse to trace the glomerular layer in each anatomical image. It should also allow users to set the center point of the bulb in each image and to define the values of some experimental parameters so that the images from adjacent fMRI slices can be properly aligned. Each trace corresponds to a band in the odor map (*see* **Fig. 1C and D**). The width of the band is calculated from the thickness of the fMRI slices and the magnification scale of the images. Assuming that the center point is defined as (0, 0), an odor map from the dorsal perspective requires that the point ($x = 0$ and $y > 0$) in the trace be first plotted in the middle of the map. The trace is then programmatically "unrolled" onto the band in both directions until it reaches the point ($x = 0$ and $y < 0$) if such a point exists. To be consistent for

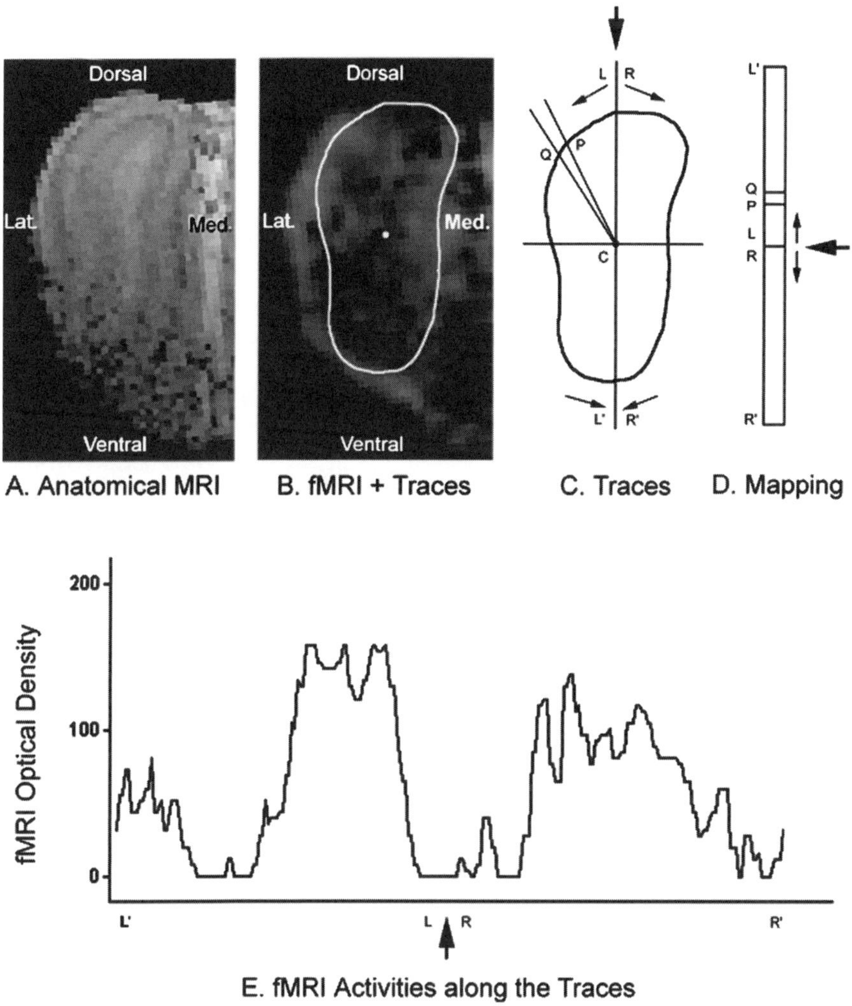

Fig. 1. Strategy of functional magnetic resonance imaging (fMRI) brain odor mapping. The anatomical (**A**) and functional (**B**) MRI images show the right half of a coronal slice of a rat olfactory bulb. The roughly spherical glomerular layer can be identified in the anatomical images as a continuous "grey" sheet sandwiched between the mostly brighter nerve layer and the external plexiform layer. The traces of the glomerular layer can be applied to the fMRI images and are unrolled starting from the top (R, L) and ending at the bottom (R′, L′) of the layer, generating an odor map from the dorsal perspective. The intensities of fMRI signals along the traces give the pixel values of the map image (**B–E**).

all individual images, unfolding of the traces is carried out in a radial manner with an increment of α degrees (by default, $\alpha = 5$). The length of the trace corresponding to each α-degree projection is defined as the distance between the two crossings (*see* **Fig. 1**; P and Q), which is plotted as a block on the band reflecting the actual distance along the trace. Therefore, the area of any given region in the odor map closely reflects the size of the glomerular sheet.

A software program, OdorMapBuilder, implements the above mapping strategy ***(17)***. The program was written in the Java programming language ***(18)***. It contains a menu bar with a rich set of commands in the main frame allowing the users to operate (*see* **Fig. 2**). The program first loads all the anatomical images, each displayed in a separate frame (*see* **Fig. 2A**). The user-defined traces and parameters can be saved in a file and can be applied

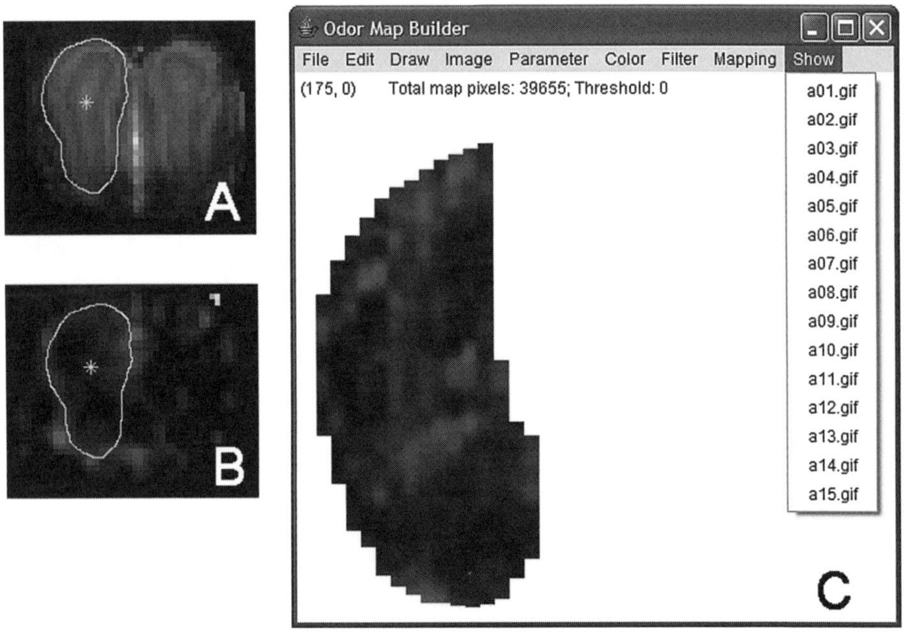

Fig. 2. OdorMapBuilder—A program to generate the functional magnetic resonance imaging (fMRI) brain odor maps. Loaded into the program are 15 consecutive anatomical MRI images from a mouse olfactory bulb, each displayed in a separate frame (**A**). It allows the user to trace the glomerular layer and set the center point of the bulb. The user-defined parameters can be applied to the corresponding fMRI images (**B**). An odor map from the dorsal perspective has been generated and shown in the main frame of the program (**C**).

to the corresponding fMRI images (*see* **Fig. 2B**). The generated odor map image is displayed in the main frame (*see* **Fig. 2C**). The raw odor maps are "red" images. They can be converted into red/green/blue (RGB) mode with an appropriate color table for better effect in presentation. The odor maps can also be smoothed to remove the jagged edges between the neighboring bands and blocks. To highlight the activity patterns, a threshold can be set so that all pixels with a value less than the threshold are set to be zero, thus decreasing the noise in the background.

In each fMRI experiment, one set of anatomical MRI images is usually matched to multiple data sets of fMRI images. The same set of traces and parameters from the anatomical images can be applied to all the data sets of fMRI images. Such a one-for-all approach allows users to generate multiple fMRI brain odor maps of the same size and shape from a single subject. These odor maps can be compared with one another without image transformation or warping. This takes full advantage of the fact that fMRI is non-invasive, allowing multiple trials in the same animals and eliminating the individual variations.

3.2. OdorMapDB—Databasing fMRI Brain Odor Maps

OdorMapDB (http://senselab.med.yale.edu/senselab/OdorMapDB/) is a Web-based database built on the Oracle relational database management system. The database interface has been developed in active server pages, which is run on the Microsoft IIS. Currently archived in the database are fMRI maps, *c-fos* maps ***(5)***, and the original 2DG maps generated in our laboratory ***(8)***. Each odor map is annotated with information about the subject, mapping technique, odor type and concentration, stimulus duration, experimenters, etc. Odor maps in the database may be browsed by the odor stimuli and mapping methods (*see* **Fig. 3**). They can also be searched by the annotating attributes. Authenticated login is required to enter or edit the content.

Using dynamic links, OdorMapDB is integrated with two other olfactory databases, that is, OdorDB (http://senselab.med.yale.edu/senselab/OdorDB/) and olfactory receptor database (ORDB) (http://senselab.med.yale.edu/senselab/ORDB/), which provide resources on the odor molecules and the olfactory receptor genes/proteins, respectively ***(19,20)***. In addition, OdorMapDB provides links to resources in the National Library of Medicine and Web sites maintained by peer laboratories. As a centralized data repository, the database not only helps to categorize the brain odor maps published in journals but also provides complementary materials, that is, the raw data and odor maps from other perspectives. This becomes increasingly significant

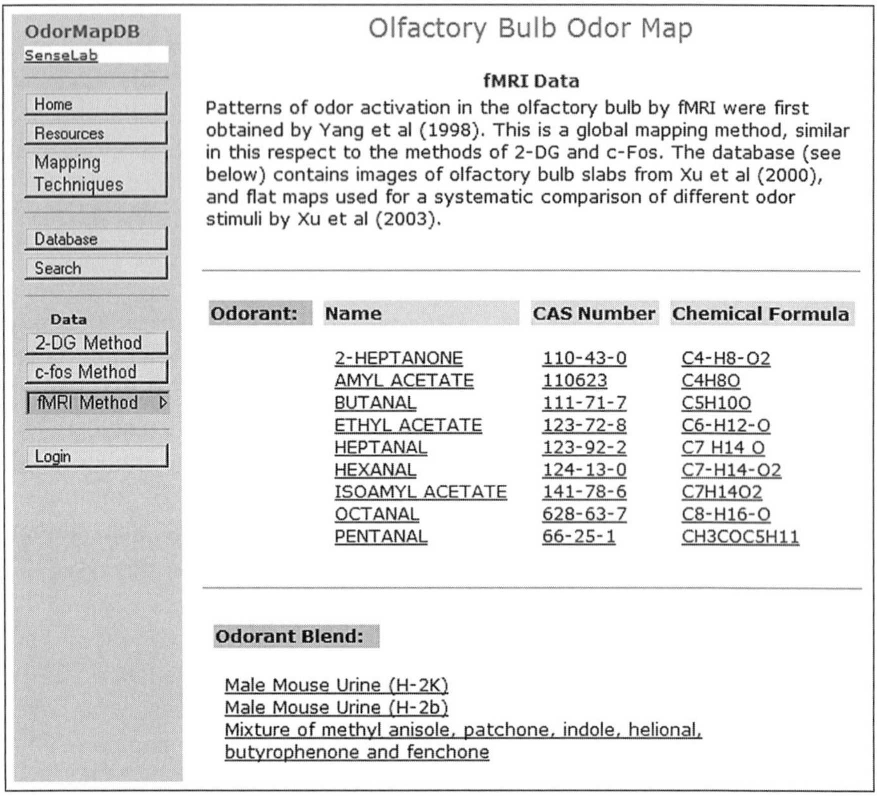

Fig. 3. OdorMapDB—A data repository for the functional magnetic resonance imaging (fMRI) odor maps. The menu items on the left bar allow users to operate, for example, search or browse the database. Displayed in this example Web page are a list of odor molecules and a list of odor blends such as urines that have fMRI odor maps in the database. The odor names, CAS numbers, and chemical formulae are hyperlinks allowing users to click on to list all the related odor maps.

when large amounts of odor maps are produced from continued experimentation with varied odor stimuli, given the existence of a large number of known odor molecules and the combinatory nature of olfactory receptor–odor interactions. With the goal to serve primarily the olfactory research community, OdorMapDB plays a unique role in the brain odor-mapping research.

3.3. OdorMapComparer—Analyzing fMRI Brain Odor Maps

Brain odor maps describe the spatial activity patterns based on the 2D images, in which the pixel values indicate the intensities of the fMRI signals in the corresponding glomerular areas. A standalone Java program, called OdorMapComparer, carries out quantitative and statistical analyses and comparisons of the fMRI odor maps. The program loads two odor maps being compared (*see* **Fig. 4**). Ideally, the loaded maps are monochrome "red" images, in which the pixel values in the "red" channel represent the signal intensities. If the loaded images are in RGB pseudo-color, they may be standardized into "red" with appropriate color tables (built-in or imported). A copy of the "red" image is automatically generated and can be converted into pseudo-color for better visual effect. All comparative analyses are performed with the underlying standardized "red" images. The program allows users to transform, that is, to scale, flip, and rotate, each of the images. It allows users to use the computer mouse to draw a pair of corresponding landmarks in the two maps, for example, the boundary of the odor maps, so that one image can be warped against the other based on the thin-plate splines algorithm ***(21)***. Users may move and properly align/overlap the two images.

For odor map warping, individual points along the two markers are pair-wise matched based on two identical reference points, for example, the centroids of

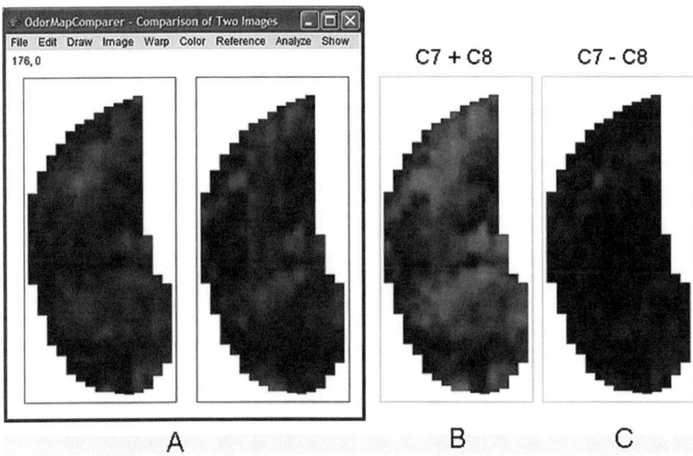

Fig. 4. OdorMapComparer—A program for quantitative and statistical comparisons of the functional magnetic resonance imaging brain odor maps. The program loads two images being compared (**A**, C7, left and C8, right), allowing image transformation and warping so that they can be appropriately aligned. Simple addition (**B**), subtraction (**C**), and averaging of the signal intensities in the two images can be performed. The program also computes the normalized correlation and spatial correlation coefficient between the images. Odor maps, C7-heptanal and C8-octanal.

all pixels within the outlines of the odor maps. For instance, when the outlines of the two odor maps are defined as markers, the program will identify a total of 72 points with a 5-degree radial separation in each odor map. Thus, two corresponding coordinate vectors **P** (original) and **V** (warped) are generated. The warped points from the original image may be defined by the warping function *(21)*:

$$f(x, y) = a_1 + a_x x + a_y y + \sum_{i=1}^{n} w_i U\left(|P_i - (x, y)|\right), \tag{1}$$

where, $U(r_{ij}) = r_{ij}^2 \log r_{ij}^2$ and $r_{ij} = \sqrt{(x_i - x_j)^2 + (y_i - y_j)^2}$ is the distance between points i and j. Two data matrices, K and P, may be defined as follows:

$$K = \begin{bmatrix} 0 & U(r_{12}) \ldots U(r_{1n}) \\ U(r_{21}) & 0 \quad \ldots U(r_{n2}) \\ \ldots & \ldots \quad \ldots \quad \ldots \\ U(r_{n1}) & U(r_{n2}) \ldots \quad 0 \end{bmatrix}, \quad (n \times n) \tag{2}$$

and

$$P = \begin{bmatrix} 1 \; x_1 \; y_1 \\ 1 \; x_2 \; y_2 \\ \ldots\ldots\ldots \\ 1 \; x_n \; y_n \end{bmatrix}, \quad (n \times 3). \tag{3}$$

A new matrix L can be constructed as follows:

$$L = \begin{bmatrix} K & P \\ P^T & O \end{bmatrix}, \quad (n+3) \times (n+3), \tag{4}$$

where T is the matrix transpose operator and O is a 3×3 matrix of zeros. Data matrix V with the corresponding points in warped landmark is defined as follows:

$$V = \begin{bmatrix} x'_1 \; x'_2 \ldots x'_n \; 0\,0\,0 \\ y'_1 \; y'_2 \ldots y'_n \; 0\,0\,0 \end{bmatrix}, \quad 2 \times (n+3). \tag{5}$$

The coefficients W $(w_1, w_2, \ldots w_3)$, a_1, a_x, a_y are solved as $L^{-1}V^T$ and then can be applied to Eq. (1) to obtain the target coordinates for every pixel in the warped image. The matrix computations are automatically performed with a dynamic script written in the R language. Execution of the R script is invoked from Java code during runtime. This warping algorithm has been used in warping MRI images against histostained tissue sections *(22)*.

After proper alignment, the two odor maps can be subjected to simple addition, subtraction, and average on a pixel-by-pixel basis upon user command from the "Analyze" menu. The result from each operation is presented as a new image, which allows users to determine visually the similarity or difference of the spatial activity patterns between the two odor maps (*see* **Fig. 4B and C**).

Two comparative statistics, normalized correlation (NC) and spatial correlation coefficient (SCC), have been implemented in the program. NC may be used to determine the degree of similarity between two data vectors. The formula for the NC is defined as follows:

$$\mathrm{NC} = \frac{\sum_{i=1}^{S} (x_i - \overline{x})(y_i - \overline{y})}{\sqrt{\sum_{i=1}^{S} (x_i - \overline{x})^2 \times \sum_{i=1}^{S} (y_i - \overline{y})^2}}, \tag{6}$$

where S is the number of pixels, x_i and y_i are the corresponding signal intensities for pixel i in odor maps x and y, respectively, and $\overline{x}$ and $\overline{y}$ are the mean intensities of x and y over the pixels within the odor map boundaries. NC has a value between -1 and 1, with 1, 0, and -1 indicating the identical, random, and reversed spatial activity patterns, respectively, between the two odor maps.

SCC is used to determine the significance of domain overlap between two data matrices or images ***(12,23,24)***. A signal threshold is set to separate all the pixels (not the entire image but those within the outline of the odor map) into two groups with a value of 1 or 0. A pixel has a value of 1 when the signal is above threshold and a value of 0 when the signal is below threshold. By default, the threshold is automatically set so that the counts of pixels in the two groups are in a 1:1 ratio. When two odor maps, A and B, are compared, a given pixel belongs to one of the four categories, [1, 1], [1, 0], [0, 1], and [0, 0], where the values in each square bracket represent the defined values of that pixel in odor maps A and B. $N_{[1,1]}$, $N_{[1,0]}$, $N_{[0,1]}$, and $N_{[0,0]}$ are the total counts of the pixels in the corresponding categories. The formulae for the SCC are defined as follows:

$$\mathrm{SCC} = \frac{N_{[1,1]}N_{[0,0]} - N_{[1,0]}N_{[0,1]}}{\left(N_{[1,0]} + N_{[1,1]}\right)\left(N_{[1,0]} + N_{[0,0]}\right)}, \tag{7}$$

when $N_{[0,0]}N_{[1,1]} > N_{[1,0]}N_{[0,1]}$ and

$$\mathrm{SCC} = \frac{N_{[1,1]}N_{[0,0]} - N_{[1,0]}N_{[0,1]}}{\left(N_{[1,0]} + N_{[1,1]}\right)\left(N_{[0,1]} + N_{[1,1]}\right)}, \tag{8}$$

when $N_{[1,1]}N_{[0,0]} < N_{[1,0]}N_{[0,1]}$ and $N_{[1,1]} < N_{[0,0]}$. If $0.2 < \text{SCC} < 1.0$, a significant degree of domain overlap occurs ($P < 0.01$). If $-1.0 < \text{SCC} < -0.2$, a significant degree of domain segregation occurs ($P < 0.01$). If $-0.2 < \text{SCC} < 0.2$, no significant correlation occurs.

In addition, OdorMapComparer can be run in batch mode. This allows users to compare odor maps without opening the program window. Running the program in this mode, however, requires that the odor maps have already been properly aligned, such as those generated from the same test animal. The batch mode facilitates comparisons of a large number of odor maps. OdorMapComparer can be used as a generic tool to analyze and compare other types of functional imaging data.

4. Discussion

This chapter has described the use of high-resolution fMRI technology and a suite of informatics tools in brain odor mapping. The fMRI method is able to produce 3D imaging data showing the odor-induced neural activities in the OB. OdorMapBuilder allows researchers to extract olfactory signals from the entire glomerular layer and generate 2D flat odor maps. Odor maps can be properly annotated and archived in the public accessible database OdorMapDB. OdorMapComparer provides a means to quantitatively and statistically analyze the fMRI odor maps. Taken together, these tools constitute a pipeline of data processing and analysis in the fMRI brain odor mapping (*see* **Fig. 5**). Merging high-resolution fMRI technology and informatics approaches undoubtedly facilitates and enhances experimental research on the global brain mapping, an important step that may lead to our understanding of the encoding mechanisms of sensory signals in the brain.

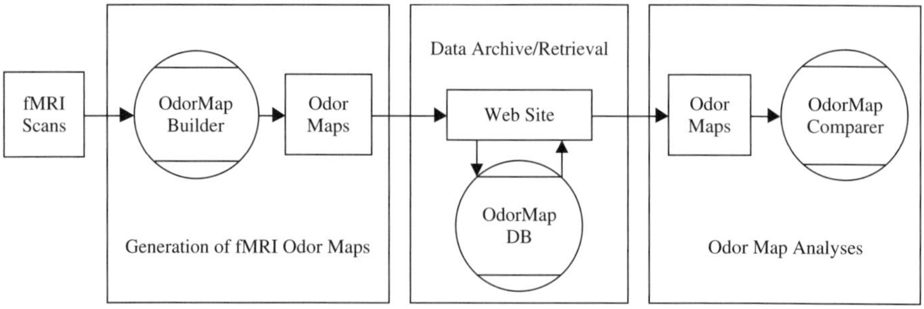

Fig. 5. A diagram showing the dataflow in functional magnetic resonance imaging (fMRI) brain odor mapping. The informatics tools, that is, OdorMapBuilder, OdorMapDB, and OdorMapComparer, serve as a pipeline for data processing and analysis of the fMRI brain odor maps.

High-resolution fMRI technique provides a mean to detect the subtle difference between the spatial patterns in response to different stimuli. Traditional MRI techniques with a resolution of millimeters have been widely used in experimental research and clinical diagnosis *(25,26)*. The converged and divergent projections of individual axons within the nerve bundle from the peripheral receptors to discrete clusters of the target neurons in the brain are part of the mechanisms employed to differentiate the identities of odor stimuli in the olfactory system *(27)*. It is impossible to use the traditional fMRI instrumentations to study the signal processing of stimulus information, which occurs at the microscopic level. Therefore, the high-resolution fMRI technique with an in-plate resolution of 0.1–0.2 mm per pixel is the key to the brain odor mapping. Using this new technique, it has been found that the similarity of spatial activity patterns in the odor maps among the aliphatic aldehyde homologs is closely related to the difference in the number of carbon atoms in the molecules *(12)*. This is further supported by behavioral tests using these aldehyde odorants *(28)*. The high-resolution fMRI technique may also be used for functional mapping in other brain regions, thus providing a generic tool for the study of the space-encoding mechanisms of sensory information in the nervous systems.

Informatics tools are indispensable components in the fMRI brain odor mapping. Although it is believed that the identities of odor stimuli are encoded in space in the glomerular layer of the OB *(29)*, the traditional method to localize the signals by overlaying the fMRI images to the anatomical images is cumbersome and non-quantitative. OdorMapBuilder allows researchers to generate single flat fMRI odor maps that intuitively describe the neural activities that occurred in any single layer of the OB. Therefore, the fMRI odor maps help to interpret the experimental results from different stimuli. They also provide materials for quantitative and statistical comparisons of the spatial activity patterns using OdorMapComparer. In the Internet era, the Web-based OdorMapDB provides a tool for data sharing in the field, facilitating the olfactory research as a whole.

5. Conclusion and Challenges

High-resolution fMRI technology has been successfully used in global brain odor mapping. Odor maps generated with OdorMapBuilder from the 3D fMRI imaging data indicate that odor-specific spatial activity patterns occurred in the glomerular layer of the OB in rodents. OdorMapDB provides a useful Web database tool to archive the odor maps and annotate the data with information about the odor stimuli and the olfactory receptors. OdorMapComparer provides statistical measures of similarity between two odor maps being compared. These software tools provide a means to process fMRI imaging data at different stages of experimental research on the brain odor mapping. Coupled with informatics tools, high-resolution fMRI plays an important role in understanding the signal-encoding mechanisms in the olfactory system.

One challenge that fMRI and all other global mapping methods face is to increase the temporal resolution in signal detection. Although the high-resolution fMRI, 2DG, and *c-fos* methods are able to describe the odor-induced spatial activity patterns at the microscopic level, it takes half a minute to tens of minutes to detect signals. This does not reconcile with our experiences and the behavior in animals. For instance, when we are exposed to odors in the environment, we may instantly smell and recognize these odors. When an animal perceives the approaching of a predator, it must react in a short period of time to survive. Therefore, a higher time resolution will certainly help better understand how the functional brain maps are related to the sensory perceptions and the responsive behaviors. In addition, a direct correlation between the fMRI signal and the transient neuronal signals such as changes in the membrane potentials of the neurons, if experimentally established, will further our understanding of brain mapping not only at the cellular but also at the molecular levels, providing a broader picture of mechanisms underlying the perception of odor stimuli.

Another fundamental and scientific challenge is to reconcile the difference among the global mapping techniques including fMRI, 2DG labeling, and *c-fos* in situ hybridization. These methods detect different kinds of neural signals in the OB in animals exposed to odor stimuli. They also produce different shapes of odor maps, which reflect the mapping projections used, just as Mollweide and Mercator world maps of the globe give different shapes and coordinates to the continents reflecting those different projections. To fully understand the encoding mechanisms of olfactory signals in the brain, it is necessary to carry out comparative analyses of the spatial activity patterns detected with different mapping methods. This requires collaborative research among involved laboratories so that appropriate algorithms can be developed to match the physical locations of each region among the different types of odor maps.

Acknowledgments

This study was supported by NIH grants K22 LM008422 and T15 LM07056 from the National Library of Medicine and P01 DC04732 under the Human Brain Project.

References

1. Kandel, E. R., Schwartz, J. H. and Jessell, T. M. (1991) *Principles of Neural Science*, 3rd ed. New York: Elsevier Science Publishing Co., Inc.
2. Conn, P. M. (1995) *Neuroscience in Medicine*. Philadelphia, PA: J. B. Lippincott Company.
3. Schoppa, N. E. and Westbrook, G. L. (2001) Glomerulus-specific synchronization of mitral cells in the olfactory bulb. *Neuron* **31,** 639–651.
4. Adelsberger, H., Garaschuk, O. and Konnerth, A. (2005) Cortical calcium waves in resting newborn mice. *Nat Neurosci* **8,** 988–990.
5. Schaefer, M. L., Young, D. A. and Restrepo, D. (2001) Olfactory fingerprints for major histocompatibility complex-determined body odors. *J Neurosci* **21,** 2481–2487.
6. Guthrie, K. M., Anderson, A. J., Leon, M. and Gall, C. (1993) Odor-induced increases in *c-fos* mRNA expression reveal an anatomical "unit" for odor processing in olfactory bulb. *Proc Natl Acad Sci USA* **90,** 3329–3333.
7. Jourdan, F. (1982) Spatial dimension in olfactory coding: a representation of the 2-deoxyglucose patterns of glomerular labeling in the olfactory bulb. *Brain Res* **240,** 341–344.
8. Stewart, W. B., Kauer, J. S. and Shepherd, G. M. (1979) Functional organization of rat olfactory bulb analysed by the 2-deoxyglucose method. *J Comp Neurol* **185,** 715–734.
9. Johnson, B. A., Ho, S. L., Xu, Z., et al. (2002) Functional mapping of the rat olfactory bulb using diverse odorants reveals modular responses to functional groups and hydrocarbon structural features. *J Comp Neurol* **449,** 180–194.
10. Rubin, B. D. and Katz, L. C. (1999) Optical imaging of odorant representations in the mammalian olfactory bulb. *Neuron* **23,** 499–511.
11. Yang, X., Renken, R., Hyder, F., et al. (1998) Dynamic mapping at the laminar level of odor-elicited responses in rat olfactory bulb by functional MRI. *Proc Natl Acad Sci USA* **95,** 7715–7720.
12. Xu, F., Liu, N., Kida, I., Rothman, D. L., Hyder, F. and Shepherd, G. M. (2003) Odor maps of aldehydes and esters revealed by fMRI in the glomerular layer of the mouse olfactory bulb. *Proc Natl Acad Sci USA* **100,** 11029–11034.
13. Takahashi, Y. K., Kurosaki, M., Hirono, S. and Mori, K. (2004) Topographic representation of odorant molecular features in the rat olfactory bulb. *J Neurophysiol* 92, 2413–2427.
14. Xu, F., Kida, I., Hyder, F. and Shulman, R. G. (2000) Assessment and discrimination of odor stimuli in rat olfactory bulb by dynamic functional MRI. *Proc Natl Acad Sci USA* **97,** 10601–10606.
15. Van Essen, D. C., Lewis, J. W., Drury, H. A., et al. (2001) Mapping visual cortex in monkeys and humans using surface-based atlases. *Vision Res* **41,** 1359–1378.
16. Van Essen, D. C. and Drury, H. A. (1997) Structural and functional analyses of human cerebral cortex using a surface-based atlas. *J Neurosci* **17,** 7079–7102.

17. Liu, N., Xu, F., Marenco, L., Hyder, F., Miller, P. and Shepherd, G. M. (2004) Informatics approaches to functional MRI odor mapping of the rodent olfactory bulb: OdorMapBuilder and OdorMapDB. *Neuroinformatics* **2,** 3–18.
18. Niemeyer, P. and Peck, J. (1997) *Exploring JAVA*, 2nd ed. Sebastopol, CA: O'Reilly & Associates.
19. Crasto C., Marenco L., Miller P. and Shepherd G. (2002) Olfactory receptor database: a metadata-driven automated population from sources of gene and protein sequences. *Nucleic Acids Res* **30,** 354–360.
20. Skoufos, E., Marenco, L., Nadkarni, P. M., Miller, P. L. and Shepherd, G. M. (2000) Olfactory receptor database: a sensory chemoreceptor resource. *Nucleic Acids Res* **28,** 341–343.
21. Bookstein, F. L. (1989) Principal warps: thin-plate splines and the decomposition of deformations. *IEEE Trans Pattern Anal Machine Intell***11,** 567–585.
22. Jacobs, M. A., Windham, J. P., Soltanian-Zadeh, H., Peck, D. J. and Knight, R. A. (1999) Registration and warping of magnetic resonance images to histological sections. *Med Phys* **26,** 1568–1578.
23. Cole, L. C. (1949) The measurement of inter-specific association. *Ecology* **30,** 411–424.
24. Ramsden, B. M., Hung, C. P. and Roe, A. W. (2001) Real and illusory contour processing in area V1 of the primate: a cortical balancing act. *Cereb Cortex* **11,** 648–665.
25. Cao, Y., Sundgren, P. C., Tsien, C. I., Chenevert, T. T. and Junck, L. (2006) Physiologic and metabolic magnetic resonance imaging in gliomas. *J Clin Oncol* **24,** 1228–1235.
26. Scouten, A., Papademetris, X. and Constable, R. T. (2006) Spatial resolution, signal-to-noise ratio, and smoothing in multi-subject functional MRI studies. *Neuroimage* **30,** 787–793.
27. Mombaerts, P., Wang, F., Dulac, C., et al. (1996) Visualizing an olfactory sensory map. *Cell* **87,** 675–686.
28. Laska, M. and Teubner, P. (1999) Olfactory discrimination ability for homologous series of aliphatic alcohols and aldehydes. *Chemical Senses* **24**, 263–270.
29. Xu, F., Greer, C. A. and Shepherd, G. M. (2000) Odor maps in the olfactory bulb. *J Comp Neurol* **422,** 489–495.

13

Brain Spatial Normalization

Indexing Neuroanatomical Databases

William Bug, Carl Gustafson, Allon Shahar, Smadar Gefen, Yingli Fan, Louise Bertrand, and Jonathan Nissanov

Summary

Neuroanatomical informatics, a subspecialty of neuroinformatics, focuses on technological solutions to neuroimage database access. Its current main goal is an image-based query system that is able to retrieve imagery based on anatomical location. Here, we describe a set of tools that collectively form such a solution for sectional material and that are available to investigators to use on their own data sets. The system accepts slide images as input and yields a matrix of transformation parameters that map each point on the input image to a standardized 3D brain atlas. In essence, this spatial normalization makes the atlas a spatial indexer from which queries can be issued simply by specifying a location on the reference atlas. Our objective here is to familiarize potential users of the system with the steps required of them as well as steps that take place behind the scene. We detail the capabilities and the limitations of the current implementation and briefly describe the enhancements planned for the near future.

Key Words: Neuroanatomical informatics; neuroinformatics; spatial normalization; image registration; brain alignment; 3D reconstruction; processing pipeline; atlas; 3D viewer; image-based query; image database.

1. Introduction

An investigator seeking access to a repository of brain images typically wishes to retrieve a particular subset defined by any of a myriad of characteristics. They may well wish to restrict their search by species, experimental manipulation, gender, brain region, imaging modality, and so forth. One of the

From: *Methods in Molecular Biology, vol. 401: Neuroinformatics*
Edited by: Chiquito Joaquim Crasto © Humana Press Inc., Totowa, NJ

grand challenges in neuroinformatics is to construct neuroimage data repositories designed to anticipate these important end-user desires. In this chapter, we focus on how one element of this need, the anatomical aspect, can be met.

A critical aspect of meeting this challenge in neuroanatomical informatics is to devise an image database interface founded in the spatial nature of anatomy. The goal is to provide the investigator with efficient access to histological or radiological images based on imaged brain location. A key element in this solution is a standardized reference coordinate system. This is invariably an atlas to which all imagery has been aligned. This registration process is called brain spatial normalization, and as its outcome the atlas becomes an index: any point on it has a known mapping to the corresponding points on all data within the image repository. In a more general neuroinformatic environment, this anatomical spatial index synergistically combines with conceptual descriptors derived from a formal semantic framework, such as NeuroNames (*see* Chapter 5) to focus a search or analytical operation.

Our goal here is to provide the reader with a working framework through which brain images can be placed in a shared spatial context using the tools developed and available from our laboratory, so that this anatomical informatics solution can be of benefit when analyzing your neuroimage data. The discussion here compliments and extends a previous published guide on building image depositories ***(1)***.

2. Materials

A central component of the indexing solution is an anatomical atlas that constitutes a standardized coordinate system. Constructing an anatomical atlas is a significant effort. However, there are a variety of such atlases already available ***(2–4)***. Our own atlas, which is freely available (http://www.NeuroTerrain.org), is that of a Nissl-stained mouse brain. It was constructed from a single adult male C57BL/6J to yield a 17.9-μm isotropic 3D volume and includes 3D delineations of numerous structures. If the brain regions of particular interest have not yet been delineated, we can provide the environment for demarcation of the regions.

The overall task of brain normalization using our system involves several software systems, from user-oriented Java applications to our back-end 3D atlas server, image file server, and a relational database management system (RDBMS). These are briefly listed here and described more extensively in the **Subheading 3**. All end-user tools and services are accessible through the NeuroTerrain Web site (http://www.NeuroTerrain.org):

- The NeuroTerrain Brain Section QC Client (NrtrnBSQC) to prepare brains for the alignment process;
- The NeuroTerrain Image File Repository (NrtrnIFP) to receive incoming image sets, store intermediate files, and deliver final results to external investigators;
- A collection of runtime components implementing the various algorithms required for alignment;
- A PostgreSQL RDBMS *(5)*, NeuroTerrain Atlas Pipeline Data Server (NrtrnAPDS) containing all data related to defining algorithmic steps, assembling algorithms into pipelines, and executing alignment jobs on sets of experimental brain images;
- The NeuroTerrain Pipeline Workflow Framework currently consisting of a collection of modular, re-useable Unix shell and Ruby scripts controlling the concerted execution of specific pipeline scenarios and handling most direct access to images on the network. Some pipeline framework components have been implemented using Java 2 Enterprise Edition (J2EE) technology *(6)*;
- The NeuroTerrain Atlas Server controlling all access to the atlas data;
- The NeuroTerrain Atlas Navigation Client (NetOStat);
- The NeuroTerrain Integrated Image Query Client to manipulate results.

The normalization task operates on brain section image sets, what we refer to as the experimental sections. Those can be provided in most common formats, although lossless formats such as TIFF or JPEG2000 are recommended. The section set can be moderately sparse with substantially higher in-plane resolution relative to the inter-section distance *(7)*. Our current production-level algorithms only support mesoscopically imaged Nissl-stained material (on the order of 20 μm/pixel as in our atlas) where the full extent of each section has been imaged. Applications supporting other modalities and contrast methods are presently under development in our laboratory. In subsequent sections of this chapter, we assume such a data set has been prepared and the end-user wishes to align those. Although our overall procedure may work with specimens other than rodent, and the software subcomponents certainly do, our performance evaluations are confined to rodent brain, and at present, we provide a suitable reference atlas for the mouse only.

3. Methods

Aligning experimental data sets is a multi-stage process. To make use of our system, researchers supply acquired brain image sets to be processed through our pipeline workflow servers. Input from the end-user is required at various steps along the processing chain. In **Subheadings 3.1 and 3.2**, we describe the steps in the alignment procedure requiring user intervention as well as the automated steps run on our machines without user intervention.

3.1. The Pipeline

Our NeuroTerrain Image Processing Pipeline (NrtrnIPP) was designed to execute our alignment workflow against data supplied by the Mouse Brain Library project ***(8)*** and by our cryoplane fluorescence microscopy facility ***(9)***. The NrtrnIPP is currently the framework for alignment of material provided to us by other researchers in the field, and we are in the midst of reorganizing the architecture to generalize its utility.

A numerical- or image-processing pipeline consists of an automated framework for passing input data into and receiving output data from a collection of algorithmic processing steps. The steps may be run in strict serial order or in parallel based on the inter-dependency of the constituent algorithmic processes. Steps can be combined in different ways so as to construct pipelines tailored to distinct overall computational tasks. Some individual steps require researcher intervention, but these tasks should include moderate amounts of automation, as typical pipelines need to support a high throughput of data. Finally, a pipeline will generally maintain a data repository to define the requirements for running any specific pipeline process and to log transactional information related to each real-time execution run of a specific version of a defined pipeline process on a given set of input data.

Though these generic requirements also apply to a neuroimage-processing pipeline, the algorithms they employ and the means created to provide external investigators access can vary broadly even in the restricted field of neuroimaging ***(10,11)***. Generally, a 2D, 3D, 4D (3D time series), and even 5D (multiple resolution) image set is passed into the initial step in the pipeline and each succeeding step generates additional images and numerical data, amongst which will be the desired analytical results, both images and numerical results. The pipeline we use provides the following:

3.1.1. Network File Access

Image files used for input to or generated as output from our NrtrnIPP are acquired from a network server, either via standard network file protocols (e.g., NFS, AFP, and SAMBA) or through a more complex mechanism such as an HTTP-based file query web service or a more proprietary, GRID-based mechanism [e.g., Storage Request Broker (SRB) used within the BIRN project ***(12)***]. We refer to these collective network file services and the image files they contain as the NrtrnIFP.

3.1.2. Algorithmic Step Execution

The specific pipeline versions contained with the NrtrnIPP system are assembled from a variety of image-processing executables [public compiled software such as AIR ***(13)***, new code written in C/C++, de novo, or new code written to the widely available itk IP libraries ***(14)***] and/or macros/scripts written in a more general purpose image- or numerical-processing environments (e.g., MATLAB and ImageMagic). Some need to run on specific, distinct hardware and software operating system environments, thereby requiring us to support a heterogeneous runtime environment, where the individual computational nodes reside on a network connected to the NrtrnIFP. Pending development will add the use of GRID computation both through connections to BIRN-related computational resources through GridSphere ***(15)*** and GRID extensions to MATLAB ***(16)***.

3.1.3. Pipeline Execution Data Management

There are several distinct approaches to handling this task. Our preferred option is to employ a RDBMS (NrtrnAPDS, **Subheading 2** listing) to store all primary numerical data files, as well as associated meta-data (e.g., paths used to access files, image provenance information related to the specific tissue workup and imaging device settings), although the images themselves, again, reside on the NrtrnIFP. Our NrtrnAPDS also stores formal algorithmic definitions (required input and output data types, algorithmic dependencies, etc.), as well as the specification of distinct processing pipelines assembled using these steps. The database also contains control information defining what collections of images constitute a defined "study set" and what processing is required for that set. Finally, transactional information on each run of each individual algorithmic step is tracked in the database. This includes information on faults and exceptional conditions encountered during a given run, so as to support a quality assurance procedure and a limited amount of automatic error recovery during execution. We have constructed an image query API implemented as an Apache Axis web service to provide programmatic access to the NrtrnAPDS and the NrtrnIFP, all through the firewall safe HTTP or HTTPS protocols. An image query is accepted as an XML-encoded PostgreSQL SELECT statement to filter on study, subject details, experimental treatment, or specific image-processing manipulation(s) applied. As an example, one might wish to obtain the following for quality control review—images related to study ID "Lab 432 synaptophysin project" for all adult mice of strain C57BL/6J treated with stress but limited to the intermediate output of the course alignment image-processing step. The image set thus defined can be returned directly in compressed or uncompressed form or one at a time based on a returned set of IDs. This system is currently being upgraded to the latest version of J2EE and Apache Axis.

3.1.4. Pipeline Workflow Management Framework

Here too, our work has evolved over time through various levels of complexity to support increasing amounts of automation. Our ultimate goal is to have a complete, J2EE framework using web services and Enterprise Java Beans ***(17)*** to provide the automation and Java Servlets/Portlets ***(18)*** to provide the user interface. As mentioned above, we also intend to incorporate use of network-based, GRID computational resources using some combination of GridSphere, the Condor System ***(19)***, and the Kepler framework ***(20)***. Some components of this J2EE framework have been completed. But currently, our primary means of running the NrtrnIPP jobs is through Unix shell scripts, Ruby scripts, and the Unix timed execution utility "cron" ***(21)***. We had considered employing the OpenPBS workflow control framework (http://www.openpbs.org), but its lack of intrinsic RDBMS support and operating system platform limits led us to reject this option. Scripts running within the NrtrnIFP perform general-purpose file management tasks on a regular basis. Local scripts running on each individual computational node in our processing farm have direct access to the NrtrnAPDS and through this mechanism know which particular steps amongst pending pipeline jobs the node is capable of running. These worker node scripts pole the NrtrnIFP to determine whether jobs it can execute are ready to run and, if so, transfers them for local execution, returning the results at completion. This architecture has turned out to be very flexible, as the overall pipeline framework itself can keep running, even if one or more of the computational nodes goes off-line. The worker node script components are highly generic, can be "wrapped" around any of the algorithmic processes we run, and can generally include whatever specific error checks and recovery procedures a given step needs to support.

3.2. Image Preparation

The goal in this stage is to take images of slides of the experimental material and isolate the brain tissue sections from the slide background. In the simplest case, the experimental material consists of a single brain section on each slide; with relatively small mouse brain sections, however, each slide often includes many sections. We have developed a flexible cropping algorithm to automatically segment the individual sections and crop each into a separate TIFF file. This program, implemented in SCIL Image ***(22)***, is the first step in the NrtrnIPP. In addition to cropping the section, the application calculates and stores the bounding rectangle and centroid of the section. The centroid defines the center of the cropper boundaries and thus each section is centered in the image window (*see* **Fig. 1**).

Although generally effective at achieving its purpose, this automatic cropper is not flawless. It occasionally combines multiple sections when they are too close together on the original slide, or it may split an individual section into multiple image files because the section is torn or it is from a level along the cutting axis where the tissue is not continuous, for example, cerebellum and brainstem on coronally sliced brains. In other cases, the cropping algorithm will inappropriately segment optically dense objects that are not sections, felt tip pen annotations, primarily.

Errors can also arise during the initial mounting of the sections on the slides. The sections' order might be incorrect, or they might be mistakenly flipped about the *x*-axis or *y*-axis. To deal with all these problems, we developed the NrtrnBSQC for end-users to employ after cropping and before submitting section sets for the next step in the pipeline. This Java application is used on the collaborator's desktop and is available for MacOS, Windows, and Linux. The cropped images must currently be downloaded from our server to the local

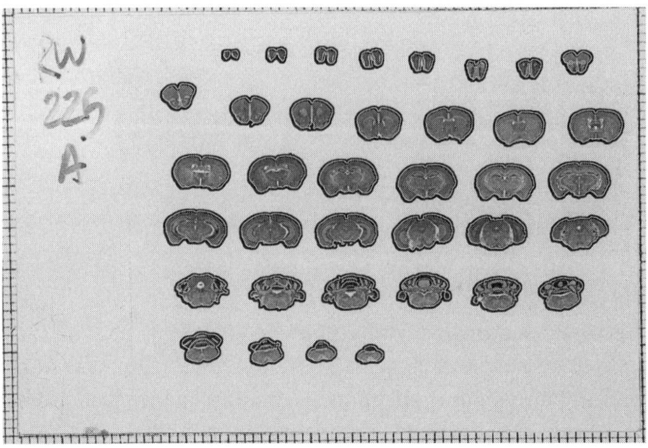

Fig. 1. A full slide image with segmentation overlay. This slide is from the Mouse Brain Library. It was automatically segmented, and the individual section masks have been dilated here for purposes of illustration. In this individual instance, the automatic cropping algorithm correctly segmented each whole, distinct section and correctly rejected all extraneous, nontissue marks.

machine where the client is running, although when we complete the upgrade to our image query web service, this will not be necessary.

NrtrnBSQC consists of three visible panels when first launched (*see* **Fig. 2**). The top-left panel is a file directory explorer providing a means to easily select the brain to be reviewed (the cropping algorithm stores its output in a distinctly labeled directory for each brain). Below it is a thumbnail view of all the sections from the current brain under view. On the right is a large view of the currently selected thumbnail image.

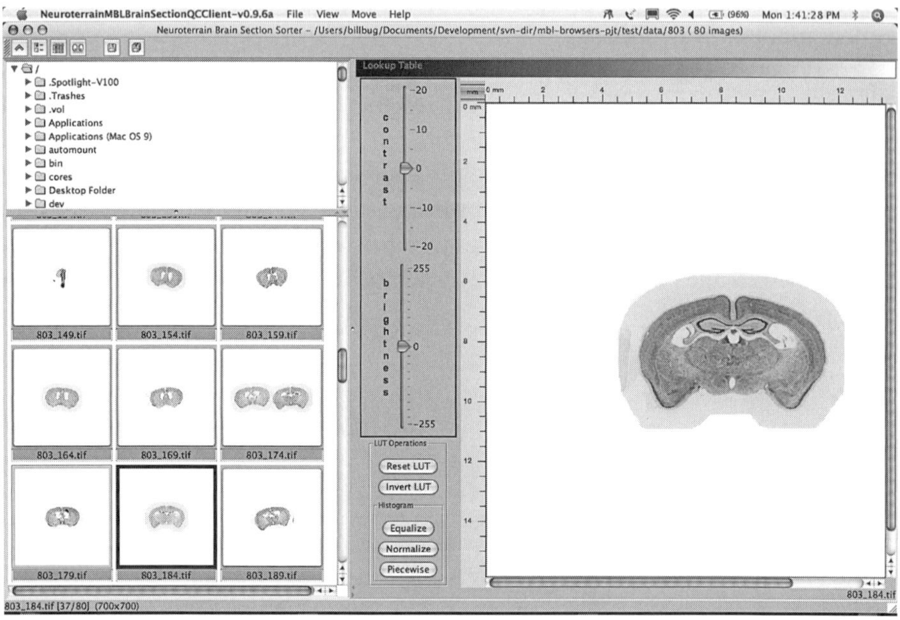

Fig. 2. Brain section QC client (main view). The three-panel view (file directory tree, image directory thumbnails, and single image view) provides for a means to quickly parse a set of sectional images for anomalies. The researcher drills down to the directory containing the collection of brain images, just as he/she would in a typical file explorer, for example, double clicking a folder expands that node in the file tree. The contents of the current directory are displayed collectively in the thumbnail view. In the single image view, the researcher can alter the image display characteristics—for example, brightness, contrast, and zoom level—to discriminate more subtle cases of horizontal transposition. Double clicking either a thumbnail or the single image view brings up the full-slide view with slide show capability (**Fig. 3**).

To correct misordering, the thumbnails can be dragged to their correct location. In the simplest case, one can choose to have only the selected section moved. Depending on how the sections were laid out on the original slide, it can be considerably more efficient to shift an entire set of images simultaneously. One has the option to either shift all subsequent or all preceding sections. Flipped sections can be easily transposed. Segmentation errors are dealt with either in the thumbnail viewer or in the single section display. A knife tool ( ) is provided to split two merged sections (*see* **Fig. 3**). A stitch tool ( ) allows the user to combine multiple separate files into a single one. In both splitting and merging operations, the centroid and bounding rectangles

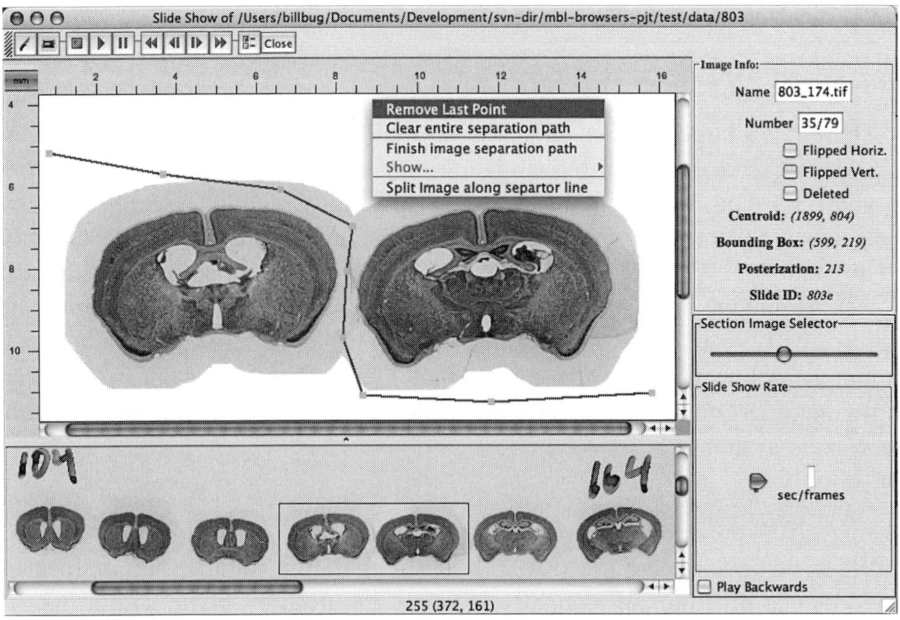

Fig. 3. Brain section registration QC client (slide view). The image of the slide below acts as a reciprocal navigation tool. Select a specific image in the viewer above, and the slide image auto scrolls to that section. Double click on a specific section on the slide image, and that section is displayed above. The "Section Image Selection" slider tool quickly jumps to any image throughout the ordered series. Finally, the "player" tool buttons at the top of the screen can be used to progressively display the sectional series forward or backward with a specified time interval so as to quickly step through the entire series. Doing so, the researcher can rapidly perform a final review of their work ensuring that sections have all been properly ordered and oriented.

are recomputed (with reference to the original slide maintained). Deletion of extraneous markings or any other nonsection material can be accomplished in the thumbnail viewer through a popup menu, as can the insertion of blanks to maintain a regular spatial frequency along the cutting axis, when a section was lost in the tissue workup.

There is an additional feature to expedite this QC task. If, for example, the sections on the original slide were placed in a regular grid pattern, extraneous markings can be automatically detected and the files marked for deletion.

The process of correcting the data is aided by a movie feature where the sections are displayed in rapid succession in accordance with all the changes introduced. This and the other available tools allow for rapid correction of the data sets. In our own work on the MBL *(8)*, a moderately experienced operator can QC about 12 brains/h.

3.3. Image Alignment

There are a large number of spatial normalization algorithms ***(23)***. The vast majority were designed for human neuroimaging using PET or MRI modalities to yield 3D data sets. Unlike this situation, in most animal research settings, the images collected are an ordered unaligned series of 2D sections that must be aligned to a 3D atlas. We have developed a multi-step solution to this 2D to 3D alignment problem. In the first step (coarse alignment), sections are automatically matched to their corresponding atlas plane ***(24)***. Next, automated or manual delineation of a limited set of geometric features is required. Finally, those features are used to warp the experimental section into the standard coordinate system.

3.3.1. Coarse Alignment

We have developed a piece-wise linear registration routine that approaches performance of a human manually locating the arbitrary plane within the 3D atlas that best matches a given 2D section. The routine first automatically reconstructs a 3D volume from the ordered sequence of 2D sections. Our approach works best with near coronal or horizontal sections, as it exploits the axis of symmetry to align the sequential sections to each other ***(25)***. One can relax the requirement of symmetry and utilize other features for alignment ***(26,27)***, although performance rapidly degrades as section spacing increases. Another available method that is insensitive to sampling frequency—image centering using moments ***(28,29)***—is not particularly robust in the presence of tears in the sectional material ***(30)***.

Typically, the experimental material consists of highly anisotropic 2D series, where the inter-section distance in the experimental material is much greater than the in-plane resolution and the result is a highly anisotropic volume. The reconstruction up to this stage does not take into account any curvature in the data perpendicular to the cutting plane. That is, if one were to cut, for example, a banana orthogonally to the long axis into a series of disks and then try to reconstruct the 3D volume using this approach, the result would be a tapered cylinder lacking the crescent curve. In addition, the physical dimensions of the atlas and the experimental volume often do not match, and therefore, scaling has to be performed to enable accurate alignment. To recover the global curvature, compensate for size differences, and bring the images into the standard coordinate system, the reconstructed volume is aligned using affine transformation and a surface-based approach ***(31)*** to the 3D atlas. This registration is guided by the outer brain surface.

As a result of this 3D–3D alignment, the experimental brain inherits the curvature of the atlas brain, and each of the experimental section is now matched to a corresponding atlas plane. The matching at this stage is still a very rough approximation, and the next series of steps refines it greatly. Although up to this point the experimental sections have been treated as rigid volumes bound to each other, in the following steps, each section is allowed to move independently within a close neighborhood to find a better matching plane. This 3D motion is aimed at finding a corresponding plane such that the image-based similarity between the section and this plane is maximized. A search is conducted within this space for a minimum difference, in terms of gray scale intensity, between the section and the atlas plane sought. The mapping is confined to affine transformation. This often improves the matching greatly over the initial attempt. Even further improvement can be obtained. To do so, the position of each plane is redefined based on a smooth interpolation of the sections' alignment parameters, and then once again each section is independently aligned to a corresponding plane in the atlas this time using a much smaller searching neighborhood.

This matching routine is quite effective despite the fact it is matching between sectional image and atlas planar image (*see* **Fig. 4**). Our current work strives to improve the process even further by seeking the best fitting curved surface in the atlas to the experimental section.

3.3.2. Fine Alignment

The above coarse alignment yields good linear matching between an experimental section and a planar image from the atlas. Applying an in-plane nonrigid

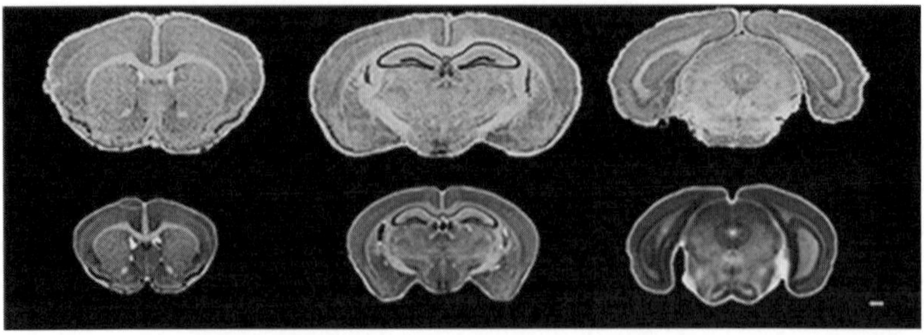

Fig. 4. Automated section matching performance. Shown in the top row are planes from the NeuroTerrain mouse atlas found to best match experimental sections (bottom row). The experimental sections are from different strains of mice (left, BXA1; middle, ABXD5F2; and right, 129X1/SVJ) within the Mouse Brain Library. The different tissue-processing procedures applied to the experimental brains and the atlas give rise to the observable size difference between matched section pairs: the former were embedded in celloidin prior to cutting, whereas the atlas was created from a fresh frozen brain. Despite the difference in shrinkage, the matching algorithm performs well. Scale bar = 500 μm.

alignment improves this registration. This feature-based registration is guided by contours of corresponding structures in the experimental section and the matched atlas plane. Some of these geometrical features, such as the external contour and the line of symmetry, can be readily located automatically. We have been developing a variety of more advanced segmentation tools to locate others, but at present, it remains the case that insufficient density of features are obtained automatically and operator delineation of some features is required. This is done quite simply using ImageJ.

Hence, a set of features are extracted from the experimental data set, referred to here as test data set, and paired with their corresponding features in the atlas, referred to here as reference data set. Once a set of guiding geometric features has been defined, each section can be further and nonrigidly deformed to better match the image overlaid on the corresponding atlas plane. We provide a detailed description of the method we employ to achieve this refinement in **ref.** ***(32)***.

Briefly, in **ref.** ***(32)***, a wavelet-based registration method with the following characterization was developed: (1) the algorithm is driven by geometrical features with the brain's external and internal contours guiding the registration; (2) the transformation model is a nonlinear deformation field where the region within the brain's image is modeled by a wavelet multi-resolution representation; (3) the distance metric used is based on the sum of squared distances where a distance is the interval between a point on the test contour and the closest point to it on the reference contour; and (4) a progressive version of the Marquardt–Levenberg optimization algorithm minimizes a functional that is the sum of two terms, namely the sum of squared distances and the elastic energy.

The resulting alignment using the algorithm in **ref.** ***(32)*** is of sufficiently high quality to permit segmentation of the experimental material using atlas anatomical templates. This is illustrated in **Fig. 5**.

3.4. Image Viewing, Navigation, and Analysis

3.4.1. NeuroTerrain Atlas Server

Recently, we have ported our Mac OS-specific MacOStat atlas navigation system ***(33,46)*** to Unix and Java. The server component now executes as a high-performance, multi-threaded Unix daemon, and the client component has been packaged as a platform-independent Java application called NetOStat. This atlas server enables a user of the Java client to login and request navigation through a specific atlas data set. As stated above, in the neuroimaging analysis environment we are describing here, the server uses our 17.9-μm isotropic 3D mouse Nissl-stained atlas. Alternative atlases can easily be substituted.

3.4.2. NeuroTerrain Aligned Image Repository

Brain Section image sets processed through the pipeline can be accessed from within our analysis tools (**Subheading 3.4.4**) either through the local file system on the computer where the tools are run or through the network from

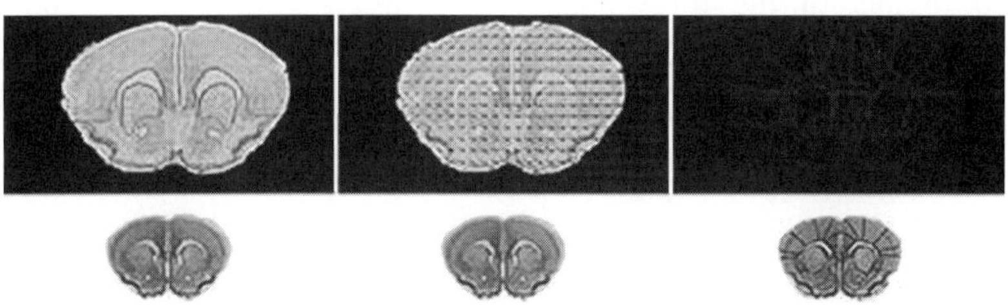

Fig. 5. Automated nonlinear registration. Shown are an experimental section (bottom) and the corresponding atlas plane (top). Shown below are the manually delineated guiding geometric features (left), the computed distortion field (middle), and the atlas segmentation layer warped onto the experimental section (right).

the NeuroTerrain Aligned Image Repository. The Java analysis tools contain a simple query form enabling the user to define the image set they wish to peruse (e.g., lab name, study name, and subject characteristics), and this collection of images is brought over the network to the local machine for viewing. As this feature requires the J2EE image query service we have created, while that service is being upgraded, aligned brain section images sets need to be downloaded manually for the time being, from which the client described in **Subheading 3.4.4** can access them.

3.4.3. NetOStat

The client portion of MacOStat now runs as a separate Java application, NetOStat *(46)*. This tool provides basic navigational capability for browsing a 3D atlas data set delivered by the NeuroTerrain Atlas Server in real-time over the network (*see* **Fig. 6**). It can be run directly from the NeuroTerrain Web site Java WebStart, or on platforms not fully supporting WebStart, downloaded, and run as a Java application.

As a part of a collaboration within the Mouse-BIRN test-bed in the BIRN project, we participated in the development of an inter-atlas communication protocol. This mechanism will make it possible for any one of the participating atlas systems [SMART ATLAS *(34)*, the LONI MAP Client SHIVA *(35)* (*see* Chapter 11), and the NeuroTerrain client–server Atlas System] to exchange atlas location, so that data sets aligned to one atlas can be viewed using any of the others. This helps to integrate the anatomical component of any data set normalized to any of the constituent atlases (within the limits of error introduced by having each of the atlases normalize to stereotaxic coordinates), thereby making it possible to analyze the superset of all data spatially mapped to any of the three atlases as if aligned to a single coordinate system. This level of integration across all three atlases is expected to mature, eventually providing higher resolution inter-atlas mapping than is currently supported.

We are adding the ability to use the NeuroNames neuroanatomical terminology (*see* Chapter 5) to specify brain region volumes of interest (VOIs). NeuroNames has recently integrated atlas terminology from several common rodent atlases making it critical resource to support in a mouse-centered neuroinformatics framework. This feature is currently under development. VOI region spatial definitions once transferred from the atlas to the aligned, experimental brain sections will also include their equivalent NeuroNames ID when available. Brain image sets aligned to the NeuroTerrain mouse atlas can thereby

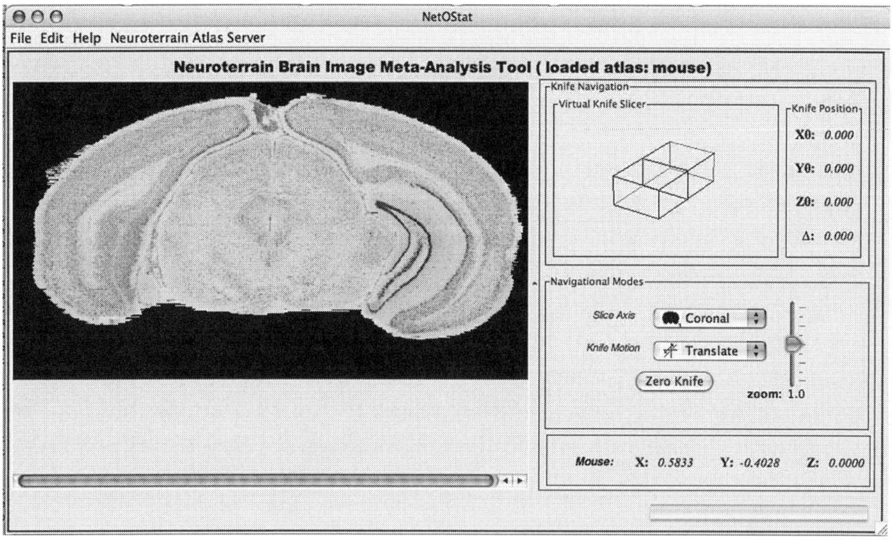

Fig. 6. NeuroTerrain atlas navigator (NetOStat). This platform-independent Java application (available through Java WebStart at the NeuroTerrain Web site) provides a wireframe "Virtual Knife" interface to enable the investigator to smoothly and intuitively "slice" through an atlas data set along a defined axis. To move the knife, the user simply clicks down and drags the mouse from left to right across the wireframe. Dragging the mouse in the reverse direction moves the knife in the opposite direction. The arrow keys can also be used to nudge the knife in small increments. In any of three traditional neuroanatomical cutting planes (coronal, horizontal, or sagittal), mouse dragging can be configured to either translate the knife along the selected axis or rotate the knife about the *x*-axis, *y*-axis or *z*-axis, thereby providing the unconstrained ability to slice the 3D volume in totally arbitrary planes. The atlas gray scale view can be zoomed in or out, and scale rulers provide an accurate measure of "true" tissue distance. The brain region volumes of interest (VOIs) defined by the neuroanatomical experts in our laboratory can be superimposed on the histological gray scale view as solid areas or outlines, and the knife controls can be constrained according to these geometric objects so that navigation is limited to planes intersecting specific VOIs. Knife positions can be saved in sets to a file and used to automatically return to the same precise location within the 3D atlas volume. Gray scale atlas image views can be saved as a TIFF image, and this ability has been automated, so an entire atlas volume can be re-sliced along an arbitrary axis and saved to disk.

link to other neuroscientific mouse data repositories on the Internet including a neuroanatomical specification ***(36–40)***. NeuroNames is rapidly becoming the *de facto* shared neuroanatomical knowledge resource and will therefore support the widest possible integration.

3.4.4. NeuroTerrain Image Query Client

Once a set of brain sections have passed through the pipeline, an "equivalent NetOStat knife position" has been calculated for each section. This means the brain sections can then be searched and analyzed within the 3D coordinate system of the atlas data set.

The NetOStat Java code base has been encapsulated with a developer framework making it easy for other programs to instantiate an embedded version of NetOStat called the NeuroTerrain Software Development Kit (NT-SDK). This has enabled us to create a fairly generic Image Query Client (*see* **Fig. 7**) supporting all the features of NetOStat for use in examining aligned brains. As stated above, the current Image Query Client requires the aligned brain section image set reside on the local computer where the tool is running, but once our Image Query web service is operational, this can be used to access aligned brain section image sets directly through the network.

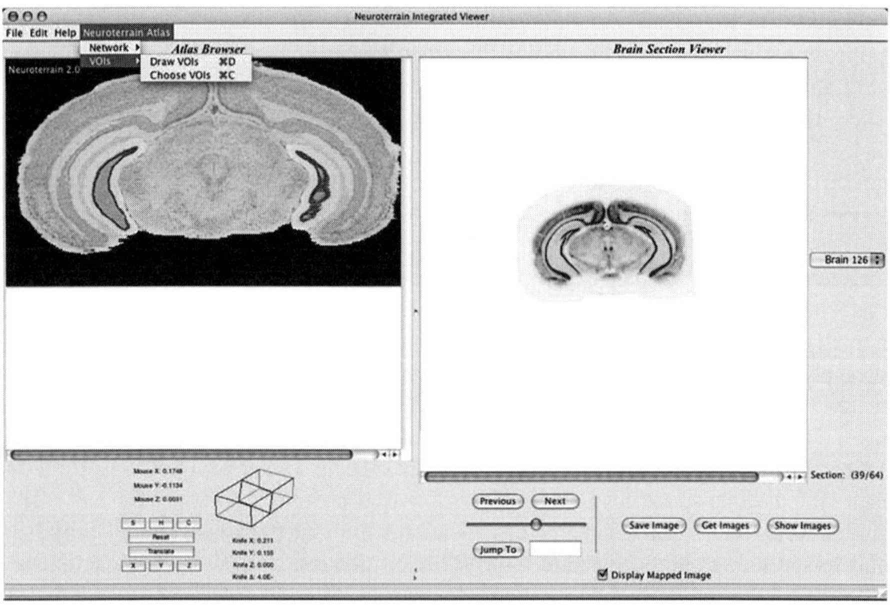

Fig. 7. NeuroTerrain Image Query Client. This tool consists of a split pane containing the NetOStat Brain Atlas Navigator on one side and an aligned brain view panel on the other. The aligned brain viewer currently allows the user to select the particular brain set for viewing through a pull-down menu and quickly navigate the experimental brain and atlas in synchrony.

3.5. Extensible Features of the NeuroTerrain Tool Set

The Image Query Client and QC Client shown in **Subheadings 3.2 and 3.4.3** respectively were developed for specific neuroinformatics projects, the Mouse Brain Library *(8)* and the cryoplane microscopy resource *(9)*. Despite this fact, all the work described here was designed and implemented with the intention of making this service more widely available to the community neuroscientists with large anatomical data sets to manage. The feature sets in the NrtrnBSQC and in NetOStat have very general applicability as was noted, but other elements in this overall architecture are also particularly suited for adaptation to broader applications. The following features highlight that fact.

3.5.1. Network File Access

The J2EE-based Image Query web service we have built can be leveraged to provide a very generic means of selecting specific image sets from the NrtrnIFP. It is firewall friendly, meaning it uses TCP network ports (80 and 443 for HTTP and HTTPS, respectively) typically not blocked by most firewalls. Typical RDBMS database connectivity ports used for direct access and the network file service ports used for remote file access are almost always blocked in most firewalls. The web service host can also provide encryption, authentication, transaction management, and load balancing, making this mechanism highly secure and scalable. The Image Query Service itself is really composed of three specific web service operations defined through Web Service Description Language (WSDL) *(41)*: one for submitting queries and receiving back the image files directly as results; one for submitting queries and receiving back a compact list of filenames for the resulting images; and a final service for accepting one or more filenames and returning the associated images. For both services returning image files, the request can specify whether they should be returned compressed (to conserve bandwidth and hasten the transfer of larger image sets). WSDL-based web services support binary MIME attachments, analogous to the standard email attachment, and this system appears robust enough to support sending very large image sets, especially in the more recent upgrades to the web service standards supported by the Web Services Interoperability Organization (WS-I, http://www.ws-i.org/). Our use of SRB also provides for more open access to the images, as the creators of SRB provide several language bindings to access files stored on an SRB infrastructure accessible to external programs.

3.5.2. NrtrnIPP Automation Wrappers

Our current pipeline is primarily composed of scripts. These have been developed as generic wrappers applied with relative ease to image-processing executable components, regardless of where the computing node resides on the network. The only requirement is that algorithmic process needs to be launchable from the command line. The script access is flexible enough to be tailored to the specific needs of most algorithm executables, and we have found, in our need to support new algorithms, this wrapper is relatively easy to adapt. The wrapper script also contains a fairly generic reporting and exception handling/error recovery capability. All these aspects make it possible for us to support the specific algorithmic processing needs of external researchers with a minimum of effort on both sides. If alignment algorithms other than the ones we provide are desired, our pipeline infrastructure can likely support that at minimal cost to you and us. The confederated nature of the computational farm—the fact worker nodes are relatively independent and can come on-line and off-line without disabling the other workers or central server—makes it conceivable that we could support running specific algorithmic steps or entire subpipelines residing on a network external to our local computational infrastructure. This capability will be significantly enhanced as we move more of the framework to J2EE, especially when we make use of GridSphere and Kepler.

3.5.3. The NeuroTerrain Atlas Server

The new version of our 3D digital atlas server implementing the highly efficient MacroVoxel 3D storage format ***(33)*** and running as a daemon on Unix provides us with a more flexible platform on which to build extended 3D data viewing and analysis services. The MacroVoxel format is particularly suited to providing over the network real-time slicing capability through a 3D volume in an arbitrary plane. It is especially effective when network bandwidth is high and latency very low as on the Internet2, where our newer server will reside as of the summer of 2006. There are other 3D anatomical viewers available ***(35,42–44)*** but none provide this unique combination of real-time slicing in arbitrary planes and spatially normalized access to experimental data sets. In addition, *any* 3D stack of images with defined resolution can be converted to MacroVoxel format and delivered through this server. It is possible for us in a more dedicated collaboration to set up both the server and our alignment pipeline to work against a different 3D atlas, should another existing atlas serve the needs of a specific biological application better than the one we use for our current applications. The new UNIX version of our atlas server has extensive, internal multiprocessing support enabling it to serve up more than one atlas data set simultaneously to the same client. This capability can be used to view aligned, full 3D MRI data sets in synchrony. If mouse MRI 3D sets are aligned to the NeuroTerrain mouse atlas, the delineated brain regions from our atlas can be used to view and analyze the MRI sets.

3.5.4. The NeuroTerrain Client Development Kit

The NetOStat Atlas Analysis Client has been significantly refactored to deliver it in a programmer-friendly package, the NT-SDK. The impetus behind investing this significant amount of programmer hours was to ensure we expend the minimum amount of effort in the long-term to support the many applications for which we need to put it to use—such as the NeuroTerrain Image Query Client (NIQC-**Subheading 3.4.4**). We realized each of the image query tools we create to support a particular scientific endeavor requires essentially identical interactions with the NeuroTerrain Atlas Server and similar client GUI capabilities. Using the new NT-SDK framework, a few dozen lines of Java code is all it took to make the stand-alone NetOStat client. A similarly minimal amount of coding was required to incorporate the NT-SDK into the NIQC. This framework provides an enclosing application methods to programmatically set the atlas ***Virtual Knife*** position, according to either the NeuroTerrain intrinsic coordinate space or using stereotaxic coordinates. Prior to creating the NT-SDK, the simple action of sending the new knife position to the server, waiting for the server to respond, and obtaining the new atlas gray scale image took many dozens of lines of code and required interacting with half a dozen different Java classes. Now an enclosing application need only write a dozen lines using 3 objects. All of the end-user atlas server interface actions—the user dragging the knife, changing slice axis or zoom level, and so on—are handled internal to these objects. The external programmer need not write any code to support these features. The programmer has the option of implementing three callbacks to receive new atlas knife position, gray scale image, or VOIs (in the form of separate vector graphic objects). The response implemented within these callbacks can be simple or complicated, as the specific application requires. All of these conveniences combine to provide a rich, easy-to-use atlas client library for a programmer to customize based on the specific analysis and visualization needed to support their laboratory's investigational approach.

4. Notes

1. What are the potential problems that may arise during alignment? The overarching issue concerns the nature of anatomical equivalency between brains. Over many decades, the issue of functional equivalency and morphological similarity has been sought at different levels of organization, from cellular level to system level. We know even in the dense and complex vertebrate CNS there are identifiable neurons ***(45)***. Yet, across individuals, even the large pair of Mauthner cells found in the fish brainstem differ significantly in dendritic pattern beyond the secondary dendrites. At the same time, functional roles of the Mauthner cell network can change. Again, the Mauthner cells provide a pithy example. They serve in coordinating a ballistic escape maneuver of the prey when attacked by a predator, but disturb the system by damaging the Mauthner cells, and functional substitution occurs as other cells take over the function ***(46)***. This conceptual issue of equivalency is, of course, not addressed by the technical solution we provide but is one that needs to be considered when interpreting results from fusing data from multiple animals.
2. A slightly more mundane extension of this problem is the technical limit on alignment. Even if there were a 1:1 matching at the cellular level between specimens, and in vertebrates there certainly is not, their location relative to each other is likely not to be deterministically defined. This obviously greatly complicates alignment. At what level does spatial normalization breakdown? Probably, registration beyond accurate routine alignment of thin mesoscopic structures such as cortical layers is not meaningful. Neither others nor we obtain this level of performance at present, but it is likely that it will be attainable in the near future. Beyond that level, other means will be required to fuse image data.
3. The further limitations of our workflow have been mentioned above. In summary, it presently requires images of whole Nissl-stained section sets. Although it supports sparse data, it does not work effectively with an incomplete set of whole sections; the whole brain is required ***(7)***. Its performance is highest when the sections are taken from the symmetrical horizontal or coronal orientation. Finally, a complete suite of tools including atlas are available for the mouse only ***(8)***. Ways to overcome these limitations are being worked on, and we hope to soon have available for the research community a more generalized solution for the rodent spatial normalization task.

Acknowledgments

This work was supported by NIH grants P20 MH62009, RR08605-08S1, RR043050-S2, and NSF DBI grant 0352421.

References

1. Bug, W., and Nissanov, J. (2003) A guide to building image centric databases. *Neuroinformatics* 1, 359–77.
2. Baldock, R. A., Bard, J. B., Burger, A., Burton, N., Christiansen, J., Feng, G., Hill, B., Houghton, D., Kaufman, M., Rao, J., Sharpe, J., Ross, A., Stevenson, P.,

Venkataraman, S., Waterhouse, A., Yang, Y., and Davidson, D. R. (2003) EMAP and EMAGE: a framework for understanding spatially organized data. *Neuroinformatics* 1, 309–25.

3. Ma, Y., Hof, P. R., Grant, S. C., Blackband, S. J., Bennett, R., Slatest, L., McGuigan, M. D., and Benveniste, H. (2005) A three-dimensional digital atlas database of the adult C57BL/6J mouse brain by magnetic resonance microscopy. *Neuroscience* 135, 1203–15.
4. MacKenzie-Graham, A., Jones, E. S., Shattuck, D. W., Dinov, I. D., Bota, M., and Toga, A. W. (2003) The informatics of a C57BL/6J mouse brain atlas. *Neuroinformatics* 1, 397–410.
5. *The PostgreSQL Relational Database Management System*. Available at http://www.postgresql.org/.
6. Shannon, B. (ed.) (2003) *Java™2 Platform Enterprise Edition Specification, v1.4 (final edition)*, Sun Microsystems, Santa Clara, CA.
7. Gefen, S., Tretiak, O., Bertrand, L., and Nissanov, J. (2003) Mouse brain spatial normalization: the challenge of sparse data, in *Biomedical Image Registration* (Gee, J. C., Maintz, J. B. A., and Vannier, M. W., eds), Springer, New York, 349–357.
8. Rosen, G. D., La Porte, N. T., Diechtiareff, B., Pung, C. J., Nissanov, J., Gustafson, C., Bertrand, L., Gefen, S., Fan, Y., Tretiak, O. J., Manly, K. F., Park, M. R., Williams, A. G., Connolly, M. T., Capra, J. A., and Williams, R. W. (2003) Informatics center for mouse genomics: the dissection of complex traits of the nervous system. *Neuroinformatics* 1, 327–42.
9. Nissanov, J., Bertrand, L., Gefen, S., Bakare, P., Kane, C., Gross, K., and Baird, D. (2006) *Cryoplane Fluorescence Microscopy*, Proceedings of the 24th IASTED International Conference on Biomedical Engineering, Innsbruck, Austria International Association Of Science And Technology for Development, 362–366.
10. Toga, A. W., Rex, D.E., and Ma, J. (2001) *A Graphical Interoperable Processing Pipeline*, 7th Annual Meeting of the Organization for Human Brain Mapping, NeuroImage Abstracts.
11. Fissell, K., Tseytlin, E., Cunningham, D., Iyer, K., Carter, C. S., Schneider, E., and Cohen, J. D. (2003) Fiswidgets: a graphical computing environment for neuroimaging analysis. *Neuroinformatics* 1, 111–26.
12. Rajasekar, A. K., and Wan, M. (2002) *SRB & SRB Rack – Components of a Virtual Data Grid Architecture*, Advanced Simulation Technologies Conference (ASTC), San Diego, CA.
13. Woods, R. P., Cherry, S. R., and Mazziotta, J. C. (1992) Rapid automated algorithm for aligning and reslicing PET images. *J Comput Assist Tomogr* 16, 620–33.
14. Ibanez, L., Schroeder, W., Ng, L., and Cates, J. (2003) *Insight Segmentation and Registration Toolkit (ITK) Software Guide*, Kitware Inc., New York, NY.
15. Novotny, J., Russell, M., and Wehrens, O. (2004) GridSphere: an advanced portal framework, *IEEE Proceedings of the 30th Euromicro Conference (EUROMICRO'04)*, 412–419.

16. Arnold, D., and Dongarra, J (2000) *The NetSolve Environment: Progressing Towards the Seamless Grid*, 2000 International Conference on Parallel Processing (ICPP), Toronto, Canada.
17. DeMichiel, L., Keith, M., and Expert Group (2006) JSR 220: Enterprise JavaBeans™ 3.0 Specification, final release.
18. Hepper, S. and Expert Group (Oct 2003) JSR 168: Portlet Specification, final.
19. Litzkow, M. J., Livny, M., Mutka, M. W. (1998) Condor-a hunter of idle workstations, *IEEE Proceedings of the 8th International Conference on Distributed Computing Systems*, 104–111.
20. Altintas, I., Berkley, C., Jaeger, M., Jones, B., Ludascher, B., and Mock, S. (2004) *Kepler: An Extensible System for Design and Execution of Scientific Workflows*, 16th International Conference on Scientific and Statistical Database Management (SSDBM), Santorini Island, Greece.
21. Powers, S., Peek, J., O'Reilly, T., and Loukides, M. (2002) *Unix Power Tools,* Third Edition, O'Reilly Media, Sebastopol, CA.
22. Van Balen, R., Koelma, D., Ten Kate, T. K., Mosterd, B., and Smeulders, A. W. M. (1994) ScilImage: a multi-layered environment for use and development of image processing software, in Experimental Environments for Computer Vision & Image Processing (Christensen, H. I. and Crowley, J. L., eds), Series in Machine Perception and Artificial Intelligence, v11, World Scientific Publishing (UK) Ltd. London: UK, 107–26.
23. Toga, A. W., and Thompson, P. M. (1999) An introduction to brain warping, in *Brain Warping* (Toga, A. W., ed.), Academic Press, New York, pp. 1–26.
24. Gefen, S., Bertrand, L., Kiryati, N., and Nissanov, J. (2005) *Localization of Sections Within the Brain via 2D to 3D Image Registration*, IEEE International Conference on Acoustics, Speech, and Signal Processing, (ICASSP), Philadelphia, PA.
25. Gefen, S., Fan, Y., Bertrand, L., and Nissanov, J. (2004) *Symmetry-Based 3D Brain Reconstruction, IEEE International Symposium on Biomedical Imaging (ISBI): Macro to Nano*, Washington DC. v1, 744–747.
26. Kim, B., Boes, J. L., Frey, K. A., and Meyer, C. R. (1997) Mutual information for automated unwarping of rat brain autoradiographs. *Neuroimage* 5, 31–40.
27. Ourselin, S., Roche, A., Subsol, G., Pennec, X., and Ayache, N. (2000) Reconstructing a 3D structure from Serial Histological Sections. *Image Vis Comput* 19, 25–31.
28. McGlone, J. S., Hibbard, L. S., and Hawkins, R. A. (1987) A computerized system for measuring cerebral metabolism. *IEEE Trans Biomed Eng* 34, 704–12.
29. Nissanov, J., and McEachron, D. L. (1991) Advances in image processing for autoradiography. *J Chem Neuroanat* 4, 329–42.
30. Shormann, T., and Zilles, K. (1997) Limitations of the principal-axes theory. *IEEE Trans Med Imaging* 16, 942–47.
31. Kozinska, D., Tretiak, O. J., Nissanov, J., and Ozturk, C. (1997) Multidimentional alignment using the euclidean distance transform. *Graphical Model Image Processing* 59, 373–87.

32. Gefen, S., Tretiak, J. O., Bertrand, L., Rosen, G. D., and Nissanov, J. (2004) Surface alignment of an elastic body using a multi-resolution wavelet representation. *IEEE Trans Biomed Eng* 51, 1230–41.
33. Gustafson, C., Tretiak, O., Bertrand, L., and Nissanov, J. (2004) Design and implementation of software for assembly and browsing of 3D brain atlases. *Comput Methods Programs Biomed* 74, 53–61.
34. Martone, M. E., Gupta, A., and Ellisman, M. H. (2004) e-neuroscience: challenges and triumphs in integrating distributed data from molecules to brains. *Nat Neurosci* 7, 467–72.
35. MacKenzie-Graham, A., Lee, E. F., Dinov, I. D., Bota, M., Shattuck, D. W., Ruffins, S., Yuan, H., Konstantinidis, F., Pitiot, A., Ding, Y., Hu, G., Jacobs, R. E., and Toga, A. W. (2004) A multimodal, multidimensional atlas of the C57BL/6J mouse brain. *J Anat* 204, 93–102.
36. Siddiqui, A. S., Khattra, J., Delaney, A. D., Zhao, Y., Astell, C., Asano, J., Babakaiff, R., Barber, S., Beland, J., Bohacec, S., Brown-John, M., Chand, S., Charest, D., Charters, A. M., Cullum, R., Dhalla, N., Featherstone, R., Gerhard, D. S., Hoffman, B., Holt, R. A., Hou, J., Kuo, B. Y., Lee, L. L., Lee, S., Leung, D., Ma, K., Matsuo, C., Mayo, M., McDonald, H., Prabhu, A. L., Pandoh, P., Riggins, G. J., de Algara, T. R., Rupert, J. L., Smailus, D., Stott, J., Tsai, M., Varhol, R., Vrljicak, P., Wong, D., Wu, M. K., Xie, Y. Y., Yang, G., Zhang, I., Hirst, M., Jones, S. J., Helgason, C. D., Simpson, E. M., Hoodless, P. A., and Marra, M. A. (2005) A mouse atlas of gene expression: large-scale digital gene-expression profiles from precisely defined developing C57BL/6J mouse tissues and cells. *Proc Natl Acad Sci USA* 102, 18485–90.
37. Lein, E. S., Zhao, X., and Gage, F. H. (2004) Defining a molecular atlas of the hippocampus using DNA microarrays and high-throughput in situ hybridization. *J Neurosci* 24, 3879–89.
38. Zapala, M. A., Hovatta, I., Ellison, J. A., Wodicka, L., Del Rio, J. A., Tennant, R., Tynan, W., Broide, R. S., Helton, R., Stoveken, B. S., Winrow, C., Lockhart, D. J., Reilly, J. F., Young, W. G., Bloom, F. E., and Barlow, C. (2005) Adult mouse brain gene expression patterns bear an embryologic imprint. *Proc Natl Acad Sci USA* 102, 10357–62.
39. Singh, R. P., Brown, V. M., Chaudhari, A., Khan, A. H., Ossadtchi, A., Sforza, D. M., Meadors, A. K., Cherry, S. R., Leahy, R. M., and Smith, D. J. (2003) High-resolution voxelation mapping of human and rodent brain gene expression. *J Neurosci Methods* 125, 93–101.
40. Gong, S., Zheng, C., Doughty, M. L., Losos, K., Didkovsky, N., Schambra, U. B., Nowak, N. J., Joyner, A., Leblanc, G., Hatten, M. E., and Heintz, N. (2003) A gene expression atlas of the central nervous system based on bacterial artificial chromosomes. *Nature* 425, 917–25.
41. Christensen, E., Curbera, F., Meredith, G., and Weerawarana, S. (2001) *Web Services Description Language (WSDL) v1.0.* Available at http://www.w3.org/TR/wsdl.

42. Feng, G., Burton, N., Hill, B., Davidson, D., Kerwin, J., Scott, M., Lindsay, S., and Baldock, R. (2005) JAtlasView: a Java atlas-viewer for browsing biomedical 3D images and atlases. *BMC Bioinformatics* 6, 47.
43. Wetzel, A. W., Ade, A., Bookstein, F., Green, W., and Athey, B. (2000) *Representation and Performance Issues in Navigating Visible Human Datasets*, The Third Visible Human Project Conference, Bethesda, MD.
44. Pieper, S., Halle, M., and Kikinis, R. (2004) 3D Slicer, *IEEE International Symposium on Biomedical Imaging (ISBI): Macro to Nano*, Washington DC, v1, 632–635.
45. Eaton, R. C., and Hackett, J. T. (1984) The role of the Mauthner cell in fast-starts involving escape in teleost fishes, in *Neural Mechanisms of Startle Behavior* (Eaton, R. C., ed.), Plenum Press, New York, pp. 213–66.
46. Gustafson, C., Bug, W. J., Nissanov, J. (2007) NeuroTerrain–a client-server system for browsing 3D biomedical image data sets. *BMC Bioinformatics*, 8, 40–55.

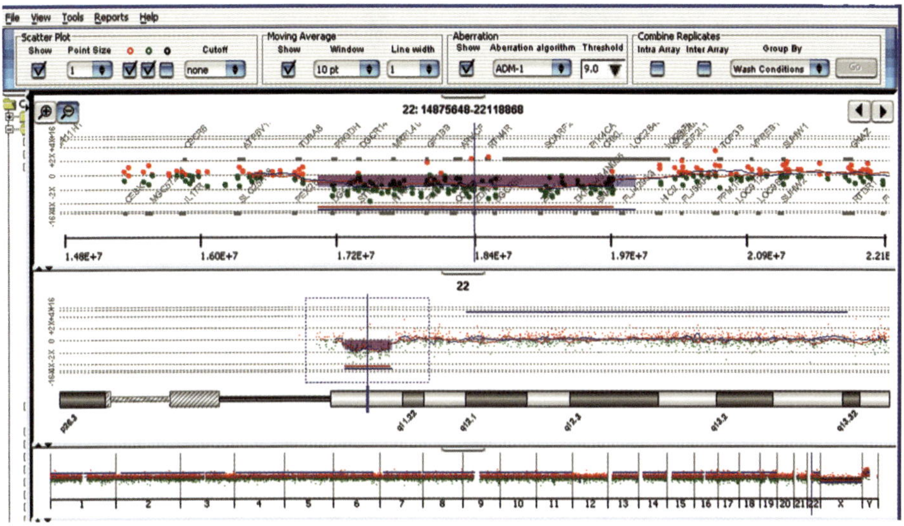

Color Plate 1. Region of the velocardiofacial syndrome (VCFS) deletion on chromosome 22 detected by comparative genome hybridization (CGH). The figure shows a screenshot of CGH Analytics software. The three windows show all chromosomes (bottom window), chromosome 22 (middle window), and the hemizygous microdeleted region with its associated genes. Red and green points are probes on the array with measured intensities. The deletion is clearly identified on the q-arm of chromosome 22, in the 17–20Mb region. A gene list can be saved from all the genes in the deleted region and imported into GeneSpring GT or GeneSpring GX. (see Fig. 1 on pg. 346)

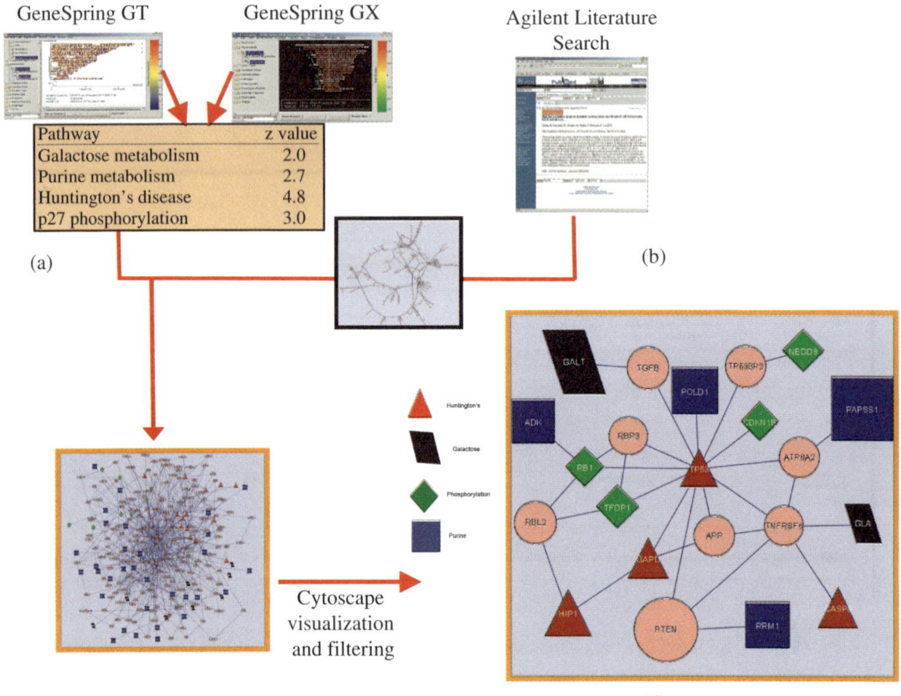

Color Plate 2. Workflow for integrating gene expression profiling and genotype analysis. **(A)** Combined gene expression and genotype gene list-based pathway enrichment analysis identifies four canonical pathways. **(B)** Agilent Literature Search tool extracts associations from literature for disease-related genes and proteins. **(C)** Literature-based extended association network for the four canonical pathways identified by combined genotype and gene expression datasets. **(D)** "Interesting" sub-network identified from the extended network in **(C)** using combined single-nucleotide polymorphism (SNP) LOD score and gene expression *P*-values. (see Fig. 3 on pg. 358)

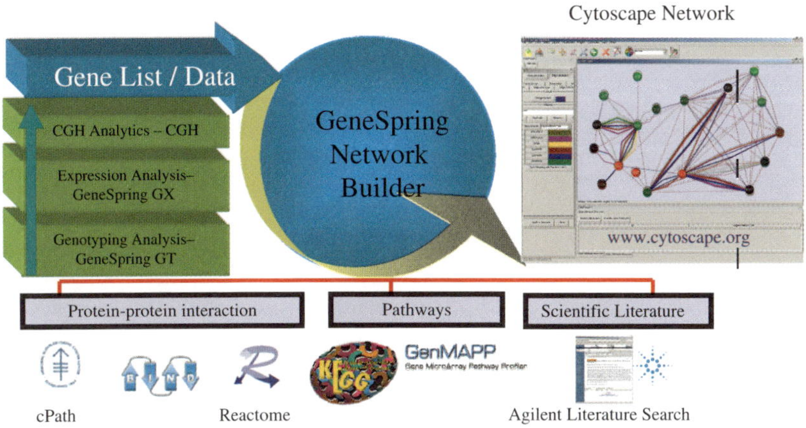

Color Plate 3. GeneSpring Network builder. Schematic representation of the workflow for the plug-in. A gene list is selected and the selected databases are searched for gene–gene interactions. A network is built using the Cytoscape viewer. (see Fig. 4 on pg. 360)

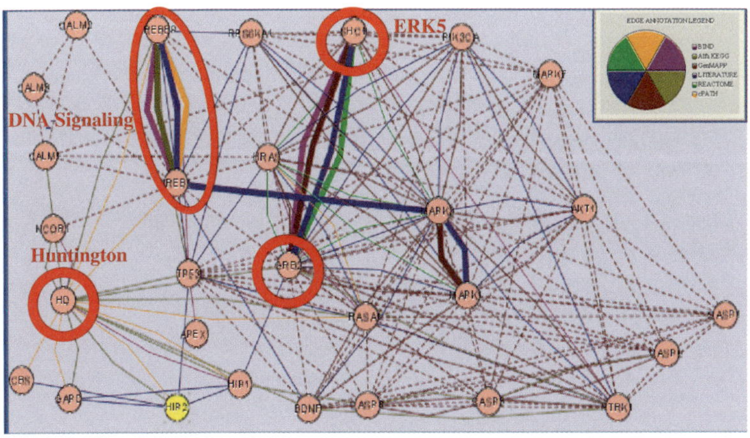

Color Plate 4. The Huntington and ERK5 gene lists from the gene expression, genotyping, and comparative genome hybridization (CGH) experiments are used to build an interaction network using the GeneSpring Network builder. New and known interactions can be identified. *GRB2* is one of the connector genes between the genes of the ERK5 pathway and the Huntington pathway. (see Fig. 5 on pg. 361)

14

Workflow-Based Approaches to Neuroimaging Analysis

Kate Fissell

Summary

Analysis of functional and structural magnetic resonance imaging (MRI) brain images requires a complex sequence of data processing steps to proceed from raw image data to the final statistical tests. Neuroimaging researchers have begun to apply workflow-based computing techniques to automate data analysis tasks. This chapter discusses eight major components of workflow management systems (WFMSs): the workflow description language, editor, task modules, data access, verification, client, engine, and provenance, and their implementation in the Fiswidgets neuroimaging workflow system. Neuroinformatics challenges involved in applying workflow techniques in the domain of neuroimaging are discussed.

Key Words: fMRI; interoperability; neuroimaging; neuroinformatics; pipeline; WFMS; workflow.

1. Introduction

Analysis of functional and structural magnetic resonance imaging (MRI) brain images requires a complex sequence of data processing steps to proceed from the raw image data to the final statistical tests. These steps include image processing, morphological operations, statistical procedures, visualization and display, validation, data format conversion, and database transactions *(1,2)*. In order to automate the analysis steps, researchers have typically relied on batch processing and scripting systems. More recently, however, workflow-based technologies have been applied to neuroimaging analysis to improve the robustness, efficiency, and reproducibility of the analysis path. This chapter describes the methods used to implement workflow technologies for the

From: *Methods in Molecular Biology, vol. 401: Neuroinformatics*
Edited by: Chiquito Joaquim Crasto © Humana Press Inc., Totowa, NJ

neuroimaging domain and the neuroinformatics challenges that arise from that effort.

A workflow is simply a series of steps that must be executed in order to complete a task. The steps are usually programs to be executed on a computer, but steps may also include manual tasks and operations performed by industrial machines or scientific instruments. Workflow management systems (WFMSs) are software systems that specify, execute, monitor, and log complex, multi-step computational procedures, very often executed over multiple computers. WFMSs consist of eight subcomponents: (1) a language in which to specify the workflow; (2) a workflow editor; (3) a set of application modules; (4) support for data access methods; (5) verification tools; (6) a job submission or "client" program; (7) the workflow engine; and (8) provenance tools. Not every WFMS supports all eight components, and in many cases, the components do not exist as distinct and independent subsystems. For example, workflow languages and engines are often tightly bound, and data access and verification procedures are sometimes merged. Scientific workflow systems emerged in the mid-nineties, drawing from technical advances in three areas (1) business and commercial workflow systems, (2) middleware and distributed computing ***(3)***, and (3) scientific visual and "workbench" computing environments ***(4–6)***.

The key advantages that workflow-based technologies bring to neuroimaging analysis are (1) interoperability across heterogeneous software packages and computing platforms; (2) transparency of data analysis methods, as processing procedures are encoded in explicit, reusable workflows; and (3) facilitated access to high-performance computing platforms (e.g., cluster, distributed, Grid) and automated computing procedures. The primary obstacles to wider acceptance and use of workflow-based technologies in neuroimaging analysis are: (1) scientific WFMSs are still an emerging technology with many issues regarding data management, neuroinformatics tools, scale, and usability still under development; (2) WFMSs can be large, complex software systems that incur system and training overhead and impose an unwanted layer of abstraction between researchers and the scientific analysis programs they are accustomed to using; and (3) the design of some software developed for neuroimaging analysis makes it difficult to integrate that software into WFMSs.

In **Subheading 2**, below, workflow systems currently available for neuroimaging analysis are listed. In **Subheading 3**, the eight components of a WFMS are discussed, and to provide a concrete example, the implementation and use of these components in the Fiswidgets *(7)* workflow system are described. **Subheading 5** (Notes) discusses issues and alternative strategies that have been used in implementing workflow systems for neuroimaging analysis.

2. Materials—Workflow Systems for Neuroimaging Analysis

A number of workflow and distributed computing tools designed specifically for neuroimaging analysis are currently available and under development. They include the following:

1. BaxGrid ***(8)***, developed under the Japanese Medical Grid project, uses the Globus ***(9)*** Toolkit and Ninf-G ***(10)*** Grid tools on Linux platforms to execute functional MRI (fMRI) analysis tasks on remote cluster-based servers across several sites in Japan and the Philippines.
2. BrainVISA/Anatomist ***(11–13)***, developed at the Institut Fédératif de Recherche, Service Hospitalier Frédéric Joliot, is an "image processing factory" for multi-modal integration and visualization of diffusion-weighted MRI, fMRI, MRI and Magneto- and Electroencephalogram (MEG/EEG) data for anatomical, morphometric, and statistical analysis techniques; the processing pipeline is implemented in the Python scripting language.
3. BAMMfx ***(14,15)***, developed at the Cambridge University Department of Psychiatry, is built on the Eclipse Java integration environment ***(16)*** and provides a visual pipeline interface to the Brain Activation and Morphological Mapping (BAMM) analysis package.
4. Fiswidgets ***(7)***, a Java-based workflow system developed at the University of Pittsburgh, provides graphical wrappers and pipelined access to modules from a number of neuroimaging packages including Automated Image Registration (AIR) ***(17)***, Analysis of Functional NeuroImages (AFNI) ***(18)***, FMRIB Software Library (FSL) ***(19)***, and tal ***(20)***, with prototype modules from the BrainVoyager ***(21)***, Nonparametric, Prediction, Activation, Influence, Reproducibility, re-Sampling (NPAIRS) ***(22)***, and Statistical Parametric Mapping (SPM2) ***(23)*** packages. Fiswidgets supports "for-each" style parameterized iteration over workflow steps, client/server-based job submission, and format conversion utilities.
5. LONI Pipeline ***(24)***, a Java-based workflow system developed at the Laboratory of Neuro Imaging, University of California, Los Angeles (UCLA), provides access to structural and functional processing modules from a number of packages including the AIR ***(17)***, BrainSuite ***(25)***, FSL ***(19)***, FreeSurfer ***(26)***, and Montreal Neurological Institute's Brain Imaging Software Toolbox. The LONI Pipeline supports client/server access to remote supercomputing facilities, management of server platforms through the Sun Grid Engine ***(27)***, and data flow parallelization of workflow tasks. Format conversions are supported by the LONI Debabeler ***(28)*** utility.
6. VoxBo ***(29)***, developed at the departments of Radiology and Neurology at the University of Pennsylvania, is a general linear model based neuroimaging analysis

package written in C++ and Interactive Data Language (IDL), with support for automated distributed and parallel processing across clustered platforms through a centralized batch scheduler. It is extensible for inclusion of modules from other software packages and includes a library of format translations.

There are a number of additional systems that address automation of neuroimaging analysis (*see* **Note 1**). Resources describing and listing WFMSs for a number of scientific fields (e.g., bioinformatics, geology) are also available (*see* **Note 2**).

3. Methods

3.1. Workflow Description Language

The first step in building a workflow is to indicate what applications need to be executed in the appropriate format or language. The workflow description language is used to specify the workflow execution logic, the tasks to be run, task parameters such as input and output datasets and algorithm options, and platform or configuration options such as required or suggested hardware platforms. The execution logic outlines the relations or constraints between the steps; these include simple sequential processing, iterated "for-each" style loops, parallel execution, conditional execution, temporal dependencies, and data dependencies.

Several different types of graphical, scripted, and structured languages can be used to encode the workflow (*see* **Note 3**). Many WFMSs use their own system-specific description language, often expressed in a machine parseable format such as Extensible Markup Language (XML), which is tightly bound to their workflow engine. For example, the workflow description language used by the Fiswidgets WFMS is a simple XML-based language. All control flow and application choices the researcher selects are stored with a set of XML tags that the workflow engine can decode in order to run the flow. **Figure 1** shows a short nine-step workflow loaded into the Fiswidgets editor; it is saved to disk in XML format through the toolbar "Save" button. The XML description for this workflow is listed in **Appendix A**. An alternative, more easily readable text version of the XML, listed in **Appendix B**, can be generated using the Fiswidgets parsing utility ("Tools/FXParse" menu item in the Fiswidgets editor). Because the XML format is too complicated for direct editing by the researcher, typically a workflow editor is used to generate workflow specifications (*see* **Subheading 3.2.**).

Regardless of the description language used, it is not trivial to express a complex neuroimaging analysis procedure in a workflow style format.

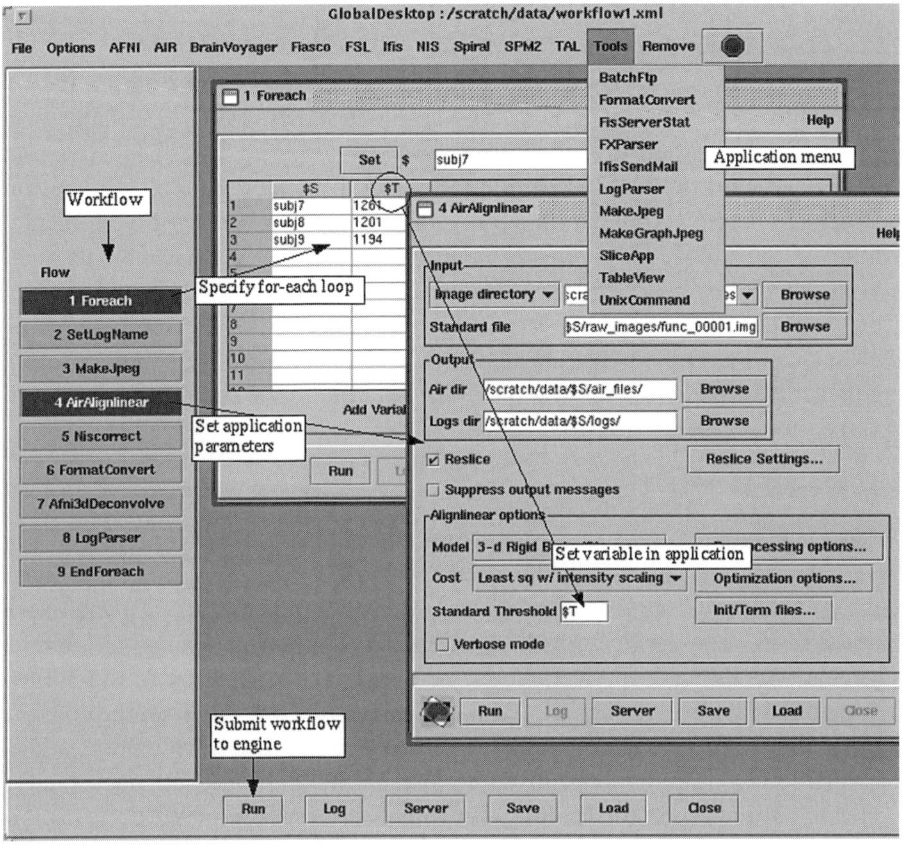

Fig. 1. Fiswidgets workflow editor. The workflow is shown in the left panel. Applications are added to the workflow by selection from the menus at the top of the editor window. Workflows are saved, loaded, and submitted to the engine using the toolbar buttons at the bottom of the editor window.

A number of technical and usability issues arise in shifting from software-package-specific interfaces to the more general workflow interface (*see* **Note 4**).

3.2. Graphical Workflow Editor

Workflow editors are a type of visual programming interface that provide graphical "drag and drop" displays to assist the researcher in workflow design and specification. The editors permit the researcher to model the workflow as a graph diagram, to simplify workflow logic by encapsulating workflow steps in a hierarchical design, and to select tasks, task parameters, and configuration choices from menu displays.

Figure 1 shows the Fiswidgets workflow editor. The menus across the top of the editor window (e.g., AIR, AFNI) list the applications available from each software package to which Fiswidgets provides an interface. Steps are added to the workflow, shown in the left panel, by selecting applications from the package menus. Application parameters are set by opening the application window in the editor work area and entering information in the text fields, radio buttons, checkboxes, etc. If the researcher needs to run an application for which a Fiswidgets interface is not available, the generic "Tools/UnixCommand" interface can be used to run any command. The workflow is submitted to the engine for execution by pressing the "Run" button on the toolbar at the bottom of the editor window.

3.3. Workflow Task Modules

The task modules are the application programs that perform the desired computations or transactions; the need to execute these programs efficiently motivated the development of the entire WFMS infrastructure. To the extent that a WFMS is an application-independent framework for managing complex computations, the task modules could be considered peripheral to the WFMS architecture. However, in practice, generic heterogeneous application programs often cannot simply be plugged into a WFMS. Programs that were not designed for workflow style interoperability may present integration challenges such as requiring particular execution environments, e.g., MATLAB ***(30)***, Java ***(31)***, or Message Passing Interface (MPI) ***(32)***, requiring runtime licenses, using interactive (e.g., dialog boxes or menu systems) rather than batch style command-line interfaces, launching additional processes (e.g., via Unix *fork* or *system* calls, or Java *Process* invocations) that are difficult for a workflow engine to monitor, generating improper exit codes, etc. When such conditions can be specified and accommodated, then it is precisely the job of the workflow engine to deal with these integration issues, by automatically locating the appropriate environment or resources. When they cannot, WFMS-compatible interfaces to application programs are obtained by (1) writing a wrapper function for each application to mediate between the application and the WFMS, (2) rewriting or revising application programs, or (3) identifying and restricting execution to, or designing the WFMS around, the set of applications with appropriate native interfaces. Because each of these approaches can require significant retooling of at least some of the task modules, and because domain-specific file format translations or visualization tools are often needed, scientific WFMSs have tended at least initially to be developed around task modules from a specific domain area.

The Fiswidgets WFMS provides interfaces to over 130 application modules from a number of widely used neuroimaging analysis packages, including AIR *(17)*, AFNI *(18)*, FSL *(19)*, and tal *(20)*. Fiswidgets application interfaces consist of customized Java graphical user interface (GUI) wrappers written using the Fiswidgets toolkit. Each wrapper consists of two parts. The front-end of the wrapper is the GUI with which the user interacts; it presents a consistent, easy way to specify application inputs, outputs, and parameters. The back-end of the wrapper (the command invocation manager) is a short Java program that collects the user-specified parameters from the front-end and builds the appropriate command-line call for the engine to execute; it provides the software interface that permits the application to be launched and monitored from within the WFMS. Both the SPM *(23)* and BrainVoyager *(21)* neuroimaging packages need specialized Fiswidgets wrappers because applications in these packages are not run through command-line calls. Fiswidgets currently provides prototype wrappers to demonstrate handling of these cases. SPM runs in the MATLAB environment, so the wrapper writes out a short MATLAB program that encodes the user's parameter choices for each application call. The WFMS engine can then execute that program through a command-line call to the MATLAB interpreter. BrainVoyager applications are typically invoked through BrainVoyager's own graphical interface. BrainVoyager, however, also provides a scripting interface for batch style execution. The Fiswidgets wrapper writes a BrainVoyager script corresponding to the user's GUI inputs; the WFMS engine can then invoke the script through a call to the BrainVoyager driver (*see* **Note 5**).

The use of these standard and specialized application wrappers permits the Fiswidgets WFMS to dynamically integrate processing steps from a number of very different neuroimaging analysis packages into a consistent workflow architecture. However, in addition to the SPM and BrainVoyager interface difficulties discussed above, there are a number of more general challenges in incorporating neuroimaging applications into WFMSs (*see* **Note 6**).

3.4. Data Access System

The input and output data for workflow task modules resides in various filesystems or databases. If the filesystems or databases are not local to the machine executing the task, the data needs to be made accessible to the local machine. In smaller systems, this can be accomplished transparently by distributed filesystems, such as Network File System (NFS) or Andrews

File System (AFS), or by network accessible data servers, such as Network Attached Storage (NAS) units. In larger systems, for example, Grid-based systems utilizing a large dynamic pool of back-end servers, more sophisticated global data systems such as the Storage Resource Broker (SRB) *(33)* are used. These systems provide a location-independent, logical naming scheme for datasets and manage the mapping of logical names to distributed and possibly replicated data stores. Alternatively, the WFMS can explicitly move and stage data as needed, using utilities such as Globus GridFTP *(34)*.

Once data are physically available to the task modules, data format issues need to be addressed. The differences in data formats required by heterogeneous computing platforms and applications are one of the major impediments to workflow integration. Increasingly automated solutions to this problem are enabled by increasingly abstract representations of data form and content. For example, in the least automated solution, the researcher building the workflow needs to explicitly insert the appropriate data conversion steps between task modules. If conversion modules are not already available for a particular domain, a significant portion of the WFMS development work may be dedicated to implementing these translations. If a degree of data structure abstraction can be obtained, for example, by the use of a self-documenting data format such as HDF5 *(35)* or logical typing *(36)*, and the applications' format requirements are well defined, a WFMS can perform some translations automatically. More sophisticated translations that depend not only on data format but also on data content require an abstraction of that content, e.g., in an ontology, to support automated translations; this approach is being developed in the semantic mediation and data registration capability in the Kepler workflow system *(37)* and the semantic typing system in the Open Microscopy Environment (OME) *(38)*.

The Fiswidgets WFMS includes a data transfer and a data conversion utility in its Tools package. The data transfer application, *BatchFtp*, was developed using the open source JFtp package *(39)*. It can be inserted as a regular step in a workflow and permits files and directories to be transferred over the Internet in an unattended, batch style use of the file transfer protocol (FTP). The format conversion application, *FormatConvert*, permits file format conversion between a number of commonly used neuroimaging formats. The user must explicitly insert *FormatConvert* steps in the workflow to ensure that applications have appropriately formatted data available. For example, in the workflow shown in **Fig. 1**, a *FormatConvert* step has been used to convert from the Analyze data format used in **steps 3–5** to the AFNI format used in **step 7**.

The data conversion and transfer utilities currently available in Fiswidgets are adequate for smaller sized workflow tasks but may not scale well to very large, widely distributed jobs. The size and complexity of neuroimaging data requires more sophisticated tools to automate larger scale data transfer, access, and conversion steps. Some of the more advanced data management approaches being developed in the neuroimaging domain are discussed below (*see* **Note 7**).

3.5. Workflow Verification

After the workflow has been constructed and the appropriate data transfer and conversion steps have been added, workflow correctness should be verified. The goal of workflow verification is that the workflow executes to completion without error. Because workflows have many more points of failure than standalone programs, can run for hours or days or more, and have complex error recovery and clean-up procedures, extensive verification checks are warranted. First, the workflow specification can be checked for syntactical correctness. Use of a workflow editor should guarantee correctness, but discrepancies may arise if the editor and engine were developed by different groups. One example of syntax verification is the Web-based Business Process Execution Language (BPEL) OnDemand Validation Service ***(40)***. Second, workflows can be checked for logical correctness (that there are not infinite loops or deadlocks in the control flow, that all necessary inputs are generated, etc.) ***(41)***. Third, an issue that is highly relevant to neuroimaging workflows is to validate that datasets of the correct data format and semantic type have been specified for each workflow step. One approach to this, proposed in **ref. *37***, is for task modules to advertise their required input and output dataset types, and for connections between one module's output and another module's input to be explicit, so that the WFMS can check for compatibility across connections. An alternative approach, proposed in **ref. *36***, is to use the XML Dataset Type and Mapping (XDTM) system ***(42)***, in which data inputs and outputs are specified by their logical structure type (e.g., Volume), independent of that structure's location or physical representation. A mapping process then automatically generates the correct location and access method at runtime. Procedures of this kind, as well as more advanced semantic based approaches, are better categorized not as verification systems but as data access systems designed to reduce or eliminate the need for data-type verification. A fourth verification is the runtime check that the system is properly configured for each step of the workflow (i.e., that the specified task modules are available, that the user has proper authentication credentials for each step).

In the Fiswidgets WFMS, syntactical correctness of the workflow is guaranteed by using the editor to construct the workflow. Because Fiswidgets currently has a very simple linear execution logic, it is not prone to logical errors such as deadlock. No tools or mechanisms are provided for dataset type checking or system configuration checking. Issues and work regarding type checking for neuroimaging applications are discussed below (*see* **Note 8**).

3.6. Workflow Engine Client

The workflow engine client is the interface used by the researcher to submit and monitor jobs on the workflow engine. This piece of software can be a standalone utility or embedded in either the workflow editor or engine. It is distinct from the editor in that the editor is an off-line process that creates workflow specifications whereas the client is in direct contact with a running workflow engine. The client is distinct from the engine because in two-tier and three-tier systems, the client runs on the researcher's desktop and the engine runs on a remote server, and multiple clients may connect to one engine. Although, in many cases, the client is customized for a particular workflow engine, in some Web services-based architectures *(3)*, a generic client such as a Netscape browser can be used for job submission and monitoring.

In the Fiswidgets system, the workflow engine client is built into the workflow editor. Once a workflow is loaded into the editor, it can be submitted to the engine through the "Run" button on the editor toolbar (*see* **Fig. 1**). The engine then returns process status information and output back to the client, where it is displayed to the researcher.

3.7. Workflow Engine

The workflow engine is the software responsible for orchestrating the execution, also known as the "enactment," of the workflow task modules. It implements the control flow logic, locates the appropriate task modules for abstract specifications, manages any required authentications, starts tasks as per the data and temporal dependencies, schedules tasks to achieve optimal workflow throughput, monitors task execution, and recovers from task failures. It plays a middleware role by encapsulating and providing uniform access to heterogeneous computing platforms (e.g., multi-processor, cluster, distributed, and Grid architectures), so that the mechanisms for this access can remain independent of both the workflow description and the individual task modules. In large workflow systems, where thousands of tasks may be running concurrently, the workflow engine functions as a kind of high-level operating system in managing the CPU, memory, disk resources, and Internet connections of multiple machines to efficiently run sets of processes (workflows).

A WFMS is typically distributed over multiple compute platforms connected over the Internet. In two-tier client/server systems, the user's interface to the WFMS runs on a local client machine (e.g., a desktop PC) and both the workflow engine and the task modules run on a remote server machine. In three- and *n*-tier systems, the workflow engine runs on a middleware server and task modules run on additional specialized servers, e.g., database servers or compute farms. Grid computing can be considered a special extension of *n*-tier systems, in which the compute-servers are high-performance or supercomputing platforms, often owned by different institutions and often very widely distributed geographically ***(43)***. The communication between tiers in a WFMS is implemented using standard distributed computing protocols: remote procedure call (RPC), remote object invocation (e.g., Java RMI, CORBA), asynchronous message or queue-based protocols [e.g., Java Message Service (JMS)], and more recently Web Services. An excellent discussion of the evolution and pros and cons of these protocols is provided in ***(3)***. Grid-enabled WFMSs typically take advantage of middleware toolkits such as the Globus Toolkit ***(9)*** that implement Grid communication and service protocols; four Grid toolkits are reviewed in ***(44)***.

The Fiswidgets engine implements sequential execution of workflow steps with optional "for-each" loops, breakpoints, and script generation. The client/server architecture is implemented using the Java Remote Method Invocation (RMI) protocol. When the workflow is submitted to the engine through the "Run" button in the editor, the engine invokes each application's wrapper, in turn, to generate the appropriate command-line call for that application. The engine then executes each command-line with a Java *Process()* call, collects the exit status of the command, and returns it to the client editor. There are three options to modify this control flow. First, the "for-each" construct can be used ("Options/Foreach loop" menu item) to iterate over a set of workflow steps with different sets of input parameters. In a spreadsheet style interface (*see* **Fig. 1**, workflow step 1), the user specifies a set of variables and lists of values for those variables for each iteration of the workflow. The user enters the variables in the input fields of the GUI wrappers; the engine automatically replaces those variables with the correct values at each iteration of the "for-each" loop. Second, the user can set breakpoints ("Options/Set Breakpoints…" menu item) between steps in the workflow. The breakpoint will delay execution of the next step in the flow until the user responds to a dialog box. This gives the user the opportunity to check and perhaps modify the outputs produced by earlier steps in a flow before proceeding with the later steps. Third, the user can run the flow in "no operation" mode and just print out a script of the commands that would have been executed ("Options/Scripting" menu item) rather than executing any applications. This mode is useful if the user just wants to check a command-line or wants to generate an initial shell script and modify it for later use.

There are a number of ways in which the Fiswidgets engine can be extended to provide a more powerful parallel and distributed computing platform. The two issues most pertinent to the development of workflow engines for neuroimaging analysis are (1) scalability across computational platforms and (2) determining which middleware tools or existing scientific workflow engines can be adapted to run neuroimaging workflows (*see* **Note 9**).

3.8. Workflow Data and Process Provenance

The final step in running a workflow is to record the procedures that were executed. Data provenance, also know as data lineage, pedigree, or derivation, is a record of how each dataset in a system was produced, i.e., what set of processes and parameters determined the final data values *(45–47)*. Provenance can be broken down into data provenance, which links to each dataset a record of all the processes that were applied to that data, and process provenance, which documents the set of processes that was executed in a workflow enactment. The two are not synonymous because there may not be a one-to-one mapping or well-defined specification of the relation between process and data. Data provenance is often used retrospectively, for example, in audit procedures, to confirm processing steps; process provenance can be used prospectively, for example, to repeat an analysis on new data.

The Fiswidgets system supports process provenance through a logging system. As each application is executed, its exact command-line and all output messages produced by that command are logged to an ASCII log file. Special tags are added to the log file by the Fiswidgets engine so that the file can be parsed and converted to HTML for easy viewing ("Tools/LogParser" menu item). For example, tags indicating the location of output JPEG images are added so that those files can be viewed from a Web browser (*see* **Fig. 2**).

Provenance is crucial in a scientific setting in order to replicate and validate results and document and disseminate procedures. There is, however, not yet widespread use of provenance systems for neuroimaging data for a number of reasons (*see* **Note 10**).

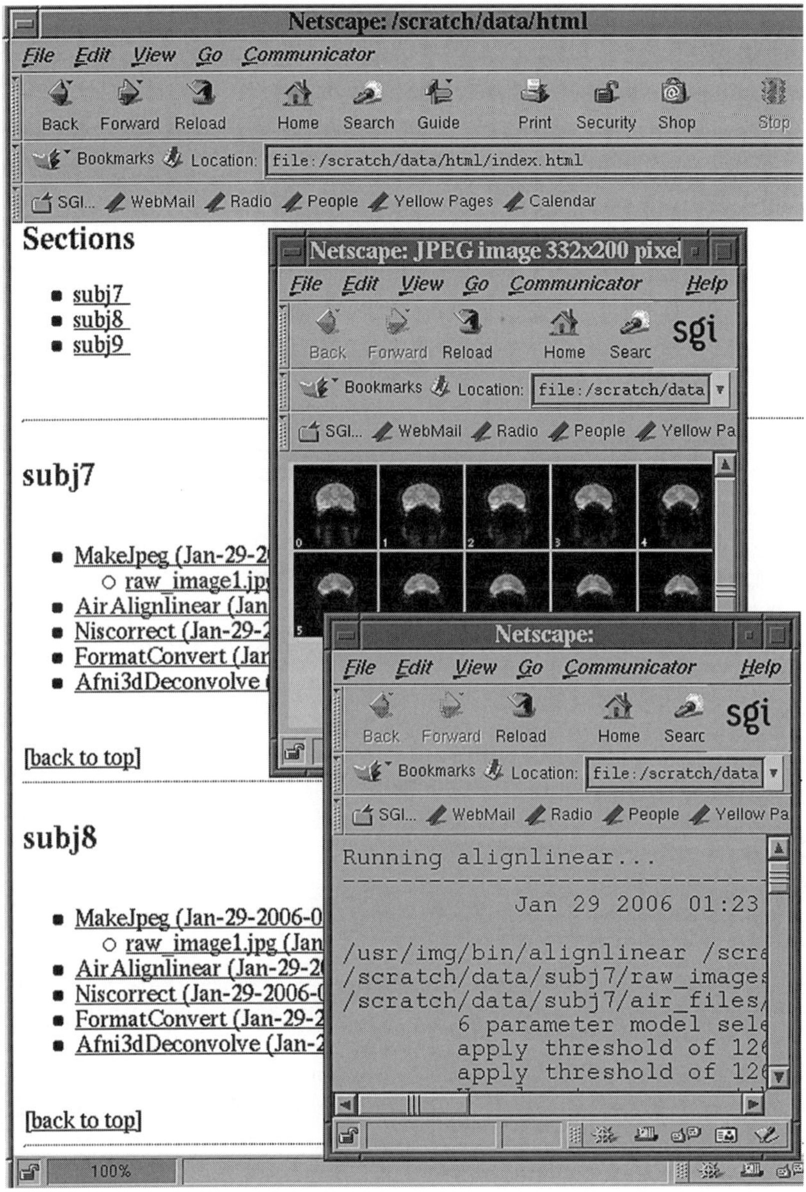

Fig. 2. HTML output from the Fiswidgets logging system. Each iteration of the for-each loop is a separate section in the HTML page. For each section, the command-line output and JPEG images (when available) from each application are saved under separate links.

4. Conclusion

Four themes emerge from this discussion of workflow-based approaches to neuroimaging analysis. First, the development of scientific workflow systems is a very active research area in computer science and in a number of scientific domains; the neuroimaging field can both benefit from and contribute to this agenda. Second, workflow systems require a significant level of abstraction in their representation and handling of data, processes, and compute platforms. Larger scale projects with a clear need for sophisticated data and process management tools recognize the role for abstraction and can promote the wider development and use of neuroinformatics tools such as databases, virtual data catalogs, and meta-data ontologies. Researchers with a more "hands-on" orientation toward data analysis may initially find these abstractions frustrating. Third, WFMSs encompass a number of diverse computing technologies, from workflow definition languages to GUI design, middleware, and powerful new models for distributed computing. Likewise, the neuroimaging research community has diverse computing requirements. Different researchers may benefit from different aspects of the WFMS: access to local or remote multi-processor compute platforms, interoperability across software packages, automated processing, or defining, archiving, and replicating analysis procedures through workflow definitions. Tailoring and scaling WFMS technologies to fit this range of research needs is a significant design criterion for neuroimaging WFMSs. Fourth, the size and complexity of neuroimaging image and meta-data, coupled with the complexity of the statistical tests through which those data are filtered, present logistical challenges that need to be addressed specifically for the neuroimaging domain.

Notes

1. There are a number of software systems developed to address automation of neuroimaging analysis that are not listed in **Subheading 2** because they do not use workflow technologies or are still in the very early stages of development. These include a number of batching systems contributed to SPM ***(48)*** that can be used to automate SPM analyses: BrainPy ***(49)***, Chiron ***(50,51)***, and "Run Pipeline in Parallel" (RPPL) ***(52)***
2. A large number of scientific WFMSs are currently in use. Ten workflow projects in geoscience, biology, and physics domains are reviewed in **ref. *53***; A WFMS taxonomy that includes workflow design, scheduling, fault tolerance, and data transfer has been used to categorize 12 WFMSs ***(54)***; the online Scientific Workflows Survey ***(55)*** currently lists 33 WFMSs. It may in future be possible to adapt some of these systems for processing neuroimaging workflows (*see* **Note 9**).

3. Workflow languages can be designed as scripting languages, graph representations such as the Directed Acyclic Graph (DAG) used in Condor DAGMan ***(56)*** or the Petri net used in ***(57)***, or as database schemas ***(38,58,59)***. Some workflow description languages have been standardized for use across multiple, independently developed workflow engines. For example, the BPEL for Web Services (BPEL4WS) ***(60)*** was developed by several corporations with a stake in commercial WFMSs. And, a language specifically addressing Grid computing requirements, the Grid Services Flow Language (GSFL) ***(61)*** has been proposed. Tasks in a workflow language can be specified concretely by designating particular versions or installations of programs and data, or through more flexible, high-level representations in what are called abstract workflows. For example, an abstract workflow step could specify a type of operation (e.g., translate from HTML to Postscript) instead of a program and could refer to virtual datasets ***(50,51,62)*** instead of specific filenames. The workflow engine translates the abstract workflow into a concrete, executable workflow at runtime by locating task modules, and locating, transferring, or generating input data ***(63,64)***.
4. One of the primary considerations in using a workflow description language to specify neuroimaging analysis procedures is not the syntax or power of the language itself, but the shift of the application logic from software package specific interfaces or scripts to a more general, transparent workflow language. (The term "application logic" refers in this context to the higher level ordering, selection, and parameterization of task modules, rather than the logic or algorithm within any specific module.) Neuroimaging analysis is inherently complex. It entails successive massaging of massive, noisy datasets to arrive at statistically valid hypothesis tests; this is done by means of computer programs containing sophisticated mixtures of image processing, morphological, statistical, visualization, and validation algorithms ***(1,2)***. In order to guide users through this process, many neuroimaging software packages provide their own application logic through a graphical or scripted interface that hides some of the underlying complexity of the data processing stream. This scaffolding embeds a valuable cache of domain knowledge and is very helpful, particularly to researchers new to neuroimaging analysis or scientific data processing. Porting this logic to a WFMS framework is difficult for two reasons. First, there are technical challenges. If the logic is not well documented, relies on non-portable shortcuts and dependencies between package modules, or includes data operations that are encoded in auxiliary scripts or user interfaces (rather than command-line programs), it may be difficult to port that logic to an automated, batch style WFMS environment. Second, there is a usability challenge because the workflow system presents the researcher with access to programs from a number of neuroimaging software packages and thus provides more choices but less guidance in navigating those choices. A number of different approaches can be taken to provide guidance. First, repositories of example or reusable template workflows ***(65)*** can be built to serve as models

for various analysis techniques. Second, the WFMS can be used to evaluate very large numbers of workflows to help determine optimal processing techniques ***(66)***. Third, application logic can be supported through specialized workflow editors that focus on particular analysis techniques (e.g., regression models) but produce workflows that can be run on the generalized workflow engine.

5. In BrainVoyager2000, it was possible to invoke a BrainVoyager script through a command-line call to the BrainVoyager driver. In early BrainVoyagerQx releases, it was only possible to invoke scripts interactively from a BrainVoyager window. In such cases, a WFMS engine would not be able to run a BrainVoyager application as a batch style workflow step without additional mechanisms.
6. There is a large amount of software available for neuroimaging analysis; for example, the Internet Analysis Tools Registry (IATR) ***(67)*** currently lists 120 packages. These programs are for the most part developed by independent academic laboratories and made available to the research community free of charge. This distributed model of software development is of great benefit to the field, because the expertise and latest research techniques from a wide range of laboratories are made quickly available to all researchers. There are, however, a number of challenges faced by a systems architect constructing an integrated, automated analysis platform. First, instead of facing the more common "legacy software" problem, in which old software programs cannot be changed, neuroimaging researchers must deal with very frequent software updates and revisions (e.g., monthly updates and major revisions every year or two). This creates a significant maintenance burden when additional middleware software has been written, for example, to handle data format translations, wrap the original programs in workflow compatible interfaces, or verify workflow correctness. Second, a concern in scientific research environments is the necessity for existing workflow specifications that have already been tested and archived to be kept up to date with software changes so that analyses can be replicated. Third, it is difficult to achieve the desired level of functional granularity in the software modules. If applications perform only very small operations, then a workflow may require very many steps and rapidly become unwieldy. If applications conglomerate together too many operations, then it becomes difficult for the researcher to construct new analysis procedures by reconfiguring and reordering subtasks. Owing to the intricate nature of neuroimaging data analysis, a very large number of quite small applications (e.g., compute a brain/background threshold, adjust a meta-data field) often need to be invoked. These steps need to be encapsulated in such a way that the user is presented with only the major or frequently adjusted processing steps and parameters but can still, if necessary, drill down to change all the detailed operations. The encapsulation can be implemented by a hierarchical arrangement of workflow steps or through scripting or merging operations at the application level.
7. Neuroimaging datasets are hard to work with in WFMSs for a number of reasons. First, the complexity of the meta-data associated with neuroimaging data makes

interoperability across software systems and hardware platforms difficult. This issue is confronted not just in developing workflow-based systems but in a number of crucial tasks including the initial import of imaging data from the MRI magnet vendor-specific format to an analysis software package, correct display of images in visualization applications ***(68)***, and storing datasets in a database ***(69)***. Neuroimaging meta-data includes information about low-level format (e.g., data dimensionality, data types, disk storage order), image acquisition (e.g., radiological parameters such as flip angle, receiver gain, k-space trajectory), clinical and administrative details (e.g., date of scan, subject age, weight), three-dimensional coordinate systems and mappings of voxel data to those coordinate systems, transformations to surface mesh representations, semantic data-types (i.e., identifying data values as gray scale intensities, mask values, statistics, indices into an external table, for example, of anatomical labels), data provenance, and links to associated datasets.

Progress has been made to standardize the meta-data interpretation and provide extensibility to one of the very simple data formats commonly used in neuroimaging software through the work of the Neuroimaging Informatics Technology Initiative (NIfTI) Data Format Working Group ***(70)***. Meta-data fields have been well defined for some image file formats, such as DICOM ***(71)*** and MINC ***(72)***; standards for XML-based interchange of neuroscience data are being developed in the BrainML system ***(73)***; and neuroscience nomenclatures and ontologies are available ***(74)***, e.g., the NeuroNames database ***(75)***. An advanced file format translation application, the LONI Debabeler ***(28)***, permits researchers to construct format converters as needed for arbitrary formats, by indicating the correct meta-data mappings between formats in a visual editor. However, bridging the gap between the low-level file format-based approaches to data access commonly used by neuroimaging researchers and the more abstract data representation and exchange systems that support interoperable, automated, and semantic based data access remains a challenge.

A second difficulty in working with neuroimaging data is the size and complexity of the image data (not meta-data). As described in ***(36)***, an fMRI neuroimaging study dataset can consist of over one million data files, hierarchically organized over group, subject, session, and image. The very large data size impacts WFMSs in three ways. First, it makes the specification of input datasets in a workflow unnecessarily fragile; the solution ***(36)*** proposed to this is a simplified, abstract dataset specification by logical structure. Second, if neuroimaging data need to be moved across a network to a remote server, the tradeoffs between data compression and replication and network bandwidth and latency need to be evaluated to obtain acceptable performance ***(8)***. Third, the large size of neuroimaging datasets motivates a need for data to be passed in memory from one step in a workflow to the next, rather than going through intermediate disk files. However, this would entail significant revision to the input/output procedures many neuroimaging applications currently use.

8. Of the workflow verification steps listed in **Subheading 3.5**, the most domain specific is the check for appropriate inputs to each task module. Because of the complexity and large amount of meta-data associated with neuroimaging datasets, this kind of verification is both sorely needed but difficult to implement for the neuroimaging domain. The primary issue is that in order to support automated data type and format checking, each of the hundreds of modules potentially accessible from a neuroimaging WFMS needs to have its type specifications expressed in a machine parseable format. This information could be stored in a central database or table, or be embedded in the logic of a workflow engine or editor. From an interoperability and software architecture perspective, embedding this information in each application (or application wrapper) confers greater flexibility, allowing the application to dynamically "advertise" or respond to queries about its data formats. An example of this approach is the work of Zhao et al. ***(36)***, who implemented logical data-type mappings for a complex brain image registration workflow consisting of AIR ***(17)*** applications. Regardless of the mechanism used however, the effort to implement such a system is considerable given the number of modules involved, the complexity of data format specifications, and the rapid pace at which neuroimaging applications and data formats change. The effort clearly warrants using a public protocol or mechanism that could be used by multiple WFMSs, but although newer protocols such as XDTM ***(42)*** or Web services ***(3)*** may be good contenders, a standard has not yet emerged. A related issue is that verification checks that go deeper than confirming data format or structure are often desirable. For example, an application may require that an input be an axial T1-weighted brain image volume, that two input datasets have the same matrix dimension, or that an input represents a region of interest (ROI) volume in some form. Progress in this direction requires further development of semantic based data access tools.
9. As has been noted in **ref. *53***, scientific workflows tend to be data intensive and thus can benefit from facilitated access to distributed and parallel compute platforms. Most neuroimaging analysis programs are written to run on a single CPU of a local machine, with no support for parallel or distributed execution. However, many analysis tasks are "embarrassingly parallel," in that the same computation is performed, without temporal or data dependencies, across multiple subjects or images, and thus can be easily parallelized at the workflow level. A key consideration in obtaining the widest possible benefit to the neuroimaging community from WFMSs is that engines scale to both low- and high-end parallel platforms. For example, a researcher working on a four CPU Linux box could benefit from workflow management tools and parallel access to those processors but might not be willing to incur the overhead of a workflow engine designed to work in a much larger clustered or Grid environment. But, workflow engines designed

for smaller platforms may not scale up successfully without changes in resource allocation and protocols ***(76)***.

Workflow engines, particularly those supporting more complex distributed compute platforms, are technically advanced, sophisticated software systems requiring a substantial development effort. Taking advantage of existing workflow engines or related distributed computing tools can reduce development costs and improve the power and robustness of the engine used in a neuroimaging WFMS. Scientific WFMSs are a new technology, dating from the mid-nineties. Although tools for distributed computing such as CORBA ***(77)*** and CONDOR ***(56)*** were available at that time, mature "off the shelf" middleware components to support scientific workflow management were not widely available. As a result, many WFMSs were initially developed using custom protocols and tools. However, the current trend is to integrate the new or enhanced public tools now available. For example, recently, the LONI pipeline ***(24)*** has incorporated the Sun Grid Engine ***(27)*** to provide virtual machine access to clustered computer platforms ***(78)***; researchers ***(79)*** have built a translator to run LONI pipelines on the CONDOR platform; the BAMMFx workflow system ***(14,15)*** is built on the Java Eclipse ***(16)*** platform; the Chiron system ***(50,51)*** uses both the GriPhyN/Chimera Virtual Data System ***(62)*** and CONDOR; and BaxGrid ***(8)*** uses Globus ***(9)*** and Ninf-G ***(10)***.

An alternative to incorporating standard middleware tools into neuroimaging specific WFMSs is to adapt neuroimaging applications, format translators, and workflows to run on the newly emerging domain-independent WFMSs. For example, the Kepler workflow system ***(37)*** was derived from the Ptolemy Project ***(80)*** to support scientific workflows across multiple domains. Researchers have reported some success in implementing a scientific workflow in the BPEL workflow description language and running it on an open source implementation of a BPEL workflow engine ***(81)***. And a medical image registration workflow ***(82)*** has been implemented on the Taverna ***(83)*** WFMS that was originally designed for the bioinformatics domain.

10. The use of provenance tools in neuroimaging analysis remains somewhat limited, particularly for researchers using multiple software packages, for a number of reasons. First, data provenance can be tracked by attaching to each dataset additional meta-data fields that store a record of each operation applied to the dataset. However, this mechanism can be difficult to implement in heterogeneous computing environments, as it requires that both the processes and the data formats support the convention. Whereas some neuroimaging software and data formats, e.g., the AFNI ***(18)*** and MINC ***(72)*** systems, record provenance information in dataset headers, several of the widely used data formats (e.g., the Analyze™ 7.5 ***(84)*** format) do not have provenance fields. Second, process provenance can be documented through the workflow specification. However, the specification may not be a complete record, as it does not capture key information available only at runtime (e.g., selection of conditional branches in the workflow, dynamic

binding to task modules). One prototype for a data provenance system in the neuroimaging domain, the Chiron portal ***(50,51)***, has tackled these issues, but it has made use of a complete virtual data access system to capture provenance information. Third, use of provenance systems is limited because comprehensive provenance collection requires a level of data surveillance that is not in tune with a research laboratory environment in which investigators have unconstrained access to relocate data on filesystems (as opposed to more controlled database access) and frequently retry, repeat, undo, abort, and revise processing steps. And finally, in addition to collecting provenance information and maintaining its links to active datasets, provenance systems also need to store and format the information so that it is accessible for easy retrieval and querying both by humans (e.g., researchers who want to understand the content of a dataset) and by computer programs (e.g., programs to re-build workflows from provenance records or search records for particular procedures).

References

1. Frackowiak, R. S. J., Friston, K. J., Frith, C., Dolan, R., Price, C. J., Zeki, S., Ashburner, J., and Penny, W. D. (2003) *Human Brain Function*. Academic Press, San Diego, CA.
2. Jezzard, P., Matthews, P. M., and Smith, S. M. (Eds.) (2001) *Functional MRI: An Introduction to Methods*. Oxford University Press, Oxford; New York.
3. Alonso, G., Casati, F., Kuno, H., and Machiraju, V. (2004) *Web Services: Concepts, Architectures and Applications*. Springer-Verlag, Berlin.
4. Maer, A., Saunders, B., Unwin, R., and Subramaniam, S. (2005) Biology workbenches, in *Databasing the Brain: From Data to Knowledge (Neuroinformatics)* (Koslow, S. H., and Subramanian, S., eds.), Wiley, Hoboken, NJ, pp. 153–65.
5. Parker, S. G., Weinstein, D. W., and Johnson, C. R. (1997) The SCIRun computational steering software system, in *Modern Software Tools for Scientific Computing* (Arge, E., Brauset, A. M., and Langtangen, H. D., eds.), Birkhauser, Boston, MA, pp. 5–44.
6. von Laszewski, G., Hui Su, M. H., Insley, J. A., Foster, I., Bresnahan, J., Kesselman, C., Thiebaux, M., Rivers, M. L., Wang, S., Tieman, B., and McNulty, I. (1999) Real-time analysis, visualization, and steering of microtomography experiments at photon sources. Ninth SIAM Conference on Parallel Processing for Scientific Computing, March 22–24, San Antonio, TX.
7. Fissell, K., Tseytlin, E., Cunningham, D., Carter, C. S., Schneider, W., and Cohen, J. D. (2003) Fiswidgets: a graphical computing environment for neuroimaging analysis. *Neuroinformatics* **1**(1), 111–25.
8. Bagarinao, E., Sarmenta, L., Tanaka, Y., Matsuo, K., and Nakai, T. (2004) The application of grid computing to real-time functional MRI analysis, in *Parallel and Distributed Processing and Applications* (Cao, J., Yang, L. T., Guo, M., and

Lau, F., eds.), Lecture Notes in Computer Science vol. 3358, Springer-Verlag, Berlin, pp. 290–302.

9. Foster, I. (2005) Globus Toolkit Version 4: software for service-oriented systems, in *Network and Parallel Computing* (Jin, H., Reed, D., and Jiang, W., eds.), Lecture Notes in Computer Science vol. 3779, Springer-Verlag, Berlin, pp. 2–13.
10. Tanaka, Y., Nakada, H., Sekiguchi, S., Suzumura, T., and Matsuoka, S. (2003) Ninf-G: a reference implementation of RPC-based programming middleware for grid computing. *Journal of Grid Computing* **1**(1), 41–51.
11. Cointepas, Y., Mangin, J. F., Garnero, L., Poline, J. B., and Benali, H. (2001) BrainVISA: software platform for visualization and analysis of multi-modality brain data. *NeuroImage* **13**(6), S98.
12. Cointepas, Y., Poupon, C., Maroy, R., Rivière, D., Le Bihan, D., and Mangin, J. F. (2003) A freely available Anatomist/Brain VISA package for analysis of diffusion MR images. *NeuroImage* **19**(2, Supplement 1), e1552–e54.
13. Rivière, D., Régis, J., Cointepas, Y., Papadopoulos-Orfanos, D., Cachia, A., and Mangin, J.-F. (2003) A freely available Anatomist/BrainVISA package for structural morphometry of the cortical sulci. *NeuroImage* **19**(2, Supplement 1), e1825–e26.
14. Ooi, C., Suckling, J., and Bullmore, E. (2004) Using Eclipse to develop a modular image processing pipeline and GUI for functional MRI data analysis. Abstract, 10th Annual Meeting of the Organization for Human Brain Mapping, June 13–17, 2004, Budapest, Hungary.
15. Suckling, J., and Bullmore, E. (2004) Permutation tests for factorially designed neuroimaging experiments. *Human Brain Mapping* **22**(3), 193–205.
16. D'Anjou, J., Fairbrother, S., and Kehn, D. (2004) *The Java Developer's Guide to Eclipse*. Addison-Wesley, Boston, MA.
17. Woods, R. P., Grafton, S. T., Holmes, C. J., Cherry, S. R., and Mazziotta, J. C. (1998) Automated image registration: general methods and intrasubject, intramodality validation. *Journal of Computer Assisted Tomography* **22**, 141–54.
18. Cox, R. W., and Hyde, J. S. (1997) Software tools for analysis and visualization of fMRI data. *NMR in Biomedicine* **10**, 171–78.
19. Smith, S. M., Jenkinson, M., Woolrich, M. W., Beckmann, C. F., Behrens, T. E. J., Johansen-Berg, H., Bannister, P. R., Luca, M. D., Drobnjak, I., Flitney, D. E., Niazy, R., Saunders, J., Vickers, J., Zhang, Y., Stefano, N. D., Brady, J. M., and Matthews, P. M. (2004) Advances in functional and structural MR image analysis and implementation as FSL. *NeuroImage* **23**(S1), 208–19.
20. Frank, R. J., Damasio, H., and Grabowski, T. J. (1997) Brainvox: an interactive, multimodal visualization and analysis system for neuroanatomical imaging. *NeuroImage* **5**(1), 13–30.
21. Goebel, R. (1997) BrainVoyager 2.0: from 2D to 3D fMRI analysis and visualization. *NeuroImage* **5**, S635.

22. Strother, S. C., Anderson, J., Hansen, L. K., Kjerns, U., Kustra, R., Sidtis, J., Frutiger, S., Muley, S., LaConte, S., and Rottenberg, D. (2002) The quantitative evaluation of functional neuroimaging experiments: the NPAIRS data analysis framework. *NeuroImage* **15**(4), 747–71.
23. Friston, K. J., Holmes, A. P., Worsley, K. J., Poline, J. P., Frith, C. D., and Frackowiak, R. S. J. (1995) Statistical parametric maps in functional imaging: a general linear approach. *Human Brain Mapping* **2**, 189–210.
24. Rex, D. E., Ma, J. Q., and Toga, A. W. (2003) The LONI pipeline processing environment. *NeuroImage* **19**(3), 1033–48.
25. Shattuck, D. W., and Leahy, R. M. (2002) BrainSuite: an automated cortical surface identification tool. *Medical Image Analysis* **8**(2), 129–42.
26. Dale, A. M., Fischl, B., and Sereno, M. I. (1999) Cortical surface-based analysis: I. Segmentation and surface reconstruction. *NeuroImage* **9**(2), 179–94.
27. Grid Engine Project. Sun Grid Engine. Electronic resource at http://gridengine.sunsource.net/.
28. Neu, S. C., Valentino, D. J., and Toga, A. W. (2005) The LONI debabeler: a mediator for neuroimaging software. *NeuroImage* **24**(4), 1170–79.
29. Kimberg, D. Y., and Aguirre, G. K. (2003) VoxBo: a flexible architecture for functional neuroimaging. The Human Brain Project Annual Conference, May 12–13, Bethesda, MD.
30. The MathWorks Inc. MATLAB. Electronic resource at http://www.mathworks.com/.
31. Gosling, J., Joy, B., Steele, G., and Bracha, G. (2005) *The Java Language Specification*, 3rd edition. Addison-Wesley, Upper Saddle River, N.J.
32. Snir, M., Otto, S., Huss-Lederman, S., Walker, D., and Dongarra, J. (1998) *MPI-The Complete Reference*, 2nd edition. MIT Press, Cambridge, MA.
33. Baru, C., Moore, R., Rajasekar, A., and Wan, M. (1998) The SDSC storage resource broker. Proceedings of the CASCON'98 Conference, November 30–December 3, Toronto, Canada.
34. Allcock, B., Bester, J., Bresnahan, J., Chervenak, A. L., Foster, I., Kesselman, C., Meder, S., Nefedova, V., Quesnal, D., and Tuecke, S. (2002) Data management and transfer in high performance computational grid environments. *Parallel Computing* **28**(5), 749–71.
35. The National Center for Supercomputing Applications (NCSA). Hierarchical Data Format 5 (HDF5). Electronic resource at http://hdf.ncsa.uiuc.edu/HDF5/.
36. Zhao, Y., Dobson, J., Foster, I., Moreau, L., and Wilde, M. (2005) A notation and system for expressing and executing cleanly typed workflows on messy scientific data. *SIGMOD Record* **34**(3), 37–43.
37. Ludäscher, B., Altintas, I., Berkley, C., Higgins, D., Jaeger-Frank, E., Jones, M., Lee, E., Tao, J., and Zhao, Y. (2006) Scientific workflow management and the Kepler system. *Concurrency and Computation: Practice and Experience* **18**(10), 1039–1065.

38. Swedlow, J. R., Goldberg, I., Brauner, E., and Sorger, P. K. (2003) Informatics and quantitative analysis in biological imaging. *Science* **300**, 100–02.
39. JFtp FTP Client. Electronic resource at http://sourceforge.net/projects/gftp/.
40. ActiveBPEL, LLC. BPEL OnDemand Validation Service. Electronic resource at http://www.activebpel.org/code/validator/.
41. Karamanolis, C. T., Giannakopoulou, D., Magee, J., and Wheater, S. M. (2000) Model checking of workflow schemas. Proceedings of the 4th international Conference on Enterprise Distributed Object Computing (EDOC). IEEE Computer Society, September 25–28, Washington, DC.
42. Moreau, L., Zhao, Y., Foster, I., Voeckler, J., and Wilde, M. (2005) XDTM: the XML dataset typing and mapping for specifying datasets. *Proceedings of the 2005 European Grid Conference (EGC'05)*, February 2005, Amsterdam, Netherlands.
43. Foster, I., and Kesselman, C. (Eds.) (1999) *The Grid, Blueprint for a New Computing Infrastructure*. Morgan Kaufmann Publishers, San Francisco.
44. Asadzadeh, P., Buyya, R., Kei, C. L., Nayar, D., and Venugopal, S. (2005) Global grids and software toolkits: a study of four grid middleware technologies, in *High Performance Computing: Paradigm and Infrastructure* (Yang, L., and Guo, M., eds.), Wiley Press, New Jersey.
45. Bose, R., and Frew, J. (2005) Lineage retrieval for scientific data processing: a survey. *ACM Computing Surveys* **37**, 1–28.
46. Miles, S., Groth, P., Branco, M., and Moreau, L. (2005) The requirements of recording and using provenance in e-Science experiments. Technical Report Electronics and Computer Science, University of Southampton.
47. Simmhan, Y. L., Plale, B., and Gannon, D. (2005) A survey of data provenance in e-science. *SIGMOD Record* **34**(3), 31–36.
48. Wellcome Department of Imaging Neuroscience. SPM Extensions, Batch Utilities. Electronic resource at http://www.fil.ion.ucl.ac.uk/spm/ext/#batch_utils/.
49. Taylor, J., Worsley, K., Brett, M., Cointepas, Y., Hunter, J., Millman, J., Poline, J. B., and Perez, F. (2005) BrainPy: an open source environment for the analysis and visualization of human brain data. Abstract, 11th Annual Meeting of the Organization for Human Brain Mapping, June 12–16, Toronto, Canada.
50. Zhao, Y., Wilde, M., Foster, I., Voeckler, J., Dobson, J., Glibert, E., Jordan, T., and Quigg, E. (2000) Virtual data grid middleware services for data-intensive science. *Concurrency and Computation: Practical Experience* **00**, 1–7.
51. Zhao, Y., Wilde, M., Foster, I., Voeckler, J., Jordan, T., Quigg, E., and Dobson, J. (2004) Grid middleware services for virtual data discovery, composition, and integration. 2nd Workshop on Middleware for Grid Computing, Oct. 18–22, Toronto, Canada.
52. Montreal Neurological Institute. Run Pipeline in Parallel (RPPL): Brain Imaging Centre Pipelining System. Electronic resource at http://www.bic.mni.mcgill.ca/~jason/rppl/rppl.html/.

53. Meyer, L. A. V. C., and Mattoso, M. L. Q. (2004) Parallel Strategies for Processing Scientific Workflows. Technical Report ES-646/04 Graduate School and Research in Engineering, Alberto Luiz Coimbra Intitute, Federal University of Rio de Janeiro.
54. Yu, J., and Buyya, R. (2005) A taxonomy of scientific workflow systems for grid computing. *SIGMOD Record* **34**(3), 44–49.
55. Slominski, A., and Laszewski, G. V. Scientific Workflows Survey. Electronic resource at http://www.extreme.indiana.edu/swf-survey/.
56. Tannenbaum, T., Wright, D., Miller, K., and Livny, M. (2002) Condor: a distributed job scheduler, in *Beowulf Cluster Computing with Linux* (Sterling, T., ed.), MIT Press, Cambridge, MA, pp. 307–50.
57. Hoheisel, A. (2006) User tools and languages for graph-based grid workflows. *Concurrency and Computation: Practice and Experience* **18**(10), 1101–1113.
58. Ailamaki, A., Ioannidis, Y. E., and Livny, M. (1998) Scientific workflow management by database management. 10th International Conference on Scientific and Statistical Database Management, July 1–3, Capri, Italy.
59. Shankar, S., Kini, A., DeWitt, D. J., and Naughton, J. (2005) Integrating databases and workflow systems. *SIGMOD Record* **34**(3), 5–11.
60. Andrews, T., Curbera, R., Dholakia, H., Goland, Y., Klein, J., Leymann, F., Liu, K., Roller, D., Smith, D., Thatte, S., Trickovic, I., and Sanjiva, W. S. (2003) Business Process Execution Language for Web Services (BPEL4WS), Version 1.1 Technical specification. Electronic resource at ftp://www6.software.ibm.com/software/developer/library/ws-bpel.pdf.
61. Krishnan, S., Wagstrom, P., and von Laszewski, G. (2002) GSFL: a workflow framework for grid services. Technical Report ANL/MCS-P980-0802 Argonne National Laboratory.
62. Foster, I., Voeckler, J., Wilde, M., and Zhao, Y. (2002) Chimera: a virtual data system for representing, querying, and automating data derivation. 14th Conference on Scientific and Statistical Database Management, July 24–26, Edinburgh, Scotland, UK.
63. Deelman, E., Blythe, J., Gil, Y., Kesselman, C., Mehta, G., Vahi, K., Arbree, A., Cavanaugh, R., Blackburn, K., Lazzarini, A., and Koranda, S. (2003) Mapping abstract complex workflows onto grid environments. *Journal of Grid Computing* **1**, 25–39.
64. Ludäscher, B., Altintas, I., and Gupta, A. (2003) Compiling abstract scientific workflows into web service workflows. 15th Annual Conference on Scientific and Statistical Database Management (SSDBM), July 9–11, Cambridge, MA.
65. Medeiros, C. B., Perez-Alcazar, J., Digiampietri, L., Pastorello Jr., G. Z., Santanche, A., Torres, R. S., Madeira, E., and Bacarin, E. (2005) WOODSS and the Web: annotating and reusing scientific workflows. *SIGMOD Record* **34**(3), 18–23.

66. Strother, S. C. (invited paper, 2006) Evaluating fMRI preprocessing pipelines, *IEEE Engineering in Medicine and Biology Magazine* **25**(2), 27–41.
67. Internet Analysis Tools Registry (IATR). Electronic resource at http://www.cma.mgh.harvard.edu/iatr/.
68. Rorden, C., and Brett, M. (2000) Stereotaxic display of brain lesions. *Behavioural Neurology* **12**, 191–200.
69. Van Horn, J. D., Grafton, S. T., Rockmore, D., and Gazzaniga, M. S. (2004) Sharing neuroimaging studies of human cognition. *Nature Neuroscience* **7**, 473–81.
70. Cox, R. W., Ashburner, J., Breman, H., Fissell, K., Haselgrove, C., Holmes, C. J., Lancaster, J. L., Rex, D. E., Smith, S. M., Woodward, J. B., and Strother, S. C. (2004) A (sort of) new image data format standard: NifTI-1. Abstract, 10th Annual Meeting of the Organization for Human Brain Mapping, June 13–17, Budapest, Hungary.
71. National Electrical Manufacturers Association (NEMA). Digital Imaging and Communications in Medicine (DICOM). Electronic resource at http://medical.nema.org/.
72. Vincent, R. D., Janke, A., Sled, J. G., Baghdadi, L., Neelin, P., and Evans, A. C. (2005) MINC 2.0: a modality independent format for multidimensional medical images. 11th Annual Meeting of the Organization for Human Brain Mapping, June 12–16, Toronto, Canada.
73. Gardner, D., Knuth, K. H., Abato, M., Erde, S. M., White, T., DeBellis, R., and Gardner, E. P. (2001) Common data model for neuroscience data and data model interchange. *Journal of the American Medical Informatics Association: JAMIA*. **8**, 17–31.
74. Bowden, D. M., and Dubach, M. (2005) Neuroanatomical nomenclature and ontology, in *Databasing the brain: From Data To Knowledge (Neuroinformatics)* (Koslow, S. H., and Subramanian, S., eds.), Wiley, Hoboken, NJ, pp. 27–45.
75. Bowden, D. M., and Dubach, M. F. (2003) NeuroNames 2000. *Neuroinformatics* **1**(1), 43–60.
76. Frank, S., Moore, J., and Eils, R. (2004) A question of scale. Bringing an existing bio-science and workflow engine to the grid. Global Grid Forum (GGF) 10th Workflow Workshop, March 9, 2004, Berlin, Germany.
77. Pope, A. (1998) *The CORBA Reference Guide: Understanding the Common Object Request Broker Architecture*. Addison-Wesley, Reading, MA.
78. Pan, M. J., Rex, D. E., and Toga, A. W. (2005) The LONI pipeline processing environment: improvements for neuroimaging analysis research. Abstract, 11th Annual Meeting of the Organization for Human Brain Mapping, June 12–16, Toronto, Canada.
79. Fox, A. S., Farrellee, M., Roy, A., Davidson, R. J., and Oakes, T. R. (2005) Condor: managing computationally intensive jobs from a computer science perspective.

Abstract, 11th Annual Meeting of the Organization for Human Brain Mapping, June 12–16, Toronto, Canada.
80. Lee, E. A. (2003) Overview of the Ptolemy Project. Technical Report UCB/ERL M03/25 University of California, Berkeley.
81. Emmerich, W., Butchart, B., Chen, L., Wassermann, B., and Price, S. L. (2005) Grid Service Orchestration using the Business Process Execution Language (BPEL). Technical Report Department of Computer Science, University College London.
82. Glatard, T., Montagnat, J., and Pennec, X. (2005) Grid-Enabled Workflows for Data Intensive Medical Applications. 18th IEEE Symposium on Computer-Based Medical Systems (CBMS'05), Dublin, Ireland.
83. Oinn, T., Addis, M., Ferris, J., Marvin, D., Senger, M., Greenwood, M., Carver, T., Glover, K., Pocock, M. R., Wipat, A., and Li, P. (2004) Taverna: a tool for the composition and enactment of bioinformatics workflows. *Bioinformatics* **20**(17), 3045–54.
84. Mayo Clinic. Software: Analyze Program. Electronic resource at http://www.mayo.edu/bir/Software/Analyze/AnalyzeTechInfo.html/.

Appendix

A. Fiswidgets Workflow Description Language, XML Version

Listed below is an example of the Fiswidgets XML workflow description language. This example has been taken from the XML specification of the workflow shown in **Fig. 1**; the XML file has been excerpted for brevity. Shown here are the XML header section, the section for the MakeJpeg application, and the start of the AirAlignlinear section.

```
<?xml version="1.0" encoding="UTF-8"?>
<!DOCTYPE FisDesktop PUBLIC "-//Fiswidgets//XML Serialized
Document 1.0//EN" "fiswidgets.dtd">
<FisDesktop name=" GlobalDesktop" date="Jan 29, 2006"
user="fissell" platform="Irix 6.5 mips" postit=""
 logging="true"logfile="Desk.log" timing="false"
 interrupt="true" winmode="1"minimized="true"
 usebreaks="false" use_remote="false"
remote_server="localhost" remote_user="fissell"
script_file="/scratch/data/script.csh" script_mode="off"
shell="/bin/csh" shell_flags="-ef" comment_char="#"
line_char="\ " quote_char="'"protected_chars="{}[]"
 add_header="false" author=""script_date="" comment="">
   <Titles>
```

```
Foreach,SetLogName,MakeJpeg,AirAlignlinear,Niscorrect,Format
Convert,Afni3dDeconvolve,LogParser,EndForeach
   </Titles>
   <Breaks>
      false,false,false,false,false,false,false,false,false
   </Breaks>
   <Visible>
      false,false,false,false,false,false,false,false,false
   </Visible>

   <FisBase name=" MakeJpeg" postit="">
      <JLabel enabled="true" postit="">
         Anz Image file:
      </JLabel>
      <JTextField name="Anz Image file:" enabled="true"
editable="true"postit="">
         /scratch/data/$S/raw_images/func_00001.img
      </JTextField>
      <JButton enabled="true" postit="">
         Browse
      </JButton>
      <JLabel enabled="true" postit="">
         Output jpg file:
      </JLabel>
      <JTextField name="Output jpg file:" enabled="true"
editable="true"postit="">
         /scratch/data/$S/logs/raw_image1.jpg
      </JTextField>
      <JButton enabled="true" postit="">
         Browse
      </JButton>
      <FisAdvancedButton enabled="true" selected="false"
name="AdvancedImage Settings..." postit="">
      <JLabel enabled="true" postit="">
         Image Quality (0-100)%
      </JLabel>
      <JSlider enabled="true" postit="">
         100
      </JSlider>
      <JLabel enabled="true" postit="">
         Image Magnification (0-200)%
      </JLabel>
      <JSlider enabled="true" postit="">
         100
      </JSlider>
      <JCheckBox name="Add overlay image" enabled="true"
```

```
postit="">
        false
     </JCheckBox>
     <JTextField name="Add overlay image" enabled="false"
editable="false" postit="">
        /scratch/data
     </JTextField>
     <JButton enabled="false" postit="">
        Browse
     </JButton>
        <JLabel enabled="true" postit="">
         Lower threshold
     </JLabel>
     <JTextField name="    Lower threshold    "
enabled="false"editable="false" postit="">
        0
     </JTextField>
     <JLabel enabled="true" postit="">
        Upper threshold
     </JLabel>
     <JTextField name=" Upper threshold " enabled="false"
editable="false" postit="">
        32767
     </JTextField>
     <JCheckBox name="Overwrite existing files"
enabled="true"postit="">
        false
     </JCheckBox>
     <JCheckBox name="View results" enabled="true"
postit="">
        false
        </JCheckBox>
     </FisAdvancedButton>
  </FisBase>

  <FisBase name=" AirAlignlinear" postit="">
     <JComboBox name="View results" enabled="true"
editable="false"selection="Image directory" selected="1"
postit="">
        Image file(s);Image directory;Image filelist
     </JComboBox>
     <JComboBox name="View results" enabled="true"
editable="true"selection="/scratch/data/$S/raw_images"
selected="0"postit="">
        /scratch/data/$S/raw_images
     </JComboBox>
```

```
      <JButton enabled="true" postit="">
         Browse
      </JButton>
      <JLabel enabled="true" postit="">
         Standard file
      </JLabel>
      <JTextField name="Standard file" enabled="true"
editable="true"postit="">
         /scratch/data/$S/raw_images/func_00001.img
      </JTextField>
      <JButton enabled="true" postit="">
         Browse
      </JButton>
      <JLabel enabled="true" postit="">
         Air dir
      </JLabel>
      <JTextField name="Air dir" enabled="true"
editable="true"postit="">
         /scratch/data/$S/air_files/
      </JTextField>
      <JButton enabled="true" postit="">
         Browse
      </JButton>
      <JLabel enabled="true" postit="">
         Logs dir
      </JLabel>
      <JTextField name="Logs dir" enabled="true"
editable="true"postit="">
         /scratch/data/$S/logs/
      </JTextField>
      <JButton enabled="true" postit="">
         Browse
      </JButton>
      <JCheckBox name="Reslice" enabled="true" postit="">
         true
      </JCheckBox>
      <FisAdvancedButton enabled="true" selected="false"
name="ResliceSettings..." postit="">
      <JComboBox name="Reslice" enabled="false"
editable="false"selection="Air file(s)" selected="0"
postit="">
            Air file(s);Air directory;Air filelist
         </JComboBox>
         <JComboBox name="Reslice" enabled="false"
editable="true"selection="/scratch/data" selected="0"
postit="">
```

```
            /scratch/data
        </JComboBox>

---- remaining XML not listed ----

</FisDesktop>
```

B. Fiswidgets Workflow Description Language, Text Display

Listed below is an example of the Fiswidgets XML workflow description language formatted as text for easier reading. This example has been taken from the XML specification of the workflow shown in **Fig. 1** and listed in **Appendix A** above; the text file has been excerpted for brevity. Shown here are the header section and the sections for the MakeJpeg and AirAlignlinear applications. The Fiswidgets *Tools/FXParser* utility translates from XML to this text output.

```
GlobalDesktop

Created on Jan 29, 2006
Created by user fissell
Created on Irix 6.5 mips platform
  Titles : Foreach, SetLogName, MakeJpeg, AirAlignlinear,
  Niscorrect, FormatConvert, Afni3dDeconvolve,
  LogParser, EndForeach

MakeJpeg
    Anz Image file: :
      /scratch/data/$S/raw_images/func_00001.img,
    Output jpg file: :
      /scratch/data/$S/logs/raw_image1.jpg,

    Advanced Image Settings...
      Image Quality (0-100)% :      100,
      Image Magnification (0-200)%  :   100,
      Add overlay image :              is not checked,
 /scratch/data,
      Lower threshold :          0,
      Upper threshold :          32767,
      Overwrite existing files :   is not checked,
      View results :                   is not checked,

AirAlignlinear
    Image directory, /scratch/data/$S/raw_images
```

```
    Standard file :
      /scratch/data/$S/raw_images/func_00001.img,
    Air dir :       /scratch/data/$S/air_files/,
    Logs dir :       /scratch/data/$S/logs/,
    Reslice :       is checked,

    Reslice Settings...
    Air file(s),      /scratch/data
      Image directory/file : /scratch/data/$S/air_images,
      File prefix :       is not checked, r,
      Overwrite existing files :   is not checked,

      Options...

         Use intensity scale factor :    is not checked, 1.0,
         JCheckBox :    is not checked,
      JComboBox :        Image file(s),
               /scratch/data
Trilinear,
         with half-window width (voxels) XYZ :
         JTextField :       , , ,
         Keep voxel dimensions same :   is selected,
         Use cubic voxels :                is not selected,
         Specify image attributes :   is not selected,
      X :            is not checked,
      Y :            is not checked,
      Z :            is not checked,
      New matrix dim(voxels) :     , , ,
      Voxel size(mm) :     , , ,
      Origin shift(voxels) :     , , ,

  Suppress output messages :   is not checked,
  Model :               3-d Rigid Body (6),
  Cost :                Least sq w/ intensity scaling,

  Preprocessing options...

    Standard file mask :           ,
    Standard threshold :           $T,
    Standard file partitions :  1,
    Standard file FWHM :
    X :            0.0,
    Y :            0.0,
    Z :            0.0,
    Use standard file options for reslice file :
              is checked,
  Mask file(s),    /scratch/data
```

```
    Reslice threshold :         500,
    Reslice file partitions :          1,
    Reslice file FWHM :
    X :             0.0,
    Y :             0.0,
    Z :             0.0,
    Assume noninteraction of spatial parameter derivatives
:            is not checked,
    Enable pre-alignment interpolation :
             is not checked,

  Optimization options...

    Sampling init/final/dec :    81, 1, 3,
    Convergence threshold : 9.999999747378752E-6,
    Repeated iterations :          25,
    Halt after n iterations :    5,
    Alt strategy after m iterations : 6,


  Init/Term files...

   Initialization File :   is not checked,
                /scratch/data,
    Termination File : is not checked,
                /scratch/data,
    Warp Format Termination File : is not checked,
                /scratch/data,
  Scaling Initialization File : is not checked,
                /scratch/data,
  Scaling Termination File : is not checked,
                /scratch/data,
  Warp Format Scaling Termination File :
                is not checked,
                /scratch/data,
  Permit overwrite of termination files :
                is not checked,

  Standard Threshold : $T,
  Verbose mode : is not checked,

---- remaining text not listed ----
```

15

Databasing Receptor Distributions in the Brain

Rolf Kötter, Jürgen Maier, Wojciech Margas, Karl Zilles,
Axel Schleicher, and Ahmet Bozkurt

Summary

Receptor distributions in the brain are studied by autoradiographic mapping in brain slices, which is a labor-intensive and expensive procedure. To keep track of the results of such studies, we have designed CoReDat, a multi-user relational database system that is available for download from www.cocomac.org/coredat. Here, we describe the data model and provide an architectural overview of CoReDat for the neuroscientist who wants to use this database, adapt it for related purposes, or build a new one.

Key Words: Autoradiography; experimental data; ligands; mapping; publications; receptors; relational database; server client architecture.

1. Introduction

Databases are essential tools for organizing and analyzing the increasing amounts of data generated in neuroscientific experiments *(1)*. Neuroscience databases can be broadly divided into databases of experimental data, knowledge databases, and software tools (see Neuroscience Database Gateway: http://www.sfn.org/ndg), where experimental databases may record unfiltered raw data (as measured by some detector), corrected or converted data (preprocessed before analysis), results of data analysis (from descriptive to inferential statistics), or selected results for publication (e.g., text-based databases). The multitude of descriptive levels and the method dependence of neuroscience data are not satisfactorily accommodated by general database designs

From: *Methods in Molecular Biology, vol. 401: Neuroinformatics*
Edited by: Chiquito Joaquim Crasto © Humana Press Inc., Totowa, NJ

but require specific developments or adaptations tailored to the respective type of data and the conditions of their acquisition.

In this chapter, we describe the design and implementation of a database for organizing quantitative values of spatially distributed markers within the brain whose density is noted with reference to some brain map. We refer specifically to the experimental results from autoradiographic ligand-binding experiments on brain slices for the analysis of receptor distributions in the mammalian brain (*see* **Note 1**).

We have implemented our database design as a prototype system entitled CoReDat (Collation of Receptor Data) in analogy to previous databases from our group: CoCoDat *(2)* and CoCoMac *(3,4)*. The purpose and the contents of CoReDat, however, differ significantly from our previous ones. The specific focus of the present database arises from the time-consuming performance of autoradiographic experiments in several mammalian species at the Vogt Brain Research Institute in Düsseldorf, which makes it desirable to keep track of animal details, experimental conditions, acquired data, and published results over years and decades. Thus, CoReDat is oriented toward the laboratory use and the collation of experimental data during the course of the experiments until publication. By contrast, both CoCoDat and CoCoMac were designed for extraction and meta-analysis of data from the published literature.

The structure and the data model of CoReDat are largely determined by the experimental approach and procedures. A typical experimental project aims at the characterization of a set of adjacent and functionally related brain regions by the differential binding of multiple radioactively labeled receptor ligands *(5–9)*. For a typical study, up to ten brain hemispheres are freshly obtained at post-mortem, cut into regional blocks, and frozen in the unfixed state. Selected blocks are consecutively cryosectioned in a standard plane of orientation, and alternating slices are exposed to the different radioactively labeled receptor ligands with or without a surplus of unlabeled displacer for the determination of total and unspecific binding, respectively (*see* **Fig. 1**). Apposition of the labeled brain sections to films for periods of weeks to months delivers gray scale film images with a spatial resolution of 50 μm, which corresponds roughly to the size of layers and columns. Further slices undergo standard histological processing for the manual delineation of brain regions such as (sub-) nuclei or cortical areas and layers (cyto-, myeloarchitectonics). Through overlay of the autoradiographic film image with the histological masks, we determine regional intensity values that are converted into receptor concentrations in units of fmol/mg protein by alignment with co-exposed scales of standard radioactivity. For each ligand and hemisphere, receptor concentrations from the same

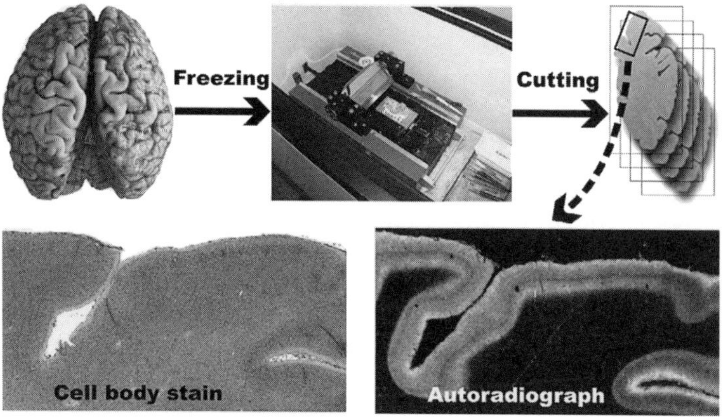

Fig. 1. Overview of tissue-processing steps for autoradiographic assessment of regional receptor distributions in the brain (see text for details).

histological region are averaged. Analyses include a typical multifactorial and multivariate analysis of variance to examine the differences in receptor densities between areas over and above the variation between individual hemispheres. Similarities between areas and ligands would also be visualized as characteristic 'fingerprints' ***(10)*** or neighborhood relationships in plots with reduced dimensionality ***(7,9)***.

The purpose of this chapter is to expose the design concepts and specific implementation of a receptor database for the neuroscientist who wants to use this database, adapt it for related purposes, or build a new one. In **Subheading 2**, we describe how the database and related material can be accessed. The **Subheading 3** exposes the data model and systems architecture demonstrating its applicability with reference to specific examples from CoReDat.

2. Materials

The current version of the CoReDat database implementation and related information can be accessed from http://www.cocomac.org/coredat. This page makes available for download the backend server database implemented in MySQL and MS Access, as well as the client frontend implemented in MS Access only.

In addition, we provide some documentation for download.

Further relevant information is contained in the extensive documents relating to the CoCoDat and CoCoMac databases available from http://www.cocomac.org/.

3. Methods

3.1. General Rationale

The general goal in the design of CoReDat was to implement a flexible tool that would provide a systematic laboratory record of material collected, experiments performed, and data generated. In addition, it should provide a basis for systematic analysis and comparison of data from different experiments, individuals, and species with the option of comparison and integration with further data and other data modalities (e.g., connectivity and imaging data) at a later stage.

The database design followed four specific objectives: (1) applicability, (2) transparency, (3) flexibility, and (4) usability. We pursued these objectives in the following ways:

1. The tables constituting the database were derived in a comprehensive survey of existing notes, procedures, and publications of the lab as well as published data from other labs. This was a prerequisite for an adequate representation of procedures and data. It also proved advantageous for the data organization and description to reflect biological concepts and mechanistic principles. Relationships between the tables were set according to logical interdependences and practical considerations.
2. Transparency was ensured by clear attribution of experimental data to identified projects in the lab with unambiguous references to lists of corresponding methods and materials. We considered the averaged data from several sections of the same brain region in one hemisphere as the smallest useful unit of information for further analysis. Thus, we abstain from storing unfiltered raw data but still maintain a level that permits re-analyses of the pre-processed data for a variety of scientific questions and types of analyses.
3. Our database can be flexibly extended or altered without having to re-design the entire database or to re-code the data contained. The tables are grouped in modular themes, such as literature, brain maps, methods, and data. Most of these have been used previously in other database systems with different contexts but very similar designs. The client-server concept is useful to evolve independently the frontend for user interaction and the backend for data management; therefore, the system is easily adjusted to different hardware and organizational environments. The connection to other independent databases is made possible through an XML-based self-description of the database, which can be automatically processed by other systems.
4. The day-to-day usability is perhaps the most critical component for the success of a laboratory-oriented database. At a basic level, this requires intuitive interfaces and a data organization that fits with experimental procedures, particularly where certain data sets are acquired together and others are obtained in successive steps.

This intuitive aspect is supplemented by adequate documentation of the database features, interfaces, and so on. We tested the usability of the database storing a significant amount of data from recent autoradiographic experiments. Additional practical aspects are the definition of access rules, the implementation of checks for data integrity, and the establishing of backup procedures, which are not immediately visible to the user but of clear long-term importance.

3.2. Data Model

To fulfill the needs of our design objectives, we had to create a correct and precise data model that conformed to the rules of relational data modeling *(11)*. It was based on an entity relationship model of the data and the procedures carried out in the experimental settings, essentially forming an adequate representation of the "real-world" problem. Therefore, the database consists of sets of related tables to map the entities and the relationships between them. While the tables contain the data (*see* **Table 1**), adding relationships between the tables is useful not only for logical reasons but also to gain access to some of the functionality implemented in relational database systems, such as cascaded updates and deletes. Major tables and parts of the relational model are shown in **Table 1** and **Fig. 2**.

As a first and crucial step in the procedure of designing the database, it is necessary to develop a so-called "semantic data model," which provides the conceptual framework for the implementation process. From this point, we devised the formal objects and integrity rules of the model as well as the formal operations on the data. The objects or entity types were characterized by their attributes with specific domains and relationships to other entity types. Formally, an entity type E with its attributes A and domains D is given by:

$$E(A_1 : D_1, \ldots, A_m : D_m).$$

These objects are mapped by database tables consisting of columns with defined properties. Examples for the implementation of this principle are given in **Subheadings 3.3 and 3.4.** The physical implementation of the whole set of entity types in a relational database system constitutes our working model.

The physical implementation of our data model was carried out with both MySQL® 4.1.10 and MS Access XP®. The tables were cross-referenced using single or combined identifiers that have unique values for each record.

3.3. Database Organization

The database architecture was organized around three general themes (*see* **Fig. 2** and **Table 1**): literature, brain maps, and project-related data. The

Table 1
Tables Implemented in the CoReDat Database and Their Contents

Table	Content
Literature	
Literature	ID, title, year of publication, publication type, abstract, keywords, comments and ID of data collator, URL for additional data accessible on the Internet
Literature Abbreviations Journals	Predefined list of journal name abbreviations
Literature Authors	Authors' initials and last names
Literature Book Chapters	Page numbers, editors, book title, publisher, place of publishing
Literature Books	Publisher, place of publishing
Literature Journal Articles	Journal name, volume and page numbers, PubMed ID for external link
Literature Link Table	Links between literature IDs and authors (implementing m-n relationship)
Mapping	
BrainMaps	Entered BrainMaps (General Map and other user-specified parcellation schemes)
BrainMaps BrainSites BrainSiteTypes	Defined BrainSiteTypes
BrainMaps BrainSiteAcronyms	BrainsiteAcronyms with their full description
BrainMaps BrainSites	BrainSites given by a combination of BrainMap, BrainSiteAcronym, and BrainSiteType
Project	
Project	Details of a project as the top level for data
Institution	Laboratory details
Contact Person	Experimentator
Project-Literature Link Table	Links between projects and literature IDs (implementing m-n relationship)
Methods	
Methods_Animal	Details of the individual animal used
Methods_Autoradiography	Details of the autoradiography method

Methods_Displacer	List of displacer substance, parameter value
Methods_Fixation	List of fixation methods, parameter value
Methods_Histology	Details of the histological procedure
Methods_Ligand	List of ligand substances, parameter value
Methods_Species	List of species, parameter value
Methods_Receptor	List of receptors, parameter value
Methods_SamplingMethods	List of experimental methods for data sampling, parameter value
Methods_TissueBlock	Details of the specific tissue where the experimental data are measured, identified by a specific ID
Methods_TissueGroup	Details of slides and substances used in a specific tissue block
Experimental Data	
ExpData	Details of the experimental findings for one set of data, identified by a specific ID
Miscellaneous	
dbCollators	Persons, entering data
Administration_Proofreading_MappingData	Status of proofreading entered mapping data
Administration_Proofreading_ReceptorData	Status of proofreading entered experimental data

concepts for representing the themes of brain maps and literature are almost identical to those implemented in the extensively documented CoCoMac database *(3)*.

Literature-related tables provide full bibliographic references of journal articles, books, and book chapters supported by additional tables for parameters such as a list of authors and link tables that implement many-to-many (m-n) relationships between database tables (*see* **Table 1** and **Fig. 3**). In the present database, the literature entries have two conceptual aspects. The first aspect is similar to the use in our previous databases and lists publications that contain information either about brain maps or about actual experimental data. The second aspect is new and concerns the housekeeping of publications that arise from data contained in the experimental database tables. The significance of

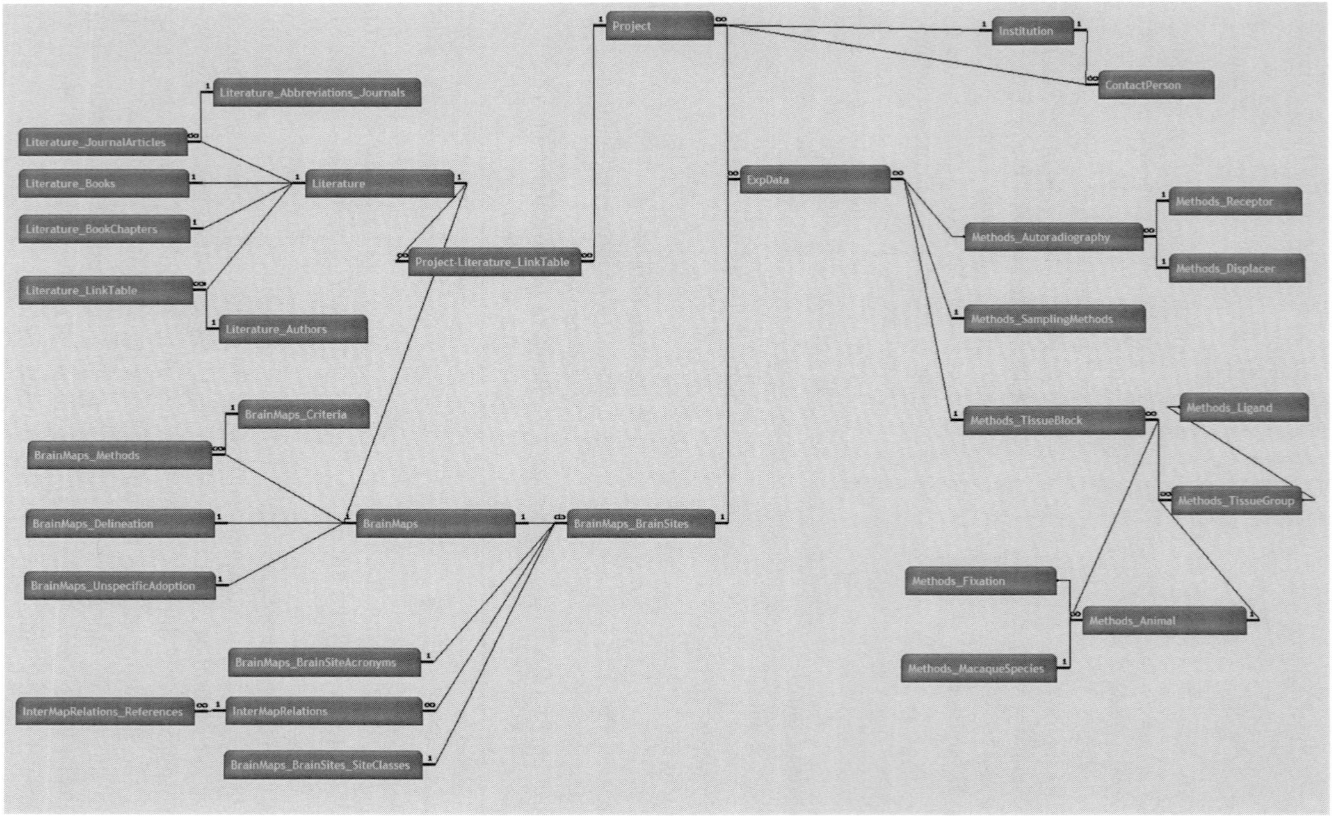

Fig. 2. Diagram of relationships between the tables of the database. The details of the tables are excluded to focus on the relational structure.

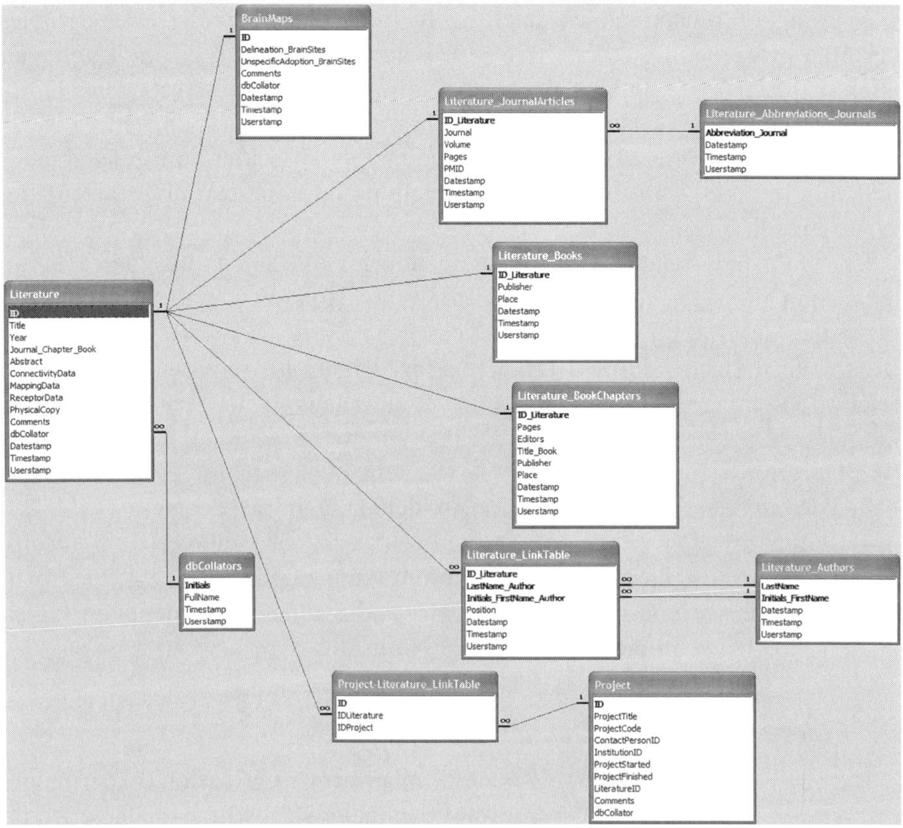

Fig. 3. Diagram with detailed view on the direct relations of the literature table (minor tables such as *Administration_Proofreading* are omitted). The primary key fields are set in bold and the types of the foreign key relations are marked on the lines (∞ represents many entries). Project-related tables are discussed in the experimental data section.

this distinction becomes clear when we consider the fundamental organizing units in a publication-based versus an experiment-based database: the former is centered on individual publications from which all data are derived and the latter is based on projects that lead to the generation of data, which may give rise to one or more publications. In the present database, therefore, the literature tables link both to the brain map-related tables and, through the project, to the group of experimental data tables.

A group of related tables, such as literature- or mapping-related tables, are often processed together. Joint processing is facilitated by the implementation of forms that combine the related information from separate tables in a common view. For instance, the form *Literature* provides a unified view and intuitive access to related information in the majority of literature-related tables including indirectly linked author information and administrative information on proofreading (*see* **Fig. 4**).

The brain map-related tables contain information about the nomenclature, delineation methods, and references for brain structures that form the units of spatial description in brain sections (*see* **Note 2**). CoReDat relies on cytoarchitectonically defined brain regions, which are referenced to specific descriptions in the published literature or alternatively to user-defined partitioning schemes as previously described for the CoCoMac database. Thus, a relationship enforces that any measurement of a receptor concentration is related to an identified brain region, which may be as coarse-grained as a major brain division or as fine-grained as a cortical sublayer or a specific cell population. Relations between different brain regions are kept track of in additional tables where the user may enter and evaluate statements on logical relationships between brain regions (*see* **Note 3**).

3.4. Experiment-Related Data

The project-related data tables comprise project information, experimental methods, experimental data, and several parameter sets, which characterize the experiments and are therefore largely specific to the autoradiographic receptor-binding studies.

Following the description of the data model above, the entity type "Project" can be described as (*see* "Project" table in **Fig. 5**):

Project (*ID*, ProjectTitle, ProjectCode, *ContactPersonID*, *InstitutionID*, ProjectStarted, ProjectFinished, Comments, *dbCollator*).

Some attributes of an entity type have special qualities as they serve for unique identification of an entity or for linking the entity type to others by means of relationships. In the nomenclature of the database system, these special attributes are columns with the properties of primary keys or foreign keys. In our example, the "Project" table has the primary key *ID* and the foreign keys *ContactPersonID*, *InstitutionID*, and *dbCollator*.

The "Experimental Data" table is the central hub pulling together specific method-related information from several other tables and containing the concentration value as the most important experimental datum (*see* **Fig. 6**). The table

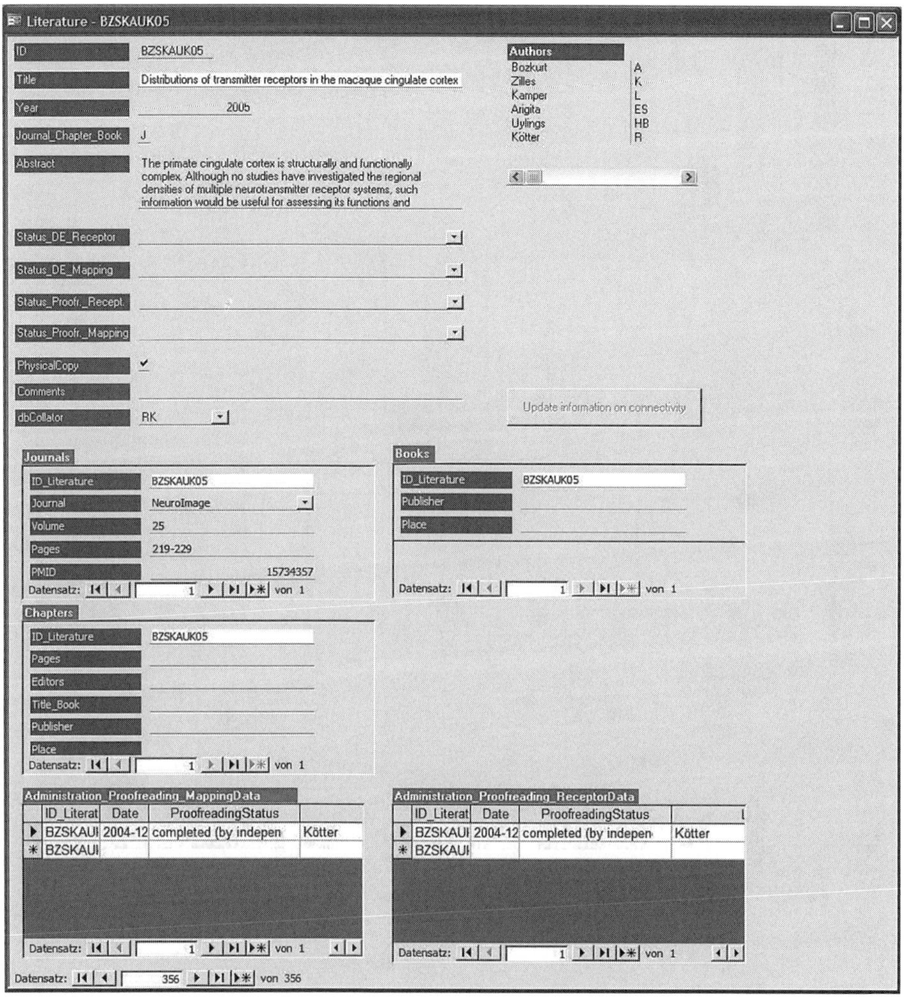

Fig. 4. Form *Literature* providing unified view of and access to the majority of literature-related data tables listed in **Fig.** 3. The main form and its subforms show related data for a single literature item.

combines references to the ID of the tissue block (IDBlock), of the autoradiographic details (IDAutoradiography), of the brain region under investigation (IDBrainSite), of the Sampling Method, and the person who entered the data (dbCollator). Further fields (table columns) contain the primary key (*ID*), the concentration value, the size of the area to which the value applies (Size of

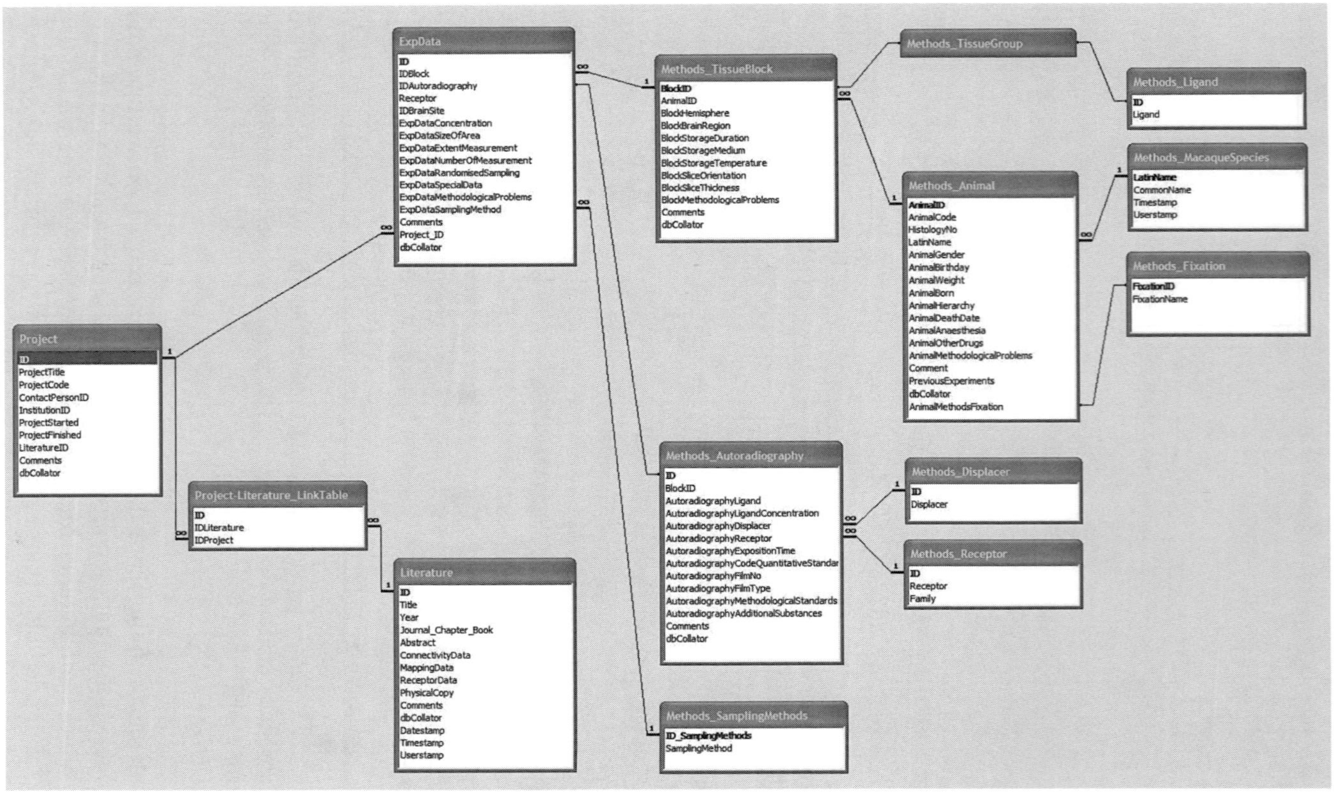

Fig. 5. Detailed snapshot of the experiment-related database tables and their relational organization. The fields in the tables and the key variables with their relationships in the group of experimental data tables are visible (primary keys in bold).

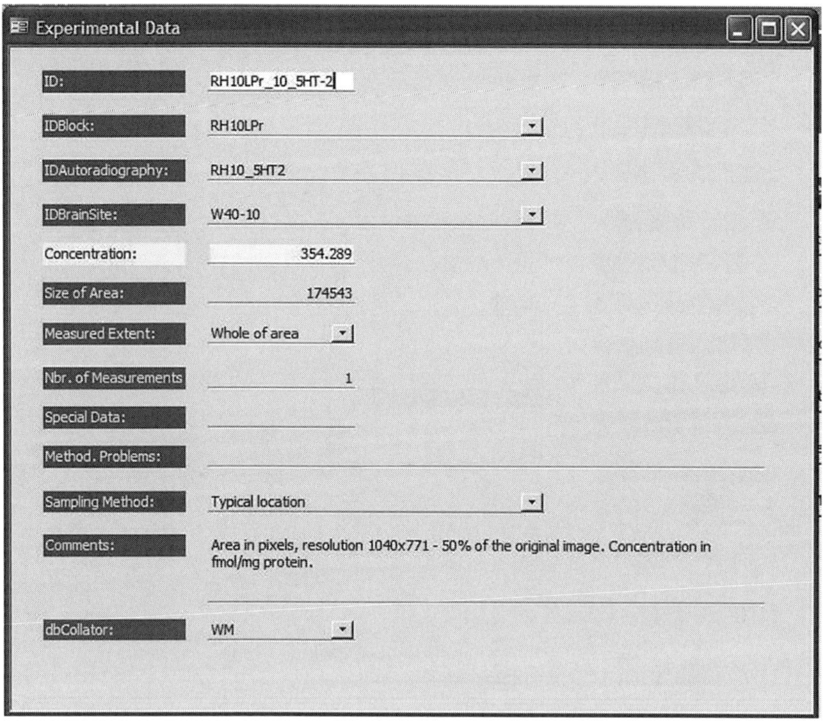

Fig. 6. Form *Experimental Data* showing a typical experimental dataset and values for the input fields. The most important concentration value is highlighted.

Area), the measured extent, number of measurements, special data, methodological problems, and comments (e.g., about the units of measurement and further methodological details).

Details about the autoradiographic procedures are listed in the "Autoradiography" table (*see* **Fig. 7**). Here, we place information on the ligand, its concentration, further substances, the displacer, the presumptive receptor, details about the detection procedure including exposition time, standards, and film media. As in previously described tables, further details are kept in the comments field, and the person who collated the data is identified.

Note that several fields in the forms (*see* **Figs 6** and **7**) are equipped with drop-down boxes to quickly select among standard field entries from linked parameter tables. These parameter tables have the same form as the contents of the drop-down boxes indicate, i.e., they contain lists of unique values for a specific parameter. For the field "Receptor" from the "Autoradiography"

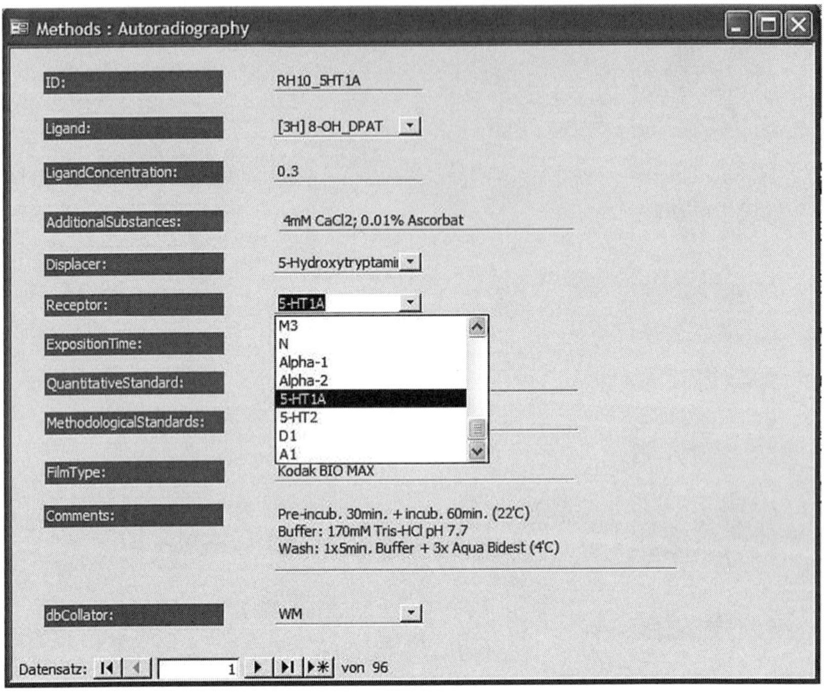

Fig. 7. Form *Methods Autoradiography* with an opened selection box in the input field "Receptor" providing in the form convenient access to predefined parameter values.

table, the corresponding "Receptor" table contains the list with the receptor classification as shown in **Fig. 8**. Note that the ID of the receptor entry is used to maintain the unique relationship with the "Autoradiography" table (*see* **Fig. 5**) but that the ID is not shown in the drop-down list because it is not informative to the user.

3.5. Forms and User Interface

The tables are the essential building blocks of the database organization, and their relations specify their interdependencies, thus providing an understanding of the database design. Although this organization delivers a thematic grouping as indicated in **Fig. 2**, it does not fully coincide with the view that a user takes when thinking about data organization, which may look more like the organization chart shown in **Fig. 9**. Here, the experimental procedures are grouped according to what can be regarded as the varying experimental data

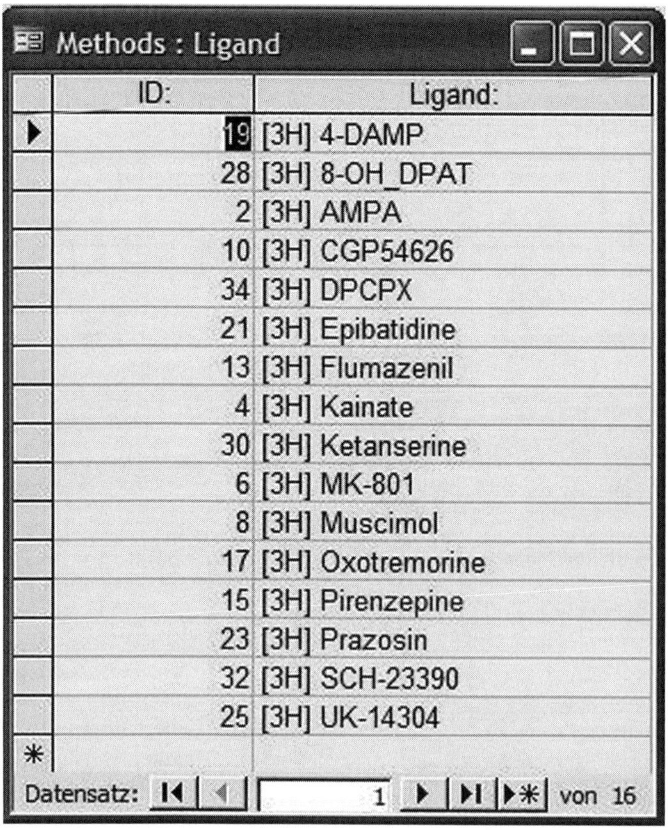
Methods : Ligand

ID:	Ligand:
19	[3H] 4-DAMP
28	[3H] 8-OH_DPAT
2	[3H] AMPA
10	[3H] CGP54626
34	[3H] DPCPX
21	[3H] Epibatidine
13	[3H] Flumazenil
4	[3H] Kainate
30	[3H] Ketanserine
6	[3H] MK-801
8	[3H] Muscimol
17	[3H] Oxotremorine
15	[3H] Pirenzepine
23	[3H] Prazosin
32	[3H] SCH-23390
25	[3H] UK-14304

Datensatz: 1 von 16

Fig. 8. Form *Methods Ligand* in the datasheet view showing a list of all predefined parameter values for the ligand.

and distinguished from the usually constant predefined parameters. Thus, this organigram may provide a good starting point for a start menu that could lead a user through the database components.

When actually browsing or entering data, users may prefer again a different layout that pulls together the data in the way as they appear in the experimental procedures. This pragmatic arrangement is at the basis of the masks designed for building a convenient user interface. Such masks are exemplified in **Figs** 4, 6, and 7, where a single mask assembles data from several tables, which may have to be accessed together or which provide parameter values for convenient data entering.

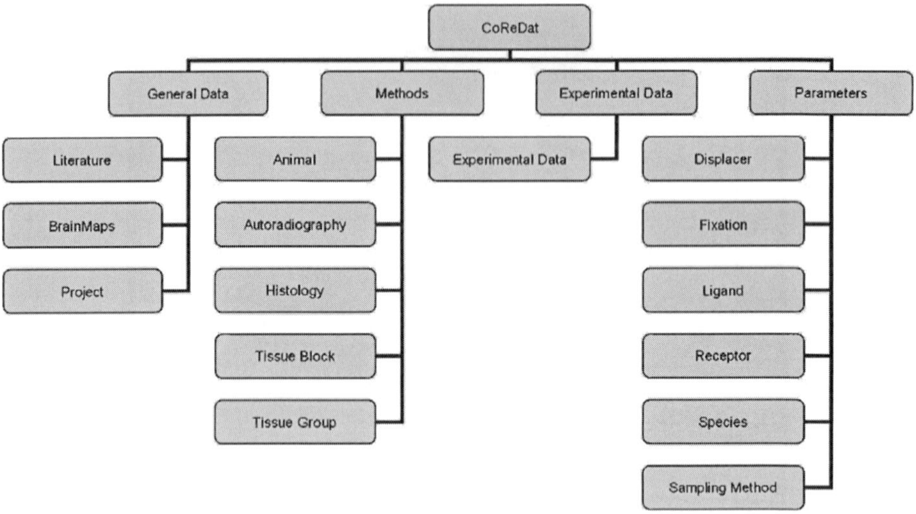

Fig. 9. Organization chart of the CoReDat database system as reflected in the user interface.

Notes

1. Although some tables in our database are specific to the autoradiographic method, the principal design applies also to several other data sets, including recent large-scale efforts to map the gene expression patterns in the mouse brain (*see* **ref. *12***; Allen Brain Atlas: www.brainatlas.org; LONI mouse brain atlas: http://www.loni.ucla.edu/MAP/; Gensat: http://www.gensat.org). Specific features of the present system are the methodological details (autoradiography), the detail of brain parcellation (histological delineation of areas), the variety of species (rodents, primates, humans), and the publication policy (free availability of the database system).
2. We have previously discussed the fundamental differences between coordinate-dependent and parcellation-based approaches ***(13)***. The strong variability of cortical folding particularly in human brains does not permit a direct comparison of voxel-based data between different individuals, and further limitations apply to comparisons between different species. This is not so much a problem with relatively uniformly shaped mouse brains provided that distortion is minimized and an adequate image resolution is used. Nevertheless, the mouse atlas systems aim to provide a comparative approach based on automated brain parcellation. Present parcellation algorithms—irrespective of whether they use feature or template-matching approaches—need considerable manual interaction given the inadequacies of automated parcellations of cerebral cortex and other brain structures

based on Nissl-stained sections or other common histological markers. The experiments contained in CoReDat, however, were primarily targeted at the detection of characteristic cortical receptor distribution patterns, which makes manual delineation indispensable. Thus, we included in our database a sophisticated approach to the storage of parcellation information and its relationships across brain maps.

3. Compared with CoCoMac, whose scope is limited to species of the genus Macaca, CoReDat may contain data from different mammalian genera. Although the concepts of brain region comparisons within a genus or species do not differ from species to species, it appears obligatory to introduce special markers for cross-genus comparisons that could be interpreted as homology statements. For example, the presumptive homology of primary visual cortex denoted area V1 in macaques and area 17 in cats should be denoted by a special H(homology) relation in contrast to the I(identiy) relation for intra-genus comparisons.

Acknowledgments

We thank Dr. Filip Scheperjans for kindly providing a draft of Fig. 1. This work was supported by the DFG Graduate School 320 (AG Kötter/Zilles) as well as DFG grant KO 1560 /2-1 and the James S. McDonnell Foundation.

References

1. Kötter, R. (2001) Neuroscience databases: Tools for exploring brain structure-function relationships. *Phil. Trans. R. Soc. Lond. B*. **356**, 1111–1120.
2. Dyhrfjeld-Johnsen, J., Maier, J., Schubert, D., Staiger, J. F., Luhmann, H. J., Stephan, K. E., Kötter, R. (2005) CoCoDat: A database system for organizing and selecting quantitative data on single neurons and neuronal microcircuitry. *J. Neurosci. Methods* **141**, 291–308.
3. Stephan, K. E., Kamper, L., Bozkurt, A., Burns, G. A., Young, M. P., Kötter, R. (2001) Advanced database methodology for the Collation of Connectivity data on the Macaque brain (CoCoMac). *Phil. Trans. R. Soc. Lond. B*. **356**, 1159–1186.
4. Kötter, R. (2004) Online retrieval, processing, and visualization of primate connectivity data from the CoCoMac database. *Neuroinformatics* **2**, 127–144.
5. Zilles, K., Schleicher, A., Rath, M., Bauer, A. (1988) Quantitative receptor autoradiography in the human brain. Methodical aspects. *Histochemistry* **90**, 129–137.
6. Zilles, K., Schleicher, A. (1995) Correlative imaging of transmitter receptor distributions in human cortex, in *Autoradiography and Correlative Imaging* (Stumpf, W., Solomon, H. eds.). Academic Press, San Diego, CA, pp. 277–307.
7. Kötter, R., Stephan, K. E., Palomero-Gallagher, N., Geyer, S., Schleicher, A., Zilles, K. (2001) Multimodal characterisation of cortical areas by multivariate analyses of receptor binding and connectivity data. *Anat. Embryol.* **204**, 333–350.
8. Bozkurt, A., Zilles, K., Schleicher, A., Kamper, L., Sanz Arigita, E., Uylings, H., Kötter, R. (2005) Distributions of transmitter receptors in the macaque cingulate cortex. *NeuroImage* **25**, 219–229.

9. Scheperjans, F., Grefkes, C., Palomero-Gallagher, N., Schleicher, A., Zilles, K. (2005) Subdivisions of human parietal area 5 revealed by quantitative receptor autoradiography: a parietal region between motor, somatosensory, and cingulate cortical areas. *NeuroImage* **25**, 975–992.
10. Geyer, S., Matelli, M., Luppino, G., Schleicher, A., Jansen, Y, Palomero-Gallagher, N., Zilles, K. (1998) Receptor autoradiographic mapping of the mesial motor and premotor cortex of the macaque monkey. *J. Comp. Neurol.* **397**, 231–250.
11. Date, C. J. (1990) *An Introduction to Database Systems*, 5th ed. Addison-Wesley, Reading, MA.
12. Visel, A., Ahdidan, J., Eichele, G. (2003). A gene expression map of the mouse brain: Genepaint.org – a database of gene expression patterns, in *Neuroscience Databases: A Practical Guide* (Kötter, R., ed.) Kluwer, Boston, MA, pp. 19–36.
13. Kötter, R., Wanke, E. (2005) Mapping brains without coordinates. *Phil. Trans. R. Soc. Lond. B.* **360**, 751–766.

IV

Neuroinformatics in Genetics and Neurodenegerative Disorders

16

An Informatics Approach to Systems Neurogenetics

Glenn D. Rosen, Elissa J. Chesler, Kenneth F. Manly, and Robert W. Williams

Summary

We outline the theory behind complex trait analysis and systems genetics and describe web-accessible resources including GeneNetwork (GN) that can be used for rapid exploratory analysis and hypothesis testing. GN, in particular, is a tightly integrated suite of bioinformatics tools and data sets, which supports the investigation of complex networks of gene variants, molecules, and cellular processes that modulate complex traits, including behavior and disease susceptibility. Using various statistical tools, users are able to analyze gene expression in various brain regions and tissues, map loci that modulate these traits, and explore genetic covariance among traits. Taken together, these tools enable the user to begin to assess complex interactions of gene networks, and facilitate analysis of traits using a systems approach.

Key Words: QTL; microarray; gene ontology; neurogenetics; transcript expression; systems genetics.

1. Introduction

GeneNetwork (GN) is a new web resource designed for the integrated analysis of complex networks of genes, molecules, cellular processes, and higher-order phenotypes, including behavior and disease susceptibility. Many of the data sets in GN are from massive transcriptome studies of major central nervous system subdivisions, including hippocampus, striatum, and cerebellum *(1)*. Rather than studying a single individual or strain, the great majority of data sets in GN are taken from genetic reference populations (GRPs) that consist of up to 100 diverse but inbred strains of mice and rats. The animal model in this case is not a single knockout or transgenic line. Instead,

From: *Methods in Molecular Biology, vol. 401: Neuroinformatics*
Edited by: Chiquito Joaquim Crasto © Humana Press Inc., Totowa, NJ

the model is an entire sample population; and the intent is to use an animal population model that incorporates the same level of genetic complexity as human populations.

We have generated and assembled large databases including databases of genomic sequences and single-nucleotide polymorphisms (SNPs), QTL locations, gene expression values, neuroanatomical and stereological measurements, behavioral and neuropharmacological observations, and many other phenotypes for several large and small GRPs. The value of one constituent database is not so much in the number of records it contains as it is in the number of relationships that can be drawn from among those records. In other words, a database of 100 records generated using the same GRP can be used to explore $100 \times 99/2$ possible relations. When this same logic is applied to microarray data sets with 45,000 expression measurements, one quickly appreciates both the power and the statistical hazards. Now, factor in that GN contains many different data sets of this size, all of which can be cross-correlated, and it is evident that literally billions of putative relations can be rapidly tested. By using a fixed GRP, one can relate any of the millions of observations being made about these strains to any other observations that have already been acquired and databased. Data do not evaporate after publication; instead, they can be relentlessly reused by other investigators employing the same GRP but their own new data. This allows one to effectively explore the mechanistic basis of genetic covariance and gene pleiotropy among sets of traits acquired at different scales, at different times and stages, and under different experimental conditions *(2–5)*. This approach can take the user from gene expression networks and their influence on cell types in a particular region, to the influence of that region's morphology on the brain, and ultimately to the genetic basis of behaviors and other phenotypes.

GN exploits a powerful synthetic approach called systems genetics. Systems genetics is a relatively new discipline that has grown out of complex trait analysis *(6)*, which in turn grew out of standard Mendelian genetics. Conventional Mendelian genetics asks a simple question: What one gene mutation or variant is responsible for the differences in phenotypes segregating in a specific population? What mutation causes albinism or Huntington's disease? This is essentially an one-to-one relationship between variation in genotype and variation in phenotype. In contrast, complex trait analysis asks a more difficult question: What combinations of gene variants and environmental factors contribute to variation of a phenotype that is segregating in a specific population? For example, what combination of gene variants and environmental risk factors contribute to polygenic diseases such as cancer, infectious disease susceptibility, or obesity? This analysis focuses on one complex trait—an one-to-many relationship with regard to genotype–phenotype relationships. Systems genetics is the next extension. Phenotypes and diseases are recognized as being parts of complex networks of related phenotypes. Schizophrenia (discussed further in Chapter 17) is not a single entity, nor is autism—these diseases are aggregates of phenotypes that we have reified with a single term. The same is true of cardiovascular disease and cancer. The set of phenotypes that comprise complex diseases are often influenced by different sets of genetic and molecular networks and sets of environmental factors. Genotype–phenotype relations in systems genetics involve *many-to-many* relationships.

Systems genetics and complex trait analysis are far more demanding from a computational and statistical perspective than Mendelian genetics. These more complex explanatory and predictive models of phenotypes often require large sample sizes—hundreds to tens of thousands. For a true systems approach, we also need to acquire various phenotypes from a defined population, whether that is a human population in Iceland or a GRP of rodents living in a single vivarium in Bar Harbor, Maine. Fortunately, the combination of rapid advances in computer performance and in molecular biology has made it possible to generate, process, and distribute the massive data sets that are required for these more complex models of disease susceptibility ***(7–10)***.

1.1. Genetic Reference Panels

Understanding the nature of GRPs is essential to fully exploit the tools in GN. Most of the GRPs in GN, and the ones that we will use in the examples below, are recombinant inbred (RI) strains. RI strains are isogenic lines (mouse, rat, nematode, rice, *Arabidopsis*, or *Drosophila*) that have derived from common parental inbred stock. Each “line” of an RI “set” is fully inbred and will have the same set of alleles, but the different RI lines in the set will by chance have fixed different combinations of alleles because of recombination and segregation. But in terms of utility, the simple way to think about RI strains is as an immortal population for which very large numbers of phenotypes have already been collected and that have also been very well genotyped. One can perform genetics without the impediments associated with genotyping.

RI strains have been used for several decades to map complex polygenic traits by statistically aligning the phenotypic differences between the strains with the known genetic differences between these strains with the measured phenotypes. In the examples outlined below, we will use the BXD mouse RI set, which contains approximately 80 strains originally derived from the mating of C57BL/6J and DBA/2J progenitors *(11–13)*.

1.2. Major Procedures

The goal of the current protocol is to enable the reader to investigate complex networks or genes, transcripts, and higher-order traits that are of interest. All tools described below are publicly available and accessible using a standard web browser. Using these tools, a user will be able to

1. Characterize variation in genes and their expression
2. Map genetic loci that affect trait variation
3. Quantify and explore genetic correlations among traits

2. Materials

All the materials required for the protocol are databases and statistical evaluation tools that are found on the web at www.genenetwork.org. A list of the databases follows:

2.1. Transcriptome Databases

Transcriptome databases provide estimates of mRNA expression as measured from microarrays of dissected tissue from the BXD and other GRPs. Databases of whole brain (minus olfactory bulb and cerebellum), eye, hippocampus, cerebellum, striatum, and hematopoietic stem cells are currently available in GN for the BXD GRP. Samples were processed using a number of different Affymetrix short oligomer microarrays. Array image data files were processed using various transformation methods including the position-dependent nearest neighbor method (PDNN, *see* **ref. *14***), robust multichip analysis method (RMA, *see* **ref. *15***), Affymetrix's own microarray suite 5 (MAS5), and a customized heritability-weighted method (www.genenetwork.org/dbdoc/BR_U_1203_H2.html). To simplify comparison among the transforms, the values from each of these different transforms have usually been adjusted to an average expression of 8 units and a standard deviation of 2 units. Details of these computational machinations are all provided on the INFO pages that accompany each of the individual data sets.

2.2. BXD Published Phenotypes

Published phenotypes represent data obtained through a search of all PubMed-indexed journals where GRPs were used (*see* **ref. *2*** for details). Whenever possible, exact values of graphically represented data were obtained from the authors. In other cases, published figures were measured using calipers. Additional published and unpublished phenotypes were submitted directly by investigators.

2.3. Genotype Database

Every GRP needs a companion genotype file. These files provide a simple summary of the genetic makeup of each individual strain. In the case of the BXD strains, the files indicate whether a given strain inherited both copies of its gene from the C57BL/6J and DBA/2J parental strain. One can imagine chromosomes of the BXD strains as being mosaics of segments inherited from either C57BL/6J or DBA/2J. The genotype file tells us approximately where those segments inherited from one parent or the other start and stop. The BXD genotype file used from June 2005 onward exploits a set of 3795 markers typed across 88 extant and a few extinct BXD strains (BXD1 through BXD100). This genotype file includes all markers, both SNPs and microsatellites that have unique strain distribution patterns (SDPs), as well as pairs of markers for those SDPs represented by two or more markers. In those situations where three or more markers had the same SDP, we retained only the most proximal and distal marker in genotype file. This particular file has also been smoothed to eliminate genotypes that are likely to be erroneous. We have also conservatively input a small number of missing genotypes (usually over very short intervals). Smoothing genotype data in this way reduces the total number of SDPs and also lowers the rate of false discovery. However, this procedure also may eliminate some genuine SDPs.

2.4. SNP Database

The SNP Database contains SNPs between the parental strains throughout the genome. Genotypes are from Celera Genomics, the Perlegen/NIEHS resequencing project, the Wellcome-CTC SNP Project, dnSNP, and the Mouse Phenome Database.

2.5. External Databases

The **Table 1** lists a sample of the external databases accessible through GN.

3. Methods

3.1. To Find Variation in Gene Expression

Gene transcription is determined by genetic, environmental, and gene $\times$ environment interactions such as age and sex. Genetic variation in gene expression can be dramatic. GN provides several large data sets of gene

Table 1
External Databases

Database	URL	Description
Biomolecular Interaction Network Database (BIND)	bind.ca	Molecular interactions
Search Tool for the Retrieval of Interacting Genes/Proteins (STRING)	string.embl.de	Protein–Protein interactions
SymAtlas	symatlas.gnf.org/SymAtlas/	Gene functionalization data sets
Synapse Database (SynDB)	syndb.cbi.pku.edu.cn	Proteins related to synapse or synaptic activity
PANTHER	panther.appliedbiosystems.com	
Entrez Gene, GenBank, OMIM, UniGene	www.ncbi.nlm.nih.gov	Gene sequences and ontology, gene–disease relationships
UCSC Genome Browser	genome.ucsc.edu	Sequences and genomics
Gene Ontology (GO)	www.geneontology.org	Gene ontology

expression estimates, most of them derived from RI mouse lines. Numerous genes have expression differences between strains ranging over fourfold.

3.1.1. Search and Data Retrieval

Point your browser to www.genenetwork.org. This brings you by default to the Search page, from which you can retrieve data from many GN data sets. We will focus on the default data set, defined by *Species: Mouse, Group: BXD, Type: Whole Brain, Database: INIA Brain mRNA M430 (Apr05) PDNN*

Enter "Kcnj*" into the ALL or ANY field and click the *Search* button. Note the location and annotation of available potassium channel genes in the Search Results page that opens.

Use the browser *Back* button to return to previous page. Enter "dopamine" into ALL or ANY field and again click *Search*. In the new Search Results page, note multiple probe sets related to dopamine.

Notice the action of the *Select All*, *Select Invert*, and *Select None* buttons on this page. Select several of the dopamine entries by clicking the checkbox at the left of the probe set description and/or using those buttons.

3.1.2. BXD Selections Page

Click the *Add to collection* button. The selected probe sets will appear in a new window, the BXD Selections page. This is one of two pages that serve as control centers for various analysis functions. The analysis functions controlled from this page are designed for a set of related traits. These multiple-trait analysis tools are invoked by the buttons at the bottom of the list: *Multiple Mapping*, *Compare Correlates*, *Correlation Matrix*, QTL *Cluster Map*, and *Network Graph*. We will explore these later.

Notice that the transcript descriptions on the BXD Selections page include the name of the data set from which they came. This page will accept traits from any BXD data set and from more than one data set. Gene expression traits from different tissues can be combined with classical traits on this page and compared or analyzed together.

For now, close the BXD Selections window. The information on the BXD Selections page is not lost when the page closes; the page can be re-opened with the selections by choosing Search > Trait Selections > BXD Selection from the menu on any GN page. The traits that are stored on this page are analogous to items saved in a shopping cart at an online store. Return your attention to the list of dopamine-related transcripts on the Search Results page.

3.1.3. Trait Data and Analysis Form

Click anywhere on a description of one of the dopamine-related probe sets (this text will highlight as you mouse over it). This will bring up the Trait Data and Analysis Form. This is the other page that serves as a control panel for various analysis functions.

Notice that this page has three major sections–an upper section with trait information and links, a middle section with analysis tools, and a lower section with phenotypic values for the trait you have chosen.

3.2. Characterize Variation in Genes and Their Expression

3.2.1. Gene Information Links

Examine the upper part of the Trait Data and Analysis Form. For gene expression traits, this section provides basic information about the gene represented by the probe set, and it provides links to information at other sites. Explore the links to Entrez Gene, Entrez Nucleotide, OMIM, and UniGene through the links marked *GeneID*, *Genbank*, *OMIM*, and *UniGene*, respectively.

3.2.2. Probe Information

Most of the gene expression data available in GN is derived from hybridization with Affymetrix microarrays. These arrays use one or more probe sets, each of which includes 11–16 25-base oligonucleotide probe sequences to represent each gene. Each probe set also includes, for each probe sequences, a mismatch probe sequence with one altered base as a control for nonspecific hybridization. Click the *Probe Tool* button for access to information about the probe sequences in this probe set.

The Probe Information window that opens has a list of information about the probe in the probe set. Odd-numbered probes, shown in green, are the probes that match sequences in the target gene. Even-numbered probes are the mismatch control probes. This table provides a wide range of information about the properties of each probe and the data derived from it. Click on the title of each column for more information about the column.

Click the *Verify UCSC* button to open a window at the UCSC Genome Browser with the positions of the probe sequences marked in comparison with other features of the mouse genome.

Click the *Verify Ensembl* button to open a similar window at Ensembl.

Click the *Select PM* button and then the *Correlation Matrix* button to display a correlation matrix of the expression estimates provided by individual probes. The correlation of individual probes may be poor. Among other things, the binding affinities of individual probes will differ, making them most sensitive in different ranges to changes in concentration of their target sequences.

3.2.3. Basic Statistics

Click the *Basic Statistics* button. This opens a Basic Statistics window with a table of trait values (mean values for each RI line), a box plot summarizing the range of trait values across lines, two bar charts showing the strain means and standard errors (when available), and a probit or Q-Q plot that displays to what extent the trait distribution differs from normal.

Notice the range of expression among RI lines. For expression data, values are $\log_2$ of the normalized fluorescence with mean values set at 8 with a standard deviation of 2; each unit therefore corresponds to a twofold change in hybridization.

In the Normal Probability Plot, the ranked trait values are plotted on the ordinate against the expected values for each rank on the abscissa. If the data are normally distributed, the plotted points will approximate a straight line. If the plotted points create a symmetric S-shape, the distribution is narrower than normal. If they are symmetric but appear to have a plateau in the center, the distribution is wider than normal. If the plotted points form a non-symmetric curve, the data are skewed.

Close the Basic Statistics window by clicking the close button and return to the Trait Data and Analysis Form.

3.3. To Find Genetic Loci that Affect Trait Variation

3.3.1. Simple Interval Mapping

The major function of WebQTL module of GN is QTL mapping ***(16)***. QTL mapping is, in essence, the search for correlation, among related individuals, between values of a trait and alleles of a marker locus somewhere in the genome. A high correlation suggests that there is a gene near the marker locus that affects the trait value. The estimated location of that hypothetical gene is a quantitative trait locus or QTL.

At the Search page, search for transcripts of the gene *Fprl1* in the BXD data set *INIA Brain M430 (Apr05) PDNN*. Click on the entry for probe set *1428589_at_A* to open a Trait Data and Analysis Form for that transcript.

Find the *Interval Mapping* button and click it (after scrolling down until the button is visible). After a short wait, a new Map Viewer window will open with a graph. The horizontal axis of this graph represents the mouse genome, and it is divided into separate sections for each chromosome. The heavy blue line represents the statistical significance (likelihood ratio statistic or LRS) of a hypothetical QTL at that location. Pink and gray horizontal lines show the thresholds at which the LRS is significant or suggestive, respectively. The thinner red or green lines represent the estimated strength of the hypothetical QTL.

For *Fprl1*, you will see a strong QTL on Chr 2. This is a so-called *trans* QTL because it is not near the location of the *Fprl1* gene, which is on Chr 17. An orange triangle on the horizontal axis at Chr 17 shows the position of the *Fprl1* gene.

3.3.1.1. Interval Analyst

Click on the numeral 2 at the top of the Chr 2 section of the graph. This will open a window with a Map Viewer for chromosome 2 and, below that, the Interval Analyst table for Chr 2. The Interval Analyst displays extensive additional information about genes and genetic variation on that chromosome. Genes in the region of the QTL peak (blue line) are candidate genes that may affect expression of the transcript being mapped.

3.3.1.2. UCSC Browser

At the top of the single-chromosome Map Viewer, there are four tracks that provide location-specific links. (These tracks only appear for maps using the default physical scale for map distance.) Clicking on one of the three colored lines at the top will display an expanded view of the part of the chromosome around the location of the click. One of the lines opens an expanded view in WebQTL Map Viewer; the others open the corresponding section of the genome in the UCSC Genome Browser or the Ensembl Genome Browser. Try clicking on one or both of the genome browser lines just above the LRS peak to see a display of the (many) genes that might be responsible for this QTL.

Below these three lines is a set of staggered blocks that indicate the positions of known genes, as derived from information of the UCSC Genome Browser. If you cause the cursor to hover over genes on this track, minimal associated information (symbol, position, and exon number) will appear.

3.3.2. Pair-Scan Mapping

3.3.2.1. Introduction

Complex traits often may be controlled by more than one locus, and, in fact, it is somewhat unrealistic to fit a single locus to a trait and expect to get an accurate description of how it is controlled. A pair-scan is a step toward detecting greater complexity. It searches for pairs of loci that can explain the trait variation. An exhaustive search, testing all available pairs of loci, would be time-consuming; WebQTL tests a grid of locations and then refines the search by testing on successively finer grids in areas that successfully explain trait variation ***(17)***.

Pairs of loci can affect a trait in two ways. They may contribute independent additive effects, so that the trait value can be determined by adding constant effects representing each allele at each locus. On the contrary, alleles at one locus may affect the effect of the other, so that the final trait value cannot be expressed as a simple sum of allele effects. This situation is described as epistasis or interaction.

The results of this mapping are displayed in a square graph or "heat map" that is divided by a diagonal line. The vertical axis represents the genomic location of one locus; the horizontal axis, the location of the other. Color in the figure represents the LRS for the association of the trait value with the pair of loci represented by that location. Warm colors, especially red, represent higher LRS values and therefore represent possible locations of a pair of interacting QTLs.

The area above and to the left of the diagonal line represents interaction effects alone. The area below and to the right represents the total effect of possible QTLs, including both additive and interaction effects. A vertical bar to the right of the interaction map is the "heat map" equivalent of an interval mapping plot. It shows the results of searching for a single QTL to explain the trait.

3.3.2.2. Example

On the Search page, search for *Zc3hav1*. From the choices returned, click on *1446244_at_A* to bring up the Trait Data and Analysis Form. Scroll down to the Pair Scan section. Choose *LRS Interact* in the *Sort by* menu. Click the *Pair-Scan* button.

After a period of computation, WebQTL will open a Pair-Scan Results window with the figure previously described. For *Zc3hav1*, There is a strong QTL on Chr 4 and a weak QTL on Chr 19. These appear as lines of yellow and red in the lower right triangle of the interaction plot. In addition, there are two locations of interactions, one for Chrs 2 and 11 and another for Chrs 2 and 19. These appear as spots of red in the rectangles, representing those pairs of chromosomes, on both sides of the diagonal line. In the lower right, the interaction appears as a red dot on the yellow line, representing the weak Chr 19 QTL.

Scroll down to the table below the figure. This table lists pairs of loci in descending order of interest. If you specified that the list be ordered by interaction effect (by choosing *Sort by LRS Interact*), the top two entries will be the Chr 2–Chr 11 and Chr 2–Chr 19 locus pairs.

Scroll up to the figure again and click anywhere in the rectangle representing Chrs 2 and 11 in the upper left triangle. WebQTL will open another window with an expanded map of just that chromosome pair. This map is created without the short-cut sampling used for the whole-genome pair-scan. That is, all pairs of available loci for Chrs 2 and 11 are evaluated to produce this map.

3.3. To Find Genetic Correlations Among Traits

If a trait is measured in a set of RI lines, differences among the strains can be attributed, in part, to their genetic differences. Environmental and stochastic differences can be minimized by testing under constant environmental conditions and by averaging trait values across several individuals. Under these conditions, traits that are affected by the same genes, directly or indirectly, would show similar variation across different strains because they would be responding to the same allelic differences in the controlling genes. Thus, traits that are affected by similar sets of genes would be expected to be correlated across a set of RI lines. These traits would be expected to be functionally related in some way. GN allows search for traits that may be functionally related by searching for those whose average trait values are correlated.

Return to the GN Search page (choose Search > Search Databases) from the menu. Using the same database as before, search for *Csf2ra*. Choose 1420703_at_A, the probe set that targets *Csf2ra* on Chr 13. The probe set that targets it on Chr 2 seems to be nonspecific. In the Trait Data and Analysis Form, scroll down to the Trait Correlations section. Under *Choose Database*, choose *BXD Published Phenotypes*. Click the *Trait Correlations* button.

The Correlation Results page that opens shows a table listing published traits for the BXD RI set and their correlations with the *Csf2ra* transcript. By default, the traits are listed in order of descending *P*-value. This transcript shows a relatively high negative correlation with brain weight for one data set. In fact, this example was chosen for this correlation. It also shows a high positive correlation to an alcohol-related trait. Farther down the list, it also shows lesser negative correlations with various measures related to brain size in other data sets.

Click on the value for the *Csf2ra*-brain weight correlation (−0.6963). This action will open a Correlation Plot page in which you can examine the relationship between the two traits. Look for linearity and outliers.

3.3.1. Selection and Saving Multiple Traits

The list of traits on the Correlation Results page represents traits that may be related in some way. You may want to select a group of them for further analysis. For example, use the *checkboxes* to the left of each entry to check entries 1, 9, 10, 14, 16, 18, traits related to brain size. Click the *Add to collection* button at the top of the page. This button will add the checked traits to your BXD Selections page.

3.3.2. Multiple QTL Mapping

Close the BXD Selections Page and return to the Correlation Results page. This page also provides direct access to two multiple-trait analysis functions. With six traits still checked on this page, click the *Multiple Mapping* button at the top of the page. This button will open a Multiple Interval Mapping page that shows QTL scans for all selected traits on the same figure. In this example, the traits seem to share weak QTLs on Chr 8 and possibly also on Chr 15 and 18.

genome where several of the lines have coincident peaks; these may represent the location of a control gene common to the mapped traits.

3.3.6.2. Correlation comparison

Return to the BXD Selections page. Change the selection of traits if you want, and click the *Compare Correlates* button. A Correlation Comparison window opens with introductory explanation and opportunity to change options for the analysis. Using the default options, click the *Correlate* button in the middle of the page.

When the calculation finishes, the Correlation Comparison page redraws with two lists of results. These list groups of genes from the database whose expression is correlated with one or more traits in the submitted set. The potentially interesting groups are those in which a several database genes are correlated with several of the input genes. This is a graph connectivity-based feature that can also be used to identify the genes that have common relations to a set of physiological or behavioral traits.

3.3.6.3. Correlation Matrix

Return to the BXD Selections page. Change the selection of traits if you want and click the *Correlation Matrix* button.

The Correlation Matrix page that opens shows a simple table with all pair-wise correlation coefficients among the submitted traits. Values are color coded to help identify the more important correlations. The table cells also show the number of value pairs on which each coefficient is based, those coefficients based on few values may be unreliable.

This page also presents principal components calculated from the correlated traits if the number of data points for each trait is sufficient. If no principal components were generated, examine the table to identify the traits with fewest values. Close the Correlation Matrix window, return to the BXD Selections page, uncheck the traits with few values, and click the *Correlation Matrix* button again. Principal components can be considered to be synthetic traits that summarize the common components of a group of correlated traits. Principal component traits can be transferred to the BXD Selections page and used for further QTL mapping or correlation analysis.

3.3.6.4. Association Network

Return to the BXD Selections page. Change the selection of traits if you want, and click the *Network Graph* button. This function creates an Association Network page with a graphical representation of the pair-wise correlations among the submitted traits. In the graph, nodes represent classical or gene expression traits, and lines connecting the nodes represent correlations. Lines are color-coded to indicate the sign and strength of the correlations.

4. Final Notes

GN is a dynamic resource. New databases, analysis tools, and search methods are being added frequently. The interface features that we describe here are also a work in progress and you can expect changes in the next few years. To track additions, improvements, and interface modifications link to the "News" on the GN home page. Also be aware that there are extensive help files accessible from the home page that contain HTML and PowerPoint tutorials, a glossary, and a frequency asked questions (FAQ) page. Finally, users are encouraged to browse and experiment—you cannot break anything.

References

1. Rosen, G. D., La Porte, N. T., Diechtiareff, B., Pung, C. J., Nissanov, J., Gustafson, C., Bertrand, L., Gefen, S., Fan, Y., Tretiak, O., Manly, K. F., Park, M. R., Williams, A. G., Connolly, M. T., Capra, J. A., and Williams, R. W. (2003) Informatics center for mouse genomics: the dissection of complex traits of the nervous system. *Neuroinformatics* **1**, 327–42.
2. Chesler, E. J., Wang, J., Lu, L., Qu, Y., Manly, K. F., and Williams, R. W. (2003) Genetic correlates of gene expression in recombinant inbred strains: a relational model system to explore neurobehavioral phenotypes. *Neuroinformatics* **1**, 343–57.
3. Chesler, E. J., Lu, L., Wang, J., Williams, R. W., and Manly, K. F. (2004) WebQTL: rapid exploratory analysis of gene expression and genetic networks for brain and behavior. *Nat Neurosci* **7**, 485–86.
4. Chesler, E. J., Lu, L., Shou, S., Qu, Y., Gu, J., Wang, J., Hsu, H. C., Mountz, J. D., Baldwin, N. E., Langston, M. A., Threadgill, D. W., Manly, K. F., and Williams, R. W. (2005) Complex trait analysis of gene expression uncovers polygenic and pleiotropic networks that modulate nervous system function. *Nat Genet* **37**, 233–42.
5. Bystrykh, L., Weersing, E., Dontje, B., Sutton, S., Pletcher, M. T., Wiltshire, T., Su, A. I., Vellenga, E., Wang, J., Manly, K. F., Lu, L., Chesler, E. J., Alberts, R., Jansen, R. C., Williams, R. W., Cooke, M. P., and de Haan, G. (2005) Uncovering regulatory pathways that affect hematopoietic stem cell function using 'genetical genomics'. *Nat Genet* **37**, 225–32.
6. Moore, K. J., and Nagle, D. L. (2000) Complex trait analysis in the mouse: the strengths, the limitations and the promise yet to come. *Annu Rev Genet* **34**, 653–86.
7. Belknap, J. K., Phillips, T. J., and O'Toole, L. A. (1992) Quantitative trait loci associated with brain weight in the BXD/Ty recombinant inbred mouse strains. *Brain Res Bull* **29**, 337–44.

8. Crabbe, J. C., Belknap, J. K., Buck, K. J., and Metten, P. (1994) Use of recombinant inbred strains for studying genetic determinants of responses to alcohol. *Alcohol Alcohol Suppl* **2**, 67–71.
9. Zhou, G., and Williams, R. W. (1999) Eye1 and Eye2: gene loci that modulate eye size, lens weight, and retinal area in the mouse. *Invest Ophthalmol Vis Sci* **40**, 817–25.
10. Airey, D. C., Lu, L., and Williams, R. W. (2001) Genetic control of the mouse cerebellum: identification of quantitative trait loci modulating size and architecture. *J Neurosci* **21**, 5099–109.
11. Taylor, B. A. (1978) Recombinant inbred strains. Use in gene Mapping, in *Origins of Inbred Mice* (Morse, H., Ed.), pp. 423–38, Academic, New York.
12. Taylor, B. A. (1989) Recombinant inbred strains, in *Genetic Variants and Strains of the Laboratory Mouse* (Lyon, M. L., and Searle, A. G., Eds), pp. 773–96, Oxford University Press, Oxford.
13. Peirce, J. L., Lu, L., Gu, J., Silver, L. M., and Williams, R. W. (2004) A new set of BXD recombinant inbred lines from advanced intercross populations in mice. *BMC Genet* **5**, 7.
14. Zhang, L., Miles, M. F., and Aldape, K. D. (2003) A model of molecular interactions on short oligonucleotide microarrays. *Nat Biotechnol* **21**, 818–21.
15. Cope, L. M., Irizarry, R. A., Jaffee, H. A., Wu, Z., and Speed, T. P. (2004) A benchmark for Affymetrix GeneChip expression measures. *Bioinformatics* **20**, 323–31.
16. Wang, J., Williams, R. W., and Manly, K. F. (2003) WebQTL: web-based complex trait analysis. *Neuroinformatics* **1**, 299–308.
17. Ljungberg, K., Holmgren, S., and Carlborg, O. (2004) Simultaneous search for multiple QTL using the global optimization algorithm DIRECT. *Bioinformatics* **20**, 1887–95.
18. Zhang, B., Schmoyer, D., Kirov, S., and Snoddy, J. (2004) GOTree Machine (GOTM): a web-based platform for interpreting sets of interesting genes using Gene Ontology hierarchies. *BMC Bioinformatics* **5**, 16.

17

Computational Models of Dementia and Neurological Problems

Włodzisław Duch

Summary

A critical goal of neuroscience is to fully understand neural processes and their relations to mental processes, and cognitive, affective, and behavioral disorders. Computational modeling, although still in its infancy, continues to play a central role in this endeavor. Presented here is a review of different aspects of computational modeling that help to explain many features of neuropsychological syndromes and psychiatric disease. Recent advances in computational modeling of epilepsy, cortical reorganization after lesions, Parkinson's and Alzheimer diseases are also reviewed. Additionally, this chapter will also identify some trends in the computational modeling of brain functions.

Key Words: Neural networks; cognitive computational neuroscience; associative memory models; cortical reorganization; computational models in psychiatry; Parkinson disease; Alzheimer disease.

1. A Bit of History

There are two primary branches of Neuroinformatics. On the one hand, it provides tools for storage and analysis of information generated from neuroscience experiments. On the other, neuroinformatics is concerned with simulations and models that capture aspects of information processing in the brain. The complexity of brain dynamics may be too high for us to conceptually understand the brain's functions in detail. Computational models that take into account neuroscientific principles may capture progressively larger number of essential features of brain dynamics, eventually leading to models of the whole brain that no individual expert will ever be able to understand in detail. This

From: *Methods in Molecular Biology, vol. 401: Neuroinformatics*
Edited by: Chiquito Joaquim Crasto © Humana Press Inc., Totowa, NJ

situation is analogous to cell biology, where the sheer number of bio-molecules and their interactions typically hamper the understanding of all genetic and metabolic mechanisms that occur within a living cell. From an engineering perspective, understanding a complex system implies the ability to build a model that behaves, at least in its crucial aspects, similar to the system that is being modeled. At this point in history, computer simulations are the easiest way to build complex models, but progress in building neuromorphic devices that implement neural functions in the hardware may change this situation in future ***(1)***.

In 1986, a two-volume compendium, *Parallel Distributed Processing: Explorations in the Microstructure of Cognition* **(2)**, written mostly by psychologists, were published. The first volume of the PDP book (as it was commonly called) focused on the general properties of parallel information processing, drawing analogies with human information processing. The Parallel Distributed Processing (PDP) name did not gain popularity. It was replaced by "connectionism," a name that stressed the importance of connections that pass information between networks of nodes representing neural cell assemblies, thus increasing or inhibiting their activation. One of the chapters described what is the now famous "backpropagation of errors" algorithm that can be used to train a network of simple artificial neurons with a one-way (feedforward) signal flow to obtain desired response to incoming signals. Such artificial neural networks became very useful for medical diagnostics support, signal and image analysis and monitoring, the search for carcinogenic agents and other data analysis tasks. In these applications, neural networks faced strong competition from statistics, pattern recognition, machine learning, and other mathematical techniques.

The second volume of the PDP book contained models of psychological processes (speech perception, reading, memory, word morphology, and sentence processing) and biological mechanisms (neural plasticity, neocortex, and hippocampus models). Here, neural models have little competition, with the exception of the continuum neural field theory (CNFT). CNFT treats neural tissue as a substrate in which various excitations are propagated. This theory contributes to the understanding of global models of the cortex; it also raises the notion of interesting phenomena that may be experimentally observed ***(3)***. But the neural field theory is not yet sufficiently developed to account for properties of biological memory, not to mention more complex functions.

The two PDP volumes generated significant interest in formal neural network models, connectionist modeling in psychology inspired by the parallel, distributed, competitive, brain-style information processing, and computational neuroscience. They focused on the biologically plausible models of single neurons or small neural assemblies. These neural models are fairly detailed, up to the level of ionic channels and specific neurotransmitters, producing spikes that have very similar characteristics to those observed in vivo. The question "at which level should one model the brain" obviously depends on the purpose of such modeling and thus cannot provide comprehensive answers to every question. It would be very interesting if various artificial neural networks based on graded (non-spiking) neurons could be derived as an approximation to biologically plausible models. A common assumption is that an average number of spikes per second represents neural activity, and therefore, artificial neural network models may use neurons that are characterized by a single number reflecting their activity. Potentially important phase relations between spikes and differences between the dynamics of various types of synapses and neurotransmitters are lost in this way. There are many other approximations to biological neural processes that remain to be explored.

Despite these difficulties, simple models of neural networks are surprisingly successful, providing insight into various brain functions. One year after the PDP book appeared, Ralph Hoffmann published, in the *Archives of General Psychiatry*, the first paper that used neural network models to address psychiatric issues—the schizophrenia–mania dichotomy ***(4)***. In 1988, the National Institute of Mental Health (NIMH) initiated the *Computational, Theoretical and Mathematical Neuroscience* program. It supported the utilization of computer simulations in psychiatry. The first international workshop on neural modeling of cognitive and brain disorders, sponsored by the National Institutes of Health, was held at the University of Maryland in 1995, resulting in the first book on this topic ***(5)***. At the same time, inspirations derived from recurrent neural networks treated as dynamical systems started to infiltrate brain science ***(6,7)*** and developmental psychology ***(8,9)***.

Computational psychiatry is thus quite young. Bearing in mind the differences between the style of thinking in psychiatry and computational modeling, it may take a long time before it will become a part of the mainstream psychiatry. Only a small percentage of the authors of papers in books devoted to this topic are professional psychiatrists ***(5,6,10)***. Nonetheless, there have been computational models that have generated useful hypotheses for almost all neuropsychological syndromes and psychiatric disorders. These can be verified by experimental work.

Spiking neurons ***(11)*** and sophisticated, biologically plausible models of neurons ***(12–14)*** applied to understanding of brain dysfunctions show the overlapping interest of computational neuroscience and psychiatry. Before reviewing the current state of the art in computational psychiatry, a short discussion in **Subheading 2** introduces key concepts and problems in this field. This is then followed by the discussion of research related to applications of these models towards neurological problems and Alzheimer's disease (AD), with more emphasis on the understanding and ideas generated with the help of computational models than on technical details. Future prospects of computational psychiatry are discussed in the final section.

2. How to Model Brain Functions

Models representing brain function have had to face the tradeoff between simplicity and biological faithfulness. Simplicity implies the ability to understand, in conceptual terms, how and why the model works the way it does, leading to new ideas that may trigger the imaginations of researchers. Lateralization of brain functions (left–right hemisphere division) found its way into popular psychology books. McLean's theory of the triune brain ***(15)*** introduced in the 1950s division into archipallium, reptile-like brain stem functions, paleomammalian limbic system functions, and neocortex, the rational mammalian brain. Although the concepts of the limbic system are rather vague and have been severely criticized ***(16)***, simplified theories are still prevalent. Computer models are usually constructed to account for results of a single experiment and have always a limited range of applicability. Computer simulations may rarely lead to wide-ranging theoretical concepts, but they may provide simplified, metaphoric language that facilitates thinking about brain processes. The brain-as-a-computer information processing metaphor has been replaced by new metaphors: brain as a connectionist system, dynamical system, and self-organizing system ***(17)***. These metaphors have now become a part of the psychology-process ***(7–9)*** and have made their way into psychiatry. They can be used to describe not only the dynamics of neural cell assemblies but also global activity of the brain, for example, mood swings in bipolar disorders ***(18)***. Such simple models are quite far from biological realism but may still be quite useful, providing metaphorical language that reflects neurodynamical processes.

Biological faithfulness requires large-scale neural simulations at a sufficiently detailed level to guarantee comparison with experimental results obtained by neurophysiologists. Until recently, such simulations were possible only for very small networks composed of a few dozens of neurons. But in 2005, projects proposing simulations of whole cortical mini-columns with 10^4–10^5 neurons and 10^8 connections on teraflop supercomputers were formulated [e.g., the IBM Blue Brain project ***(19)***]. Such detailed simulations, if successful, should allow one to create in silico experiments that will be very difficult or even impossible to perform in vivo. It is possible, therefore, that answers to fundamental questions, such as the nature of the neural information codes, will be found using computational models. Once these questions have been solved, simplified architectures for brain-like computing, as well as models for different disorders, can be constructed.

2.1. Neural Models

Traditional neuropsychological theories describe information processing in the brain using box-and-arrow diagrams, treating brain dysfunctions as disturbances in the information flow. Biophysical models may be very detailed, aiming for biologically faithfulness. The majority of neural and connectionist brain function models are somewhere in between. They are usually based on inspirations of, rather than systematic approximations to, biophysical models. Successful connectionist models capture principles behind the biological realization of certain brain functions, providing proofs that postulated mechanisms may, in principle, explain some results of observations and experiments, without quantitatively predicting any directly measurable variables.

A network of nodes interacting with each other through connections of different strengths can collectively result in interesting computations. Some of these computations involve:

- transforming incoming signals: filtering irrelevant information, aggregating information to construct new features, enhancing salient features;
- categorizing elementary signals (such as phonemes or graphemes) to simplify analysis of complex signals in discretized form;
- self-organizing to reflect general topographical properties of sensory inputs and motor actions;
- memorizing associations at many levels, providing models of different types of distributed, content addressable memory;
- making decisions, setting goals, and controlling behavior.

Each network node, frequently called "an artificial neuron," may represent a single neuron of a specific type, a generic neuron, a group of a few neurons, a neural assembly, or an elementary function realized in an unspecified way by some brain area. Two key features of neural models are: (1) internal signal processing done by their nodes; and, (2) the type of interactions among these nodes. Internal signal processing mechanisms of neurons (network nodes) are a form of internal knowledge representation allowing for transformation of the incoming signals. In the simplest case, this knowledge is specified by just a single parameter, a threshold for activation of a neuron by incoming signals. This threshold is an "adaptive parameter." It means that it is adjusted by the learning algorithm of the model to improve performance of the task as required by the whole network. Real neurons send single spikes when the sum over all incoming signals in a given time window (called neuron's activation) exceeds certain threshold value. This fact, and the brain–computer analogy, gave rise to the logical neuron model, introduced by McCulloch and Pitts (for anthology of early papers on neurocomputing *see* **ref. *20***). If the activation of a logical neuron is below the threshold, its output is "0", and, if it is above the threshold, its output is "1". Networks of such logical neurons may implement arbitrary Boolean function, but training such networks is fairly difficult. Single spikes do not matter much. The frequency of spikes is approximately proportional to the activation of the neuron around the threshold activation value, going to 0 for activation value significantly below the threshold, and saturating at maximum frequency for high activation values. This gives rise to the graded response neuron models, with output defined by a non-linear tilted S-shaped (sigmoidal) function of activation. The response gain of the neuron is defined by the steepness of this function. Because the output functions are squashing linear activations to keep all outputs between zero and some maximum values, they are sometimes called "squashing functions."

It is clear that such simplified models lose key factors associated with biological networks, such as the ability to synchronize the activity of several units for specific time delays between spikes (phase relations). Models of real neurons at different abstraction levels, from simple "integrate and fire" neurons without specific spatial structure, to multi-compartmental models that take into account geometry of dendrites and axons, are available in the NEURON and GENESIS software packages ***(12–14)***. Sophisticated biophysical models of compartmental neurons provide information directly related to neurophysiological parameters measured in experiments. In 1994, Callaway and collaborators, who had modeled reaction times to different drugs, stated: "Neural network models offer a better chance of rescuing the study of human psychological responses to drugs than anything else currently available" ***(21)***. Such detailed models related to psychiatric problems are available only in rare cases (a few papers may be found in books ***(5,6,10)*** and ***(22)***).

Connectionist models, whose nodes are not related to single neurons, involve tracking different directions, representing memory states recognized as a joint activity of neural ensembles. The high activity of such nodes may represent recognition of letters, phonemes, words, or iconic images. These models provide a bridge between neural and symbolic information processing, rule-based systems for systematic reasoning, and grammatical analysis. Knowledge stored in nodes of connectionist networks is still rather simple, although many parameters may be required to represent it. Within this limit, each node may become an agent, with more internal knowledge and some methods to process this knowledge depending on the information received ***(23)***.

The second key aspect of neural models is the communication between nodes. Signals carrying information about activity of input units may flow through the network in one direction, from designated inputs, through some hidden elements that process these signals, to output elements that determine the behavior or signify decisions of the network. In logical networks, signals have only two values, True and False, and connections between two logical neurons may either leave them unchanged or negate them. As an analogy to biological neurons, these connections are sometimes called "synaptic," direct connections are "excitatory," and negated connections "inhibitory." Activity of the neuron is simply calculated as the number of true inputs received by the neuron minus the number of the false inputs. This activity is then compared with the threshold to determine whether the neuron should output the signal True or False. Although this is a very simple model, threshold logic implemented by such neurons allows for the definition of non-trivial concepts, such as "the majority" (at least half of the inputs should be True to output True value).

Graded neurons are used in the popular Multilayer Perceptron (MLP) neural networks that assume feedforward flow of information between layers of neurons. Such architecture (*see* **Fig. 1**) simplifies training of the network parameters that include thresholds and the strength of connections W_{ij} between pairs of neurons n_i and n_j . The only information that is passed through such networks is the strength of the signals X_i, and the only knowledge that neurons hold is activation thresholds θ_i. MLPs are examples of mapping networks

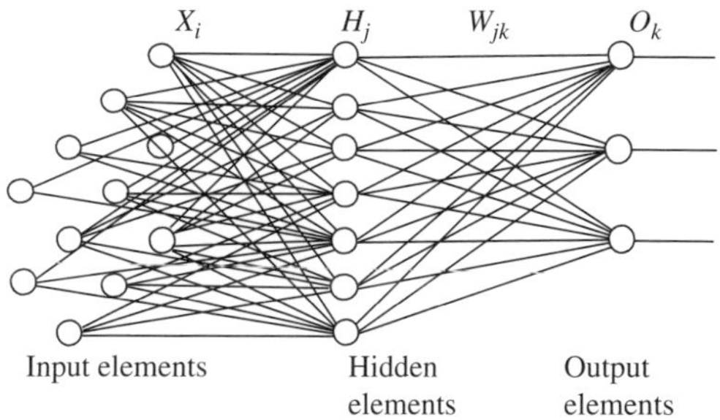

Fig. 1. Mulit-layer perceptron feedforward network with a single hidden layer.

that transform input signals into output signals and may learn, given sufficient examples of inputs and desired outputs, how to set the connections W_{ij} and the thresholds θ_I (or the interactions between the nodes and the local knowledge in the nodes) to perform heteroassociation between inputs and outputs. Learning algorithms of such networks is driven by errors that are assumed to be propagated from the output to the input layer, although such a process is difficult to justify from biological point of view. MLP networks may thus signify the presence of certain categories of inputs, reduce the amount of information needed for making decisions, or provide new set of features for further processing. Biological networks are almost never feedforward, containing many feedback loops, and assuming dynamical states that memorize or represent information received by the network. Given n logical neurons in some initial state and connecting them all together, one may observe how their mutual excitations and inhibitions evolve until a static configuration of activities is reached and no further change is possible (point attractor) or cyclic changes occur indefinitely (cyclic attractor) or the system reaches chaotic state. In 1982, John Hopfield described such fully connected network of the two-state neurons *(24)*. With additional assumptions that there are no self-connections and all weights are symmetrical, the evolution of such network always ends in one of several static configurations, depending on the initial state and on the weights that connect neurons. Such networks are commonly called "the Hopfield networks" (*see* **Fig. 2**) and are the simplest type of attractor neural networks or dynamical systems motivated by neural ideas. A version of such a

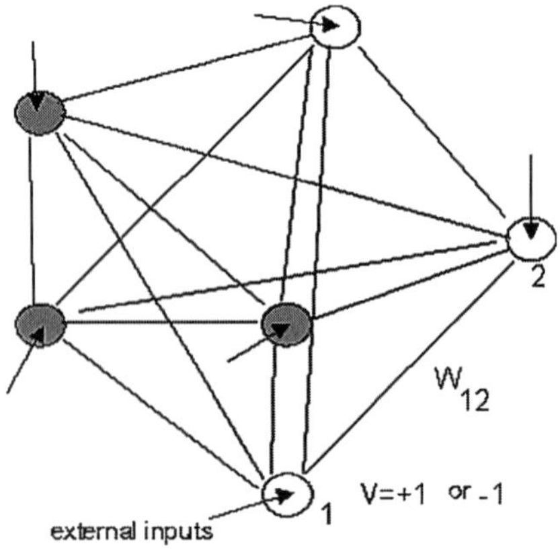

Fig. 2. Hopfield network with 6 neurons, 3 dark neurons are in an active state and 3 light neurons are in the inactive state.

network with graded neuron activities and continuous weights is very useful as a model of autoassociative memory. Learning in the Hopfield type of network may be based on the Hebb principle that neurons that show correlated activity should have stronger excitatory connections, those that show negative correlation (one active when the other is not active) should have stronger inhibitory connections, and those that do not show any correlations may have weak connections or lose them completely. Therefore, Hebbian learning is also known as the correlation-based learning.

The MLP and the Hopfield networks are frequently used in the modeling of brain functions related to perception and memory. One more basic model that is very useful in understanding the development of topographical mappings has been introduced by Teuvo Kohonen *(25)* and is known as the Self-Organizing Mappings (SOM) or Kohonen network. It is also a very simple model, with 2D grid of nodes (neurons) that do not communicate with each other directly (*see* **Fig. 3**), but tend to react in a similar way when the training signals are presented. These neurons have no internal parameters, only weights W_i that may be compared to the input signals. Evaluation of similarity, the basic operation in SOM network, may be viewed as a simplification of neural activation by a specific train of spikes. Initially, all weights are random; but after presenting training signals to the network many times, the learning mechanisms pull together those weights that are similar to typical signals (centers of clusters

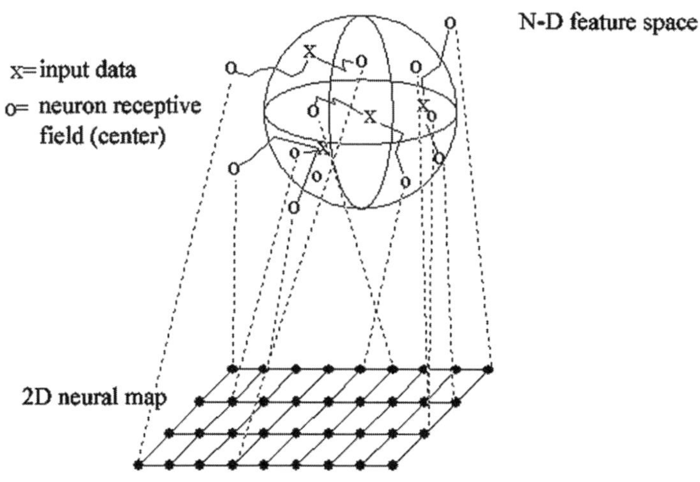

Fig. 3. Self-Organised-Mapping is a grid of processors, each with parameters that learn to respond to signals from the high-dimensional input feature space.

in the input space). In effect, different part of the neural grid show maximum activity for different types of signals, discovering, for example, the phonetic structure of speech signals. SOM networks may model spontaneous perceptual and motor learning in infancy but may also be useful as a model of reorganization of cortical functions after stroke or loss of limbs (*see* **Subheading 3**).

These three types of models and their many variants are used extensively in the *Computational Explorations in Cognitive Neuroscience* book *(26)* that contains many experiments made with accompanying PDP++ software for biologically based modeling of psychological functions. Although the software is not directed at psychiatric disorders, it has been used to set up many experiments to elucidate various brain mechanisms involved in perception, attention, recognition, semantic, episodic and working memory, phonology and speech perception, reading and dyslexia, mapping phonological to orthographical representations and executive frontal lobe functions.

2.2. Brain Simulations and Mental Functions

Large neural networks based on biologically plausible spiking neural models may be used to model many brain functions, but such models may not only be difficult to simulate, sometimes requiring supercomputing power, but also are rather difficult to analyze and understand. Although the relatively simple approaches presented above are much less faithful to biology, they may help elucidate some brain mechanisms, generating useful hypothesis. Neural simulations should capture casual relations between the activity of brain structures and their general neuroanatomical features, in particular, the influence of lesions and neuropathological changes on the modification of normal behavior and cognitive performance. It is not clear a priori, that simplified neural models will be sufficient to capture such casual relations. Convergence of the modeling processes could be too slow to make them useful; for example, some pathological effects could appear only in models based on complex multicompartmental spiking neurons.

Fortunately, there are some indications that the qualitative behavior of complex models based on spiking neurons *(27)* may also be obtained in simplified neural models *(5)*. Even the simplest neural models of associative memory show many characteristics known from psychology, such as: content-addressability—cues lead to the whole memorized patterns, even when they are imperfect (computer memory needs an address to find information), graceful degradation—damage to some connections does not erase specific facts—only increases the recall error rates; time of recall does not depend on the number of memorized items (in computers, time of recall is proportional to the number of items); similar items get more frequently mixed up; an attempt to understand too many things too quickly leads to chaotic behavior, and so on.

Thus, there is a chance that simple neural models may help us understand neurological and neuropsychological syndromes, providing some insight into the source of pathologies and some understanding of the effects of therapeutic procedures. Classical methods of psychiatry and neuropsychopharmacology are restricted to observations of correlations between behavior and physiological responses of the organism to medical treatments. They usually try to understand the mechanisms leading to neuropathological behavior at the neural level, whereas cooperative network-level effects may be quite complex and difficult to infer from experiments. Brain simulations can complement traditional techniques in several ways. They provide insights into possible causal relations, allowing for full control of every aspect of the experiments; they are inexpensive and are not restricted by ethical considerations. An early review article *(28)* and the books *(5,6,10,22,26,29)* provide many insights into the mechanisms behind the memory and language impairments, psychiatric disorders, AD and Parkinson disease, epilepsy, and other neurological problems.

The hierarchical approach to modeling of brain functions at different levels of complexity has been presented in *(23)* and *(30)*, addressing the problem of creating understandable descriptions of complex phenomena and the gap between neuroscience and psychology. Neural models may predict behavioral patterns but do not capture the inner, first-person subjective perspective; they do not offer any vision of the mind. Transitions between quasi-stable states of attractor networks lead to behaviorist rules, reducing complex neurodynamics to simple symbolic descriptions of animal and human behavior. These rules are a very rough approximation to neurodynamics; but an intermediate level of modeling, presenting the current state of the network in reduced dimensionality spaces, is possible. At the level of conscious decision-making, only some highly processed features are accessible, with active "mind objects" in psychological spaces reflecting neural dynamics. This type of modeling, while still quite rare, is in line with Roger Shepard's search for universal laws of psychology in appropriate psychological spaces *(31)*. This modeling strategy has been applied to the analysis of psychological experiments on human categorization, providing both neurodynamical and psychological perspectives of the processes responsible for puzzling human judgments *(32)*. Although such mental-level models have a chance to complement neural models providing deeper understanding of brain dysfunctions in the future, so far they have not been used in this area.

In the remaining part of the chapter, a review of two application areas of computational psychiatry—various neurological problems and AD—is presented,. This review illustrates the type of models that has been used and emphasize insights, understanding, and hypothesis generated by computational models.

3. Neurological Problems

Neurological problems, such as epilepsy, ischemic or hemorrhagic strokes, cerebral palsy or parkinsonism, are usually associated with well-defined anatomical and physiological changes in the brain. In this respect, the modeling problem is easier than in case of such psychiatric disease such as *schizophrenia*. An early review of different neurological problems has been presented in **ref.** ***33***.

The most detailed models so far have been created for *epilepsy*, which is one of the most widespread neurological disorder. Epilepsy is characterized by recurrent, unprovoked seizures that frequently lead to the loss of consciousness. Many types of epilepsy exist. They differ in the localization of origin of synchronous neural oscillations (usually the hippocampus and certain neocortical regions), with frequency that may be either quite low or extremely high (up to 600 Hz). Dynamical neural networks with strong feedback connections and little inhibition may easily be brought to a threshold of epileptic-like discharge, but this is a rather simplistic observation. Much deeper and more detailed understanding is required to understand this process. For example, it is known that small changes in the molecular structure of neuron's sodium ionic channels may be one of the causes of epilepsy, and some drugs act by modulating those currents. Therefore, detailed models that include many pyramidal neurons with several types of ion channels are particularly useful here. Many such models have been developed in the research groups of Jeffreys ***(34)*** and Traub ***(35,36)***. They have been able to generate electroencephalogram (EEG) patterns that closely resemble experimental ones, elucidating the role of various types of neurons and pharmacologic agents in epilepsy.

The generation of extremely high-frequency oscillations is not possible in model networks without electrotonic synapses, which are much faster than classical chemical synapses. Although it is known since the early 1980s that such synapses should exist (because blocking of chemical synaptic transmission does not stop epileptic seizures), direct evidence for their existence in the neocortex is still controversial. Fast, electrical, non-synaptic communication is possible through gap junctions filled with connexins, intramembranous proteins that have rapidly modifiable conductance properties. Predictions of computational models are even more specific: gap junctions should exist between the axons of pyramidal cells in hippocampus and neocortex.

These hypotheses focus on the search for therapeutic strategies on the direct manipulations of gap junctions to decrease synchrony. A better understanding of molecular processes at the gap junctions may be achieved by detailed biophysical modeling techniques based on molecular dynamics.

Parkinson's disease is connected with the loss of dopaminergic neurons in the substantia nigra that project to one of the basal ganglia's large nuclei, the putamen. As a result, motor control problems develop, including akinesia (difficulty initiating movements), tremor, rigidity, slowing of movements, and poor balance. In a substantial percentage of patients, cognitive impairment, hallucinations, and depression may occur. The origins of this disease are not yet clear, and therapy is based on dopamine precursor, L-DOPA, that slows down the progress of the disease.

Movements are smooth if the timing between activation of agonist and antagonist muscles is synchronized. Low level of dopamine in the putamen may lead to the imbalance in the cortico-basal-ganglia-thalamic-cortical loop that controls the motor system (involvement of the cerebellum in this loop is usually neglected). In computational models therefore, the dynamics of four layers of neurons was investigated, with feedback from the fourth to the first layer ***(38)***. In such attractor networks, a change of parameters leads to sudden bifurcation, changing qualitative behavior of the network from point attractor to a cyclic attractor. Borrett et al. ***(38)*** interpret this change as the sudden onset of tremors when a significant decrease in dopamine level is reached. In their model, it is preceded by a slower network response that may lead to slow movements. Edwards et al. ***(39)*** investigated richly connected inhibitory neural networks of the Hopfield type, showing that such transitions between irregular and periodic dynamics are common if synaptic connections are weakened, because the number of units that effectively drives the dynamics is reduced, leading to simpler behavior. Fixed points in the dynamics of networks correspond to akinesia, whereas irregular dynamics may correspond to the normal low-level physiological tremor. High-amplitude tremors in Parkinson's disease are regular and result from cyclic attractors in networks with simplified dynamics. Such models may produce various symptoms, relating them to different forms of cooperative behavior (see also papers in **refs *5,6***).

Cognitive aspects of Parkinson's disease, and the influence of pharmacological therapy on cognitive abilities, have been modeled by Frank ***(40)***. Complexity of brain systems involved is too high to have a good descriptive (verbal) model of interactions among basal ganglia, motor cortex, and prefrontal cortex. Although dopaminergic drugs ease the difficulties with the motor symptoms, they may increase problems with the execution of cognitive tasks.

Basal ganglia are not only involved in motor actions but also working memory updates and initiation of thought movement. This is a subtle system that normally operates with wide dynamic range of dopamine levels, but in Parkinson's patients taking dopaminergic medication, this range is reduced, leading to frontal-like and implicit learning impairments. Frank's model, implemented in the *Leabra* framework ***(26)***, provides many detailed testable predictions for neuropsychological and pharmacological studies.

Stroke and brain lesions have the longest history of computational modeling. The situation in case of focal cortical lesions in some areas is rather clear, and detailed experimental data exist from animal and human studies. For example, in the somatosensory cortex, topographical organization, reflecting spatial relationships has been studied at many levels (see the review in **ref.** ***41***). Cortical representations may reorganize as a result of lack of stimulation because of the nerve damage, limb amputation, direct lesion, or other processes. The simplest model that leads to formation of topographical maps is based on the self-organized mapping networks ***(25)*** and has been used with success to model details of visual system development ***(42)***. SOM has been used in a number of studies involving cortical reorganization after lesions and stroke ***(43,44)***. More complex networks designed for cortical map reorganization with excitatory and inhibitory cells were used to model inputs from the hand, including fingers ***(45,46)***. The model showed a number of physiologically correct responses, such as expansion and contraction of representations of the stimulated areas, effects of nerve deafferentation and gradual disappearance of the "silent" regions in the map, as seen in experiments. Self-organizing models with competitive activation and representation of thalamus (a relay of sensory inputs) account also for the "inverse magnification" rule: the size of cortical representation is inversely related to the size of the cortical cell receptive fields, leading to large cortical representations devoted to the skin areas with small receptive fields ***(43)***. In this model, thalamic and cortical areas are represented by 2D sheets of neurons, folded into a toroidal shape to avoid the border areas. Each of the network nodes should stand for a microcolumn of about 110 neurons, with hexagonal 32×32 node network. Thalamic neurons, activated by external inputs from the skin, excite cortical receptive field composed of a central neuron and 60 neurons surrounding it in four rings. These thalamo–cortical connections are trained using competitive learning, whereas cortico–cortical connections have local excitatory connections to the nearest neighbors and inhibitory connections to the further neighbors (the shape of the function describing the type of connections resembles a Mexican hat, with broad peak and a dip before leveling off).

Effects of lesions may be investigated either by cutting the thalamic connections or by removing all connections of a group of neurons, which simulates their death. The remaining neurons surrounding the lesioned region do not receive inhibitory inputs; therefore, their dynamics change quickly, and they become more responsive to weak stimulations from the thalamic area. As a result of these activations, slower synaptic learning processes lead to the reorganization of the cortex responses, with neurons close to the lesioned areas partially taking over the function of the dead neurons. This reorganization may be faster if only those skin areas that have lost their representation in the somatosensory cortex are repeatedly activated. Simple devices that provide tactile, vibration, and temperature stimuli placed over selected parts of the skin should speed up rehabilitation and prevent formation of phantom limbs.

Reorganization, because of the diffuse branching thalamic projections, is limited to no more than 1 mm; but experimental evidence shows that, after a long time, reorganization processes may extend to over 10 mm. Cortical areas devoted to a hand that has been amputated may, for example, start to demonstrate relocalization in the face. It is not yet clear how exactly the reorganization process happens, but it is quite likely that new connections because of sprouting (and may be even because of the development of new neurons) are formed. Of course, it is quite easy to put such effects into computational models, but without experimental constraints, models can provide explanations that have nothing to do with reality. On the contrary, some models are over-constrained by their assumptions; for example, the neglect of inhibition between cortical neurons ***(45)***.

Intriguing effects have been experimentally observed in the case of anesthesia: receptive fields immediately expand to neighboring neurons even before additional stimulation. This is probably because of the presence of weak connections between the thalamus and the topographically adjacent cortical areas, and the lack of inhibition from neurons that become silent. Similar effects are also observed in the visual cortex when artificial scotoma are created—the fill-in processes make it invisible after several minutes ***(33)***. In this case, the lack of inhibition, making neurons more responsive, leads to changes in the gain response of neurons.

A biologically plausible state-of-the-art model of the 3b area of the primary somatosensory cortex and thalamic nuclei was constructed by Mazza et al. ***(47)*** using the GENESIS simulator. In this model, the cortex is composed of excitatory (regular and burst spiking) and inhibitory (fast spiking) neurons. All parameters of the model were taken from literature; therefore, some aspects of the simulations may be compared with in vivo measurements. NMDA, AMPA, and GABA types of synapses were used and the synaptic changes observed during reorganization of maps after lesions. The hand mechanoreceptors give a short burst of action potentials when stimulated, adapting quickly to the pressure. A palm and four fingers containing 512 receptors were simulated. The ventral posterior lateral (VPL) part of the thalamus contains excitatory neurons and inhibitory interneurons. The thalamic relay cell model contains a soma and three dendrites, with sodium, low threshold inactivating calcium, fast calcium, voltage-dependent potassium, slow calcium-dependent potassium, and delayed rectifier potassium ionic channels. The thalamic interneuron has a soma and a dendrite with three types of ionic channels.

Cortical neurons are arranged in three layers (corresponding to the layer III, IV, and V of the area 3b), with each layer composed of a 32×32 node grid of excitatory and inhibitory neurons. Three types of neurons were included: excitatory pyramidal cells, fast spiking inhibitory basket cell GABAergic neurons, and burst spiking excitatory stellate neurons, for a total of 3072 excitatory neurons and 1536 inhibitory neurons. Although real cortex contains more kinds of of neurons, these were selected as the most common.

Equations describing currents flowing through ionic channels have many parameters, but they are fixed using experimental values. Connectivity between these neurons is matched to the statistical information about real connectivity. Simulation of 1 s of real-time processes took less than 4 h on a personal computer. Many properties of normal maps are exhibited by this model, including stable representation of the hand that has been developed in each cortical layer and in the thalamus nuclei. The most precise topographical maps occur in layer IV, and this is also shown in the model. Simulation of the amputation of a finger showed immediate reorganization of the cortical (but fewer such thalamic) representations, leading to expansion of the representation areas of the intact hand regions. This is followed by a slower improvement and consolidation of the new map. When a finger is removed, the lack of activity from this finger removes inhibition, and expansion follows. This process may be analyzed in every detail, including the dynamics of the AMPA, NMDA and GABA receptors in the dendrites of model neurons. It appears that the information in the NMDA channel increases rapidly after a lesion. This may be an important clue for pharmacological treatment of patients during rehabilitation.

Although the model is quite detailed, it does not include important features such as long-term potentiation (LTP). Of course, the complexity of real neurons is several orders of magnitude greater. Nevertheless, it is clear that models at this and higher levels of detail will be more common in future. *Phantom limbs* are experienced by some patients after amputation of arm, hand, or breast. This curious phenomenon has been studied by Ramachandran ***(48)***. In the somatosensory cortex, the representation of the hand is adjacent to that of the face. Patients who experienced phantom limbs had also rather detailed representation of the hand in several places on their face, for example, the sensation of touching a given finger could be elicited by touching three different well-localized spots on the face. In another case, sensation of touching the removed breast was elicited by touching the ear ***(49)***.

These sensations appear within hours after amputation and evidently show that large-scale reorganization and rapid expansion of the cortical receptive fields took place rapidly. Some attempts at modeling this process have been based on self-organizing networks ***(50)*** with the reorganization processes driven by noise generated by dorsal root ganglion sensory neurons that fire irregularly after an injury. Unfortunately, this model has not yet been tested. It seems that more experimental data are needed to create detailed models of this phenomenon and gain more understanding of large-scale reorganization processes. Many questions arise, for example, as to why only some patients experience it (because of stimulation of the skin?) or how to slow down the rapid reorganization (blocking NMDA receptors? avoiding stimulation of certain skin areas through local anesthesia?) to prevent appearance of the phantom limbs that create pain and discomfort for many patients.

The phantom limb phenomenon is similar to another strange phenomenon: the experience of an additional ghost arm parallel to the normal arm. Despite an apparent similarity, the mechanism in this case seems to be quite different. The unitary percept of the body after stroke or other lesions may dissociate, creating a perception of an additional arm that occupies the previous position of the real arm with a time lag of 60–90 s. An fMRI study of one such stroke patient ***(51)*** with a right frontomesial lesion showed a strong supplementary motor area (right medial wall) activity. This could be interpreted by the brain as an intention to move a hand that does not exist in this location.

Computational models of visual hemi-neglect and similar phenomena ***(52–55)*** may help to explain such experiences. Neglected patients, usually those having suffered lesions to the right parietal cortex, are not able to see objects in their left visual field if there is an object in their right field. In the particularly curious version of neglect, called object-based visual neglect, patients do not see the left half of each object spread out horizontally in the visual field. Because this is object-based visual impairment, attention mechanisms have to be involved. Deco and Rolls ***(55)*** created a model of attention that included three areas: primary visual cortex (V1), object recognition areas (inferior temporal cortex, IT), and spatial orientation areas (posterior parietal cortex, PP). The spatially local lateral inhibition in the parietal and visual cortex produces high-contrast effects at the edges of each object. If the damage in the parietal cortex increases linearly through the left visual field, local peaks in the resulting neuronal activation appear only for the right half of each object. There are many new models of attention ***(52,56)***, and by combining the results of experimental observations and computational modeling, we now have a better understanding of attention and executive function.

4. Alzheimer Disease

AD is the most common neurodegenerative disorder. AD involves a gradual deterioration of global cognitive and behavioral function. The disease is always progressive. Remissions are rare, but variability are numerous: life expectancy for sufferers ranges between 1 and 25 years. The earliest symptoms involve memory degradation, both for learning new things and recalling known facts. This is followed by degradation of language skills, poverty of thoughts and associations, intellectual rigidity, loss of initiative and interest, disturbances in motor, and executive functions. At an advanced stage, judgments are impaired, psychotic features (such as paranoid delusions) and personality disintegration may appear.

The disease usually first attacks the entorhinal cortex and the adjacent limbic areas and then spreads out to the neocortex. Prominent atrophy of the predominantly frontal and temporal cortex is observed in neuroimaging studies, and large amounts of senile plaques and neurofibrilliary tangles are found in the brain. Exact relations between cognitive decline and changes in the brain are rather complex and still not understood. Although knowledge of the possible causes AD have been over time accumulating, the real causes of pathogenesis are still unknown. There are no reliable clinical tests; definitive diagnosis is made only after autopsy. Mild cognitive impairment, involving loss of short-term memory, represents the early stage of AD ***(57)***. The few drugs available specifically for Alzheimer treatment (e.g., Cognex and Aricept) do not slow the progress of disease directly but are rather aimed at improving and stabilizing memory and cognitive state of the patient, helping to retain and utilize in a better way acetylcholine (ACh), one of the most important neurotransmitters (cholinergic neurons in the basal forebrain are destroyed in the early stages of AD, resulting in low level of acetylocholine). A better understanding of the mechanisms and development of AD are thus necessary to propose better diagnostic and therapeutic procedures. This is a challenge to the neural modeling community because too little is known about the progression of the disease.

The neurofibrillary tangles and senile plaques are clear markers of AD. Neuronal death is probably preceded by a loss of synaptic connections. Earliest changes in the brain have been observed around the enthorinal cortex and the hippocampal formation. The first symptom of the mild cognitive impairment and of AD itself concerns memory. All these facts point to the importance of the memory models ***(58,59)*** in understanding AD. The hippocampus is involved in memory consolidation and seems to be an intermediate memory storage that allows for the formation of stable long-term memories without catastrophic forgetting of contradictory information ***(58)***. Neural models should unravel the relation between changes at the cellular level and various clinical manifestations sof the disease. Three models of pathogenesis of AD have been proposed, all focusing on synaptic processes and their role in memory maintenance.

The "synaptic deletion and compensation" model of Horn et al. ***60***, developed further by Ruppin and Reggia ***(61)***, has been motivated by the following experimental observations. In the brains of AD patients, the density of synaptic connections per unit of cortical volume decreases with progress of the disease, whereas the surviving synapses increase physically in size and thus provide stronger connections. It is likely that these synapses compensate for synaptic deletion. In the feedforward neural network models, pruning of weak synaptic connections is frequently used to improve their predictive powers. Pruning allows one to forget the accidental details of mapping learned from data, whereas the essential characteristics are captured with simplified network structure using large synaptic weights of the remaining connections, necessary for the realization of strongly non-linear behavior. How do these two processes—synaptic deletion and compensation—influence memory deterioration? What are the best compensation strategies that may slow down the memory deterioration process?

The simplest associative memory models that may be used to investigate such questions are based on the Hopfield networks (*see* **Fig. 2**). Assuming that the synaptic matrix W_{ij} determines the strength of connections between neurons i and j, each of the N neurons has threshold θ_i for firing, and is in one of the two states $V_i = \pm 1$, the external inputs are E_i, the total activation of neuron n_i at the next time moment $t+1$ is a sum of all weighted activations at time t of neurons that connect to it, minus the threshold for activation, and plus the external input:

$$I_i(t+1) = \sum_{j=1}^{N} W_{ij} V_j(t) - \theta_i + E_i$$

The simplest network dynamics is defined by taking the sign of the activation:

$$V_i(t+1) = \text{sgn}[I_i(t+1)$$

This dynamic has only a point-attractor corresponding to the minima of the energy function:

$$E(V) = -\frac{1}{2}\sum_{i \neq j} W_{ij} V_i V_j$$

These stationary states are totally determined by synaptic weights and thresholds, and may be interpreted as the memory patterns of the system: starting with activation pattern $V(t_0) = V_0$ that carries partial information about memorized pattern the system will evolve, changing neural activity in a series of steps until at some step the activity $V(t)$ does not change any more, reaching one of the memorized states.

The number of patterns (V_i vectors in the stationary states) that may be correctly memorized in the fully connected Hopfield autoassociative memory model is about 0.14N. Trying to memorize more patterns leads to chaotic and random associations. Deleting a large number of synaptic connections will cause forgetting of some patterns and distortions of others. Assume that a certain percentage *d* of synaptic connections is randomly deleted (in the model, their value is set to 0). Suppose also that some biological mechanism makes the remaining connections stronger. One way to express it is W' $_{ij} = c(d, k)W_{ij}$, where the compensating factor $c(d, k) > 1$ is a multiplicative factor depending on d and a parameter *k(d)*, called a compensation-strategy parameter, that is fitted to experimental data. Horn et al. ***(62)*** proved that taking $c(d, k)\_ = k(d)d/(1-d)$ significantly slows the memory deterioration. Depending on the compensation strategy *k(d)* after the same evolution period various degrees of deterioration are obtained. Thus, failure of proper compensation for synaptic deletion may explain why patients with similar density of synaptic connections per unit of cortical volume show quite different cognitive impairments. This approach gives the modeler the freedom to fit *k(d)* function to experimental data, but because the data is missing, it simply shows what is possible rather than what really happens in nature.

There are several problems with such a simplistic model. Hopfield neural networks are not plausible from a neurobiological point of view because they require symmetric weights, have only point attractors, and are trained using non-local learning procedures. Compensation quite efficiently maintains memory capacity even when more than half of the connections and neurons are deleted, while in the latest stages of AD, no more than 10% of neurons are dead, but the density of synaptic connections may drop below 50% of the normal level. Addressing some of these questions, Ruppin and Reggia ***(61)*** and Horn et al. ***(62)*** improved the compensation model in several ways, obtaining similar conclusions from other memory models (Willshaw, Hebbian, and modified Hopfield networks), with over 1000 neurons used in simulations. Activity-dependent Hebbian models allow for studying memory acquisition, showing such effects as faster in forgetting of more recent memories. The memory recency effect, known in the psychological literature on memory as the "Ribbot gradient," has been noticed a long time ago in retrograde amnesia ***(63–65)*** and has also been observed in AD patients. Temporal gradients of memory decline and several other experimental phenomena characterizing memory degradation in AD patients have been recreated in Hebbian models. In such models, there is no global error function that is optimized, local compensatory mechanisms are sufficient to maintain high capacity of memory ***(62)***. The way deletion and compensation factors change in time has an influence on the final performance of the network. Cognitive impairments are therefore history-dependent in this model, leading to a broad variability of AD symptoms despite similar levels of structural damage of the brain.

The *Synaptic runaway model* developed by Hasselmo ***(66,67)*** is focused on a different phenomenon observed in associative memory attractor networks. To store a new pattern in the memory, such networks first explore all similar patterns to find out whether this is indeed a new pattern. If certain memory capacity is exceeded, the new pattern may interfere with existing ones, creating an exponentially large number of slightly different patterns that the system tries to store. This initiates pathological, exponential growth of synaptic connections, known as the "synaptic runaway" effect. It is not clear whether such an effect exists in biological neural networks, but if it does, it should lead to very high metabolic demands of hyperactive neurons, demands that in the longer time period cannot be satisfied. As a result, toxic products should accumulate and neurons will die because of excitotoxicity, creating senile plaques.

Synaptic runaway may arise because of excessive memory overload, reduced synaptic decay, or a low level of cortical inhibition. If external synaptic strength is sufficiently large or if internal inhibition is sufficiently strong, synaptic runaway may be prevented, but after critical storage capacity is exceeded, it is unavoidable. This model explains some intriguing experimental facts about AD:

- Enthorinal regions (involved in recognition memory) suffer greater degradation than cortical areas and are usually impaired at an earlier stage; these regions lack internal inhibition present in cortical modules.

- Cholinergic innervation in dentate gyrus, the primary afferent area to the hippocampus, is sprouting in AD patients.

ACh is a neurotransmitter that has complex functions. In the dentate gyrus, it does not influence external afferent synaptic transmission but selectively suppresses the internal excitatory transmission, effectively increasing internal inhibition. Experiments that proved this were inspired by Hasselmo's theoretical considerations ***(67)***. Thus, the sprouting of cholinergic innervation may reflect the brain's attempts to stop the synaptic runaway by increasing internal inhibition. Another way to avoid this effect could be through separation of learning and recall mechanisms in the hippocampal networks. Perhaps acetylocholine level can switch networks between the two modes ***(68)***, with high levels present during active waking facilitating encoding new information in the hippocampus without interference from previously stored information (by partially suppressing excitatory feedback connections), and lower levels of ACh during slow-wave sleep facilitating consolidation of memory traces.

Acetylocholine neuromodulation in the CA3 region of the hippocampus is also the focus of the Menschik and Finkel model ***(69–72)***. The neuroregulatory network is quite intricate and severely perturbed in AD, involving the death of neurons in several nuclei (locus coeruleus, dorsal raphe) that control norepinephrine and serotonin levels. ACh is produced ***(73)*** in the medial septal nuclei and the vertical nucleus (diagonal band of Broca). Understanding the pathological effects arising in neuroregulatory networks requires detailed, biophysical models of neurons. Menschik and Finkel used a parallel version of GENESIS (*see* Chapter 7) to construct quite detailed model of hippocampal pyramidal neuron with 385 compartments with ionic channels of several types. Although this model was too complex for network computations, some conclusions may be drawn even from single-neuron simulations. For example, switching between learning and recall, corresponding to switching between burst and regular spiking, requires high levels of acetylocholine that may not be available in hippocampus of the AD patient. Moreover, complete lack of ACh may lead to excitotoxic levels of calcium in dendrites.

Networks up to 1032 cells were constructed using 51- or 66-compartment neurons, based on the anatomy of the CA3 region of the hippocampus. The network was pre-wired in a Hopfield-like way to store some memory patterns, assuming that it has already undergone learning. This network behaves as an attractor network, with patterns presented as a series of spikes at the beginning of theta rhythm cycle, progressing to one of the stored cycles that appears as gamma bursts (100 Hz) within the theta cycle. In a large network, 40 randomly chosen patterns of 512 bits each were stored. This network helps to assess the neuromodulatory role of acetylocholine on different ionic channels. Decline in ACh slows down the intrinsic gamma rhythm, and this in turn makes memory retrieval more difficult, giving less time for the network to settle in the attractor. Unfortunately, after the initial burst of activity, this interesting model has not been developed further.

All three neural models complement rather than compete with each other. Simple models that can be analyzed in details may be source of inspirations for more precise questions that biophysical models may answer. There may be several routes to the development of AD: synaptic loss and insufficient compensation should lead to AD cases with little structural damage of the brain, whereas synaptic runaway should eventually lead to death of the hyperactive neurons and significant structural damage. Both types of AD cases are indeed known, and the great variability in the life expectancy and manifestations of clinical symptoms is probably a reflection of different underlying mechanisms.

The assumption that computational models reflect real neural mechanisms leads to several therapeutic suggestions, summarized in *(74)* and extended below. They may help to slow down the degeneration of synaptic connections and thus the development of the disease, at least in its early stages. These suggestions may be tested experimentally and in light of high variability of the AD symptoms should be matched to individual cases.

If synaptic runaway processes and the failure of proper compensation are the cause of rapid memory impairment, then one should minimize new memory load for AD patients. This should involve regular, simple daily routine, and minimization of the number of new facts or items that should be remembered. Heavy memory load may contribute to the rapid progress of synaptic deletion. Patients should not be allowed to follow visual, auditory, or printed stories, such as the TV news, soap operas, or TV series that require remembering of new facts, names, and interpersonal relations. Sedatives may have a positive effect on the memory overload; because, in the absence of strong emotions, the limbic neuromodulatory systems do not increase synaptic plasticity, preventing formation of new memories. On the contrary, activation of non-declarative memory, for example, by learning new skills, may work in a positive way. Engagement in new activity seems to benefit patients with mild dementia. Modeling this type of activity has not yet been attempted, and it is likely that patients with different types of AD will respond in different ways.

- Strengthen the old, well-established memory patterns.

A significant portion of time should be spent on recalling the stories and facts of patient's life with the help of family members. These memories form a skeleton of the concept of "self." Antonio Damasio *(75)* expressed it this way: "...the endless reactivation of updated images about our identity (a combination of the memories past and planned future) constitutes a sizable part of the state of self as I understand it." These memories are probably based on strong synaptic connections between cortical columns, with little involvement from limbic inputs required by more recent memories (cf. Murre *(63,76)*). Strengthening old memory patterns related to one's self is very much in line with the "Self-Maintenance-Therapy" (Selbst-Erhaltungs-Therapie) proposed by Romero *(77)* on quite different theoretical grounds and used in the treatment of the early stages of AD. In this therapy, patients are required to tell stories recalling various events of their life as a means to strengthen their self.

Compensation effects should selectively reinforce strong synaptic connections. This may be achieved through a combination of Self-Maintenance-Therapy (with help from family members) with drugs that allow for a short period of emotional arousal, thus increasing synaptic plasticity.

- "Cool the brain": simplify the brain dynamics to avoid memory interferences.

Formation of new memory patterns or activation of existing memories requires repetitive high-frequency reverberations in the neocortex. For example, hearing and recognizing a real word leads to a noticeable rise in the EEG frequency, in comparison with a pseudo-word, that is, a meaningless combination of phonemes *(78)*. Research on rats showed large effects of temperature on hippocampal field potentials *(79)*. Integrated electrical activity of cortical columns gives a measure of the overall activity of the brain. The power spectrum obtained from the multi-electrode EEG measurements should allow, in the limit of a large number of electrodes, to evaluate this energy. In an analogy to thermodynamics of systems far from thermal equilibrium, one could thus define the "brain temperature" and think about the synaptic runaway processes as overheating the system. Direct measurements of the brain temperature may also show interesting differences between patients. Only recently, a non-invasive technique for monitoring temperature through "Brain Temperature Tunnel" has been developed *(80)*. Monitoring blood flow to the brain is an alternative way of estimating the total energy used by the brain.

Blood pressure-lowering drugs significantly decrease the risk of dementia, including AD. This may be related to the decrease of overall brain temperature. Therefore, "cooling the brain," or reducing the average brain temperature, should decrease the effects of synaptic runaway and slow down the synaptic deletion processes. It may be achieved with the help of biofeedback, yoga, meditation, or other deep relaxation techniques. In particular, the alphabiofeedback is aimed at reducing the average EEG frequency ***(81)*** or achieving the "alpha relaxation state." Mental activities such as mantra repetition, chanting, visualization, or contemplative absorption should lower the brain temperature, stopping the background thoughts and other processes that may lead to the synaptic runaway. There is a lot of evidence in the medical literature showing various health benefits of such activities and neuroimaging studies detailing the effect of relaxation response and meditative practices ***(82)***.

Therefore, in the early stages of AD, it may be worthwhile to experiment with various relaxation techniques to slow down the development of the disease.

It should be possible to draw more detailed therapeutic suggestions from better models related to AD. The existing models should be extended in several directions. Human memory involves interactions among hippocampal formation, neocortex and neuromodulatory systems, regulating plasticity of synapses depending on the emotional contents of the situation ***(63,76,83)***. Such models have been initially created only at the conceptual level, but computational simulations followed ***(64,65)***. More realistic memory models that would allow studying the influence of different neurotransmitters on the inter-module inhibition and between-module excitation should help to evaluate potential benefits of new drugs. Models based on simplified spiking neurons are needed to make direct connections with neurophysiology. Many associative memory models based on simplified spiking neurons have been created recently and could be used in a near future to study the AD and other memory-related diseases. Other approaches to AD are discussed in **ref.** ***84***.

5. Conclusions

Neuroinformatics covers not only databases and computer software to analyze data from neuroscience experiments. Its ultimate goal should be an integration of all information about the brain (a good example here is the Visome platform ***(85)***). This should include data from neuroscience experiments and tools for analysis of such data but also tools for and results of computer simulations that provide models for various brain functions and dysfunctions. Simple neural models based on small number of assumptions allow for the qualitative understanding of experimental observations in neuroscience and may help to generate interesting ideas for new experiments. Biophysical models capture sufficient amounts of details to answer in silico precise questions that may be very difficult to answer experimentally.

A few models selected for this review show the potential of different approaches to neurological and psychiatric disorders, including Parkinson's disease and AD. An interesting area of neural modeling concerns reorganization processes following focal damages of neocortex (stroke, lesions) and damages to afferent pathways (amputation of limbs). Some therapeutic suggestions may be offered for faster recovery of sensorimotor competence after stroke ***(43)***, reduction of pain in phantom limb phenomena ***(44)*** and even such strange neuropsychological syndromes as the object-based unilateral hemi-neglect ***(55)***. Although therapeutic suggestions drawn here from AD models are speculative they are also worth testing.

Computer simulations appeared only quite recently as tool for modeling real brain processes. In view of the great complexity of the brain and lack of detailed understanding of its functions, skepticism toward such models may seem to be justified. There are many fundamental problems related to the convergence of computational models, hypothesis on which they are based (Alzforum.org in Chapter 19 provides a forum for AD hypothesis by users and experts), selection of minimal neural models that capture relevant phenomena and are still amenable to computer simulations. Surprisingly, even very simple neural models of associative memory show a number of features that reflect many properties of real biological memories known from cognitive psychology. Therefore, even rough neural models may show interesting properties, elucidating some brain mechanisms.

Neural models provide a new level of reasoning about brain diseases, level that cannot be adequately described in the language of psychiatry or psychopharmacology ***(30)***. They show how difficult it is to draw conclusions about causal mechanisms if only behavior is observed. Owing to the limited capacity of human working memory, verbal models, dominating in neuroscience, have to be relatively simple and cannot incorporate too many factors and interactions. Computational models do not have such limitations, and it is possible that such models will eventually capture all neuroscience knowledge, becoming repositories of the collective effort of many experts ***(85)***. However, so far, most computer simulations are aimed at the explanation of a single experiment or a single type of phenomena (with notable exceptions ***(86)***), and they are frequently based on in-house computer algorithms and programs. The creation of flexible simulators that provide in silico models for a wide range of phenomena is a great challenge for the future. There are already general purpose, low-level modeling systems, such as NEURON (Chapter 6) or GENESIS (Chapter 7), that provide models of specific dendrites, axons, whole neurons, or small networks. But in most cases, each new model has to be laboriously constructed from low-level elements.

There are two intermediate-level software packages designed for construction of computational models of brain functions. The PDP++ approach of O'Reilly and Munakata ***(26)*** has been used mostly to model different aspects of normal brain functions but was also used recently for cognitive deficits in Parkinson's disease ***(40)***. The Neural Simulation Language (NSL), developed by Weitzenfeld, Arbib, and Alexander ***(87)***, provides another simulation environment for modular brain modeling, including the anatomy of macroscopic brain structures. It offers object-oriented language applicable to all levels of neural simulation, providing high-level programming abstraction that use the Abstract Schema Language to create hierarchies of modules from the leaky integrator and other types of neural elements. Development tools are provided for visualization and analysis of models.

Availability of these type of neural simulators will eventually lead to the development of a community of experts who will use them as primary tools for analysis and understanding of experimental data, as well as generating new ideas about normal and pathological brain functions. On the hardware side, integrated circuits suitable for "neurophysiological" experimentations have already been constructed ***(88)***, and the whole neuromorphic sensory systems may slowly be developed ***(1)***. Computational models are certainly going to play an important role at every step of the long way leading to full understanding of the brain.

Acknowledgments

I am grateful to the Polish Committee of Scientific Research for a grant (2005–2007) that partially supported this research.

References

1. Rasche, C. (2005) *The Making of a Neuromorphic Visual System*. Springer-Verlag, New York.
2. McClelland, J.L. and Rumelhart, D.E, eds. (1986) *Parallel Distributed Processing: Explorations in the Microstructure of Cognition*, Vol. I, II. MIT Press, Cambridge, MA.
3. Taylor, J.G. (2003) Bubbles in the brain? *Trends in Cognitive Neurosci*ence **7**, 429–430.
4. Hoffmann, R.E. (1987) Computer simulations of neural information processing and the schizophrenia–mania dichotomy. *Archives of General Psychiatry* **44**, 178–188.
5. Reggia, J.A, Ruppin, E. and Berndt, R.S., eds. (1996) *Neural Modeling of Brain and Cognitive Disorders*. World Scientific, Singapore.
6. Reggia, J.A, Ruppin, E. and Glanzman, D.L., eds. (1999) *Disorders of Brain, Behavior, and Cognition: The Neurocomputational Perspective*. Elsevier, New York.

7. Scott Kelso, J.A. (1995) *Dynamic Patterns. The Self-Organization of Brain and Behavior*. Complex Adaptive Systems series. MIT Press, Bradford Book, Cambridge, MA.
8. Thelen, E. and Smith, L.B. (1994) *A Dynamic Systems Approach to the Development of Cognition and Action*. MIT Press, Cambridge, MA.
9. Smith, L.B. and Thelen, E., eds. (1994) *A Dynamic Systems Approach to the Development*. MIT Press, Cambridge, MA.
10. Parks, R.W., Levine, D.S. and Long, D., eds. (1998) *Fundamentals of Neural Network Modeling*. MIT Press, Cambridge, MA.
11. Maass, W. and Bishop, C. (1999) *Pulsed Neural Networks*. MIT Press, Bradford Book, Cambridge, MA.
12. Carnevale, N.T. and Hines, M.L. (2006) *The NEURON Book*. Cambridge University Press, Cambridge, UK.
13. Hines, M. (1993) NEURON – A program for simulation of nerve equations. In: *Neural Systems: Analysis and Modeling* (F. Eeckman, ed.). Kluwer Academic Publishers, Norwell, MA, pp. 127–136.
14. Bower, J.M. and Beeman, D. (1998) *The Book of GENESIS: Exploring Realistic Neural Models with the GEneral NEural SImulation System*, 2nd ed. Springer-Verlag, New York.
15. MacLean, P. (1990) *The Triune Brain in Evolution*. Plenum Press, New York.
16. LeDoux, J. (1999) *The Emotional Brain*. Phoenix, London.
17. Erdi, P. (2000) On the 'dynamic brain' metaphor. *Brain and Mind* **1**, 119–145.
18. Ehlers, C.L. (1995), Chaos and complexity: Can it help us to understand mood and behavior? *Archives of General Psychiatry* **52**, 960–964.
19. IBM Blue Brain Project, http://bluebrainproject.epfl.ch/
20. Anderson, J.A. and Rosenfeld, E. (1988) *Neurocomputing*. The MIT Press, Cambridge, MA.
21. Callaway, E., Halliday, R., Naylor, H., Yano, L. and Herzig, K. (1994) Drugs and human information processing. *Neuropsychopharmacology* **10**, 9–19.
22. Stein, D.J., ed. (1997) *Neural Networks and Psychopathology*. Cambridge University Press, Cambridge, UK.
23. Duch W. and Mandziuk J. (2004) Quo Vadis computational intelligence? In: *Machine Intelligence. Quo Vadis? Advances in Fuzzy Systems - Applications and Theory* (P. Sinčák, J. Vaščák, K. Hirota, Eds), Vol. 21. World Scientific, Singapore, pp. 3–28.
24. Hopfield, J.J. (1982) Neural networks and physical systems with emergent collective computational abilities. *Proceedings of the National Academy of Sciences of the United States of America* **79**, 2554–2558.
25. Kohonen, T. (1995) *Self-Organizing Maps*. Springer-Verlag, New York.
26. O'Reilly, R.C. and Munakata, Y. (2000) *Computational Explorations in Cognitive Neuroscience: Understanding the Mind by Simulating the Brain*. MIT Press, Cambridge, MA.

27. Xing, J. and Gerstein, G.L. (1996) Networks with lateral connectivity. I. Dynamic properties mediated by the balance of intrinsic excitation and inhibition. II. Development of neuronal grouping and corresponding receptive field changes. III. Plasticity and reorganisation of somatosensory cortex. *Journal of Neurophysiology* **75**, 184–232.
28. Ruppin, E. (1995) Neural modeling of psychiatric disorders. *Network* **6**, 635–656.
29. Levine, D.S. (2000) *Introduction to Neural and Cognitive Modeling*, 2nd ed. Lawrence Erlbaum Associates, Hillsdale, NJ.
30. Duch, W. (1997) Platonic model of mind as an approximation to neurodynamics. In: *Brain-Like Computing and Intelligent Information Systems* (S.-I. Amari, N. Kasabov, eds). Springer, Singapore, chap. 20, pp. 491–512.
31. Shepard, R.N. (1987) Toward a universal law of generalization for psychological science. *Science* **237**, 1317–1323.
32. Duch, W. (1996) Categorization, Prototype Theory and Neural Dynamics. Proceedings of the 4th International Conference on SoftComputing '96, Iizuka, Japan (T. Yamakawa, G. Matsumoto, eds), pp. 482–485.
33. Crystal, H. and Finkel, L. (1996) Computational approaches to neurological disease. In: *Neural Modeling of Brain and Cognitive Disorders*. World Scientific, Singapore, pp. 251–272.
34. Jefferys, J.G.R. (1998) Mechanisms and experimental models of seizure generation. *Current Opinions in Neurology* **11**, 123–127.
35. Traub, R.D., Whittington, M.A., Stanford, I.M. and Jefferys, J.G.R. (1996) A mechanism for generation of long-range synchronous fast oscillations in the cortex. *Nature* **382**, 621–624.
36. Traub, R.D., Whittington, M.A., Buhl, E.H., LeBeau, F.E., Bibbig, A., Boyd, S., Cross, H. and Baldeweg, T. (2001) A possible role for gap junctions in generation of very fast EEG oscillations preceding the onset of, and perhaps initiating, seizures. *Epilepsia* **42**, 153–170.
37. Bragin, A., Mody, I., Wilson, C.L. and Engel, J., Jr. (2002) Local generation of fast ripples in epileptic brain. *Journal of Neuroscience* **22**, 2012–2021.
38. Borrett, D.S., Yeap, T.H. and Kwan, H.C. (1993) Neural networks and Parkinson's disease. *Canadian Journal of Neurological Science* **20**, 107–113.
39. Edwards, R., Beuter, A. and Glass, L. (1999) Parkinsonian tremor and simplification in network dynamics. *Bulletin of Mathematical Biology* **61**, 157–177.
40. Frank, M.J. (2005) Dynamic dopamine modulation in the basal ganglia: a neurocomputational account of cognitive deficits in medicated and non-medicated Parkinsonism. *Journal of Cognitive Neuroscience* **17**, 51–72.
41. Buonomano, D.V. and Merzenich, M.M. (1998) Cortical plasticity: from synapses to maps. *Annual Review of Neuroscience* **21**,149–186.
42. Erwin, E., Obermayer, K. and Schulten, K. (1995) Models of orientation and ocular dominance columns in the visual cortex: a critical comparison. *Neural Computation* **7**, 425–468.

43. Reggia, J., Goodall, S., Chen, Y., Ruppin, E. and Whitney, C. (1996) Modeling post-stroke cortical map reorganization. In *Neural Modeling of Brain and Cognitive Disorders*. World Scientific, Singapore, pp. 283–302.
44. Spitzer, M. (1996) Phantom limbs, self-organizing feature maps, and noise-driven neuroplasticity. In *Neural Modeling of Brain and Cognitive Disorders*. World Scientific, Singapore, pp. 273–282.
45. Pearson, J.C., Finkel, L.H. and Edelman, G.M. (1987) Plasticity in the organization of adult cerebral cortical maps: a computer simulation based on neuronal group selection. *Journal of Neuroscience* **7**, 4209–4223.
46. Finkel, L.H., Pearson, J.C. and Edelman, G.M. (1997) Models of topographic map organization. In: *Pattern Formation in the Physical and Biological Sciences* (H.F. Nijhout, L. Nadel, D. Stein, F. Nijhout, eds). Addison-Wesley, Reading, MA.
47. Mazza, M., de Pinho, M., Piqueira, J.R.C. and Roque, A.C. (2004) A dynamical model of fast cortical reorganization. *Journal of Computational Neuroscience* **16**(2), 177–201.
48. Ramachandran, V.S. and Hirstein, W. (1998) The perception of phantom limbs. *Brain* **121**, 1603–1630.
49. Aglioti, S., Cortese, F. and Franchini, C. (1994) Rapid sensory remapping in the adult brain as inferred from phantom breast perception. *Neuroreport* **5**, 473–476.
50. Sptizer, M., Bohler, P., Weisbrod, M. and Kishka, U. (1996). A neural network model of phantom limbs. *Biological Cybernetics* **72**, 197–206.
51. McGonigle, D.J., Hänninen, R., Salenius, S., Hari, R., Frackowiak R.S.J. and Frith, C.D. (2002) Whose arm is it anyway? An fMRI case study of supernumerary phantom limb. *Brain* **125**, 1265–1274.
52. Cohen, J.D., Romero, R.D., Servan-Schreiber, D. and Farah, M.J. (1994) Mechanisms of spatial attention: the relation of macrostructure to microstructure in parietal neglect. *Journal of Cognitive Neuroscience* **6**, 377–387.
53. Thier, P. and Karnath, H.O., eds (1997) *Parietal Lobe Contribution in Orientation in 3D Space*. Springer-Verlag, New York.
54. Rao, R.P.N. and Ballard, D.H. (1997) A computational model of spatial representations that explains object-centered neglect in parietal patients. In: *Computational Neuroscience: Trends in Research* (J.M. Bower, ed.). Plenum Press, New York.
55. Deco, G. and Rolls, E.T. (2002) Object-based visual neglect: a computational hypothesis. *European Journal of Neuroscience* **16**(10), 1994–2000.
56. Deco, G. and Rolls, E.T. (2005) Neurodynamics of biased competition and cooperation for attention: a model with spiking neurons. *Journal of Neurophysiology* **94**, 295–313.
57. Morris, J.C., Storandt, M., Miller, J.P., McKeel, D.W., Price, J.L., Rubin E.H. and Berg, L. (2001) Mild cognitive impairment represents early-stage Alzheimer disease, *Archives of Neurology* **58**, 397–405.
58. McClelland, J.L., McNaughton, B.L. and O'Reilly, R.C. (1995) Why there are complementary learning systems in the hippocampus and neocortex: insights from

the successes and failures of connectionist models of learning and memory. *Psychological Review* **102**, 419–457.
59. Hasselmo, M.E. and McClelland, J.L. (1999) Neural models of memory. *Current Opinion in Neurobiology* **9**, 184–188.
60. Horn, D, Ruppin, E., Usher, M. and Herrmann, M. (1993) Neural network modeling of memory deterioration in Alzheimer's disease. *Neural Computation* **5**, 736–749.
61. Ruppin, E. and Reggia, J. (1995) A neural model of memory impairment in diffuse cerebral atrophy. *British Journal of Psychiatry* **166**(1), 19–28.
62. Horn, D., Levy, N. and Ruppin, E. (1996) Neuronal-based synaptic compensation: a computational study in Alzheimer's disease. *Neural Computation* **8**, 1227–1243.
63. Murre, J.M. (1996) TraceLink: a model of amnesia and consolidation of memory. *Hippocampus* **6**, 675–684.
64. Meeter, M. and Murre, J.M.J. (2004) Simulating episodic memory deficits in semantic dementia with the TraceLink model. *Memory* **12**, 272–287.
65. Meeter, M. and Murre, J.M.J. (2005) TraceLink: a model of consolidation and amnesia. *Cognitive Neuropsychology* **22**(5), 559–587.
66. Hasselmo, M.E. (1991) Runaway synaptic modification in models of cortex: implications for Alzheimer's disease. *Neural Networks* **7**(1), 13–40.
67. Hasselmo, M.E. (1997) A computational model of the progression of Alzheimer's disease. *MD Computing* **14**(3), 181–191.
68. Hasselmo, M.E. (1999) Neuromodulation: acetylcholine and memory consolidation. *Trends in Cognitive Sciences* **3**, 351–359.
69. Menschik, E.D. and Finkel, L.H. (1998) Neuromodulatory control of hippocampal function: towards a model of Alzheimer's disease. *Artificial Intelligence in Medicine* **13**, 99–121.
70. Menschik, E.D. and Finkel, L.H. (1999) Cholinergic neuromodulation and Alzheimer's disease: From single cells to network simulations. *Progress in Brain Research* **121**, 19–45.
71. Menschik, E.D. and Finkel, L.H. (2000) Cholinergic neuromodulation of an anatomically reconstructed hippocampal CA3 pyramidal cell. *Neurocomputing* **32–33**, 197–205.
72. Menschik, E.D., Yen S.-C. and Finkel, L.H. (1999) Model and scale independent performance of a hippocampal CA3 network architecture. *Neurocomputing* **26–27**, 443–453.
73. Mesulam, M. (2004) The cholinergic lesion of Alzheimer's disease: pivotal factor or side show? *Learning and Memory* **11**, 43–49.
74. Duch, W. (2000) Therapeutic applications of computer models of brain activity for Alzheimer disease. *Journal of Medical Informatics and Technologies* **5**, 27–34.
75. Damasio, A.R. (1996) *Descartes' Error: Emotion, Reason and the Human Brain*. Papermac, London.
76. Murre, J.M. (1997) Implicit and explicit memory in amnesia: some explanations and predictions by the TraceLink model. *Memory* **5**(1–2), 213–232.

77. Romero, B. (1998) Self-maintenance-therapy (SMT) in early Alzheimer Disease. *European Archives of Psychiatry and Clinical Neuroscience* **248**, 13–14.
78. Pulvermuller, F. (2003) *The Neuroscience of Language. On Brain Circuits of Words and Serial Order*. Cambridge University Press, Cambridge, UK.
79. Andersen, P. and Moser, E.I. (1995) Brain temperature and hippocampal function. *Hippocampus* **5**(6), 491–498.
80. Haddadin, A.S., Abreu, M.M., Silverman, D.G., Luther, M. and Hines, R.L. (2005) Noninvasive assessment of intracranial temperature via the medial canthal-brain temperature tunnel. *American Society of Anesthesiology Annual Meeting*, Abstract A38.
81. Criswell, E. (1995) *Biofeedback and Somatics*. Freeperson Press, Novato, CA.
82. Newberg, A.B. and Iversen, J. (2003) The neural basis of the complex mental task of meditation: neurotransmitter and neurochemical considerations. *Medical Hypotheses* **61**(2), 282–291.
83. Banquet, J.P, Gaussier, P., Contreras-Vidal, J.L., Gissler, A., Burnod, Y. and Long, D.L. (1998) A neural model of memory, amnesia and cortico-hippocampal interactions. In: *Disorders of Brain, Behavior, and Cognition: The Neurocomputational Perspective*. Elsevier, New York, pp. 77–120.
84. Adeli, H., Ghosh-Dastidar, S. and Dadmehr, N. (2005) Alzheimer's disease and models of computation: imaging, classification, and neural models. *Journal of Alzheimer's Disease* **7**(3), 187–199.
85. Usui, S. (2003) Visiome: neuroinformatics research in vision project. *Neural Networks* **16**, 1293–1300.
86. Wallenstein, G.V. and Hasselmo, M.E. (1997) Are there common neural mechanisms for learning, epilepsy, and Alzheimer's disease? In: *Neural Networks and Psychopathology*. Cambridge University Press, Cambridge, UK, pp. 314–346.
87. Weitzenfeld, A., Arbib, M.A. and Alexander, A. (2002) *The Neural Simulation Language. A System for Brain Modeling*. MIT Press, Cambridge, MA.
88. Fusi, S., Del Giudice, P. and Amit, D.J. (2000) Neurophysiology of a VLSI spiking neural network. Proceedings of the IEEE-INNS-ENNS International Joint Conference on Neural Networks, Vol. III, 121–126.

18

Integrating Genetic, Functional Genomic, and Bioinformatics Data in a Systems Biology Approach to Complex Diseases: Application to Schizophrenia

F. A. Middleton, C. Rosenow, A. Vailaya, A. Kuchinsky, M. T. Pato, and C. N. Pato

Summary

The search for DNA alterations that cause human disease has been an area of active research for more than 50 years, since the time that the genetic code was first solved. In the absence of data implicating chromosomal aberrations, researchers historically have performed whole genome linkage analysis or candidate gene association analysis to develop hypotheses about the genes that most likely cause a specific phenotype or disease. Whereas whole genome linkage analysis examines all chromosomal locations without a priori predictions regarding what genes underlie susceptibility, candidate gene association studies require a researcher to know in advance the genes that he or she wishes to test (based on their knowledge of a disease). To date, very few whole genome linkage studies and candidate gene studies have produced results that lead to generalizable findings about common diseases. One factor contributing to this lack of results has certainly been the previously limited resolution of the techniques. Recent technological advances, however, have made it possible to perform highly informative whole genome linkage and association analyses, as well as whole genome transcription (transcriptome) analysis. In addition, for the first time we can detect structural DNA aberrations throughout the genome on a fine scale. Each of these four approaches has its own strengths and weaknesses, but taken together, the results from an integrated analysis can implicate highly promising novel candidate genes. Here, we provide an overview of the integrated methodology that we have used to combine high-throughput genetic and functional genomic data with bioinformatics data that have produced new insights into the potential biological basis for schizophrenia. We believe that the potential of this combined approach is greater than that of a single mode of discovery, particularly for complex genetic diseases.

From: *Methods in Molecular Biology, vol. 401: Neuroinformatics*
Edited by: Chiquito Joaquim Crasto © Humana Press Inc., Totowa, NJ

Key Words: Linkage; association; family-based association; SNP; gene expression; genetic analysis; systems biology.

1. Introduction

In order to lay the groundwork for describing our combined systems biology approach, we first review the basic approaches to studying simple and complex genetic disorders, and then, using our data set developed for the study of neuropsychiatric disease as an example, we demonstrate how combinatorial methods offer a picture whose whole is greater than the sum of each individual part.

1.1. What Is a Genetic Disease?

Simply put, a genetic disease is an unhealthy condition that is heritable. If the disease is produced by inheritance of one or two copies of a single DNA sequence alteration, and shows up in a highly predictable pattern in a family or set of families, then the disease is often termed a "simple" genetic disease. If, however, more than one distinct DNA sequence alteration must be present for the disease to occur, or if the occurrence of the disease requires a strong epigenetic or environmental trigger, then the disease is described as a "complex" disease. Two clear-cut examples of simple and complex genetic diseases are Huntington's disease (HD) and schizophrenia. In HD, subjects with the disorder typically develop their first symptoms (cognitive deficits) in their third or fourth decade, followed by progressive development of frequent uncontrollable movements of the arms or legs, termed "chorea." The cause of HD is the presence of a DNA sequence abnormality on chromosome 4 in the gene *Huntingtin*, which directly alters the amino acid sequence of the protein. This particular DNA sequence abnormality is termed a trinucleotide repeat, involving three bases (CAG) that code for the amino acid glutamine. Subjects with HD have 30 or more copies (up to 100) of the trinucleotide repeat in their *Huntingtin* gene, which leads to a highly abnormal protein product containing polyglutamine repeats. The expanded CAG repeats function as an autosomal dominant allele. If a subject has HD, then their offspring have a 50% chance of inheriting the expanded trinucleotide repeat. Thus, HD is an example of a fairly simple genetic disease caused by a single DNA sequence alteration of a single gene. It is also an example of an autosomal dominant disease, because the mutation does not occur on one of the sex chromosomes and a subject only needs to inherit a single copy to show the disease. In contrast to HD, the genetics of schizophrenia are much more complex.

Schizophrenia is a chronic and debilitating mental disorder that affects 1% of the population worldwide. Patients with schizophrenia typically experience the onset of symptoms in their twenties or thirties and show a progressive deterioration of function. The symptoms shown include "positive symptoms" (delusions and hallucinations) and "negative symptoms" (apathy and impaired thought associated with cognitive deficits). Although a small number of genes have been identified that may underlie the cause of schizophrenia in small isolated families, no mutations have yet been identified that could account for the occurrence of the disease worldwide. Rather, it is now generally agreed that schizophrenia is a complex disorder influenced by genes, environmental risk factors, and their interaction ***(1)***. Examples of this fact can be found in twin studies. In monozygotic twins, if one twin has been diagnosed, the risk for the second twin developing schizophrenia is approximately 50%. This is certainly less than an autosomal dominant disease such as HD, but much greater than the 1% global risk. Likewise, if one has a first-degree relative with schizophrenia (parent or sibling), then there is approximately a 10% risk of developing the disease, or 10 times the global risk. Such findings have been reinforced in well-controlled studies of the children of schizophrenic parents who have been raised by adoptive parents. For example, when the children of schizophrenic mothers were raised by adoptive parents, they showed higher rates of schizophrenia as adults (13%), compared with control adoptees (approximately 2%) ***(2)***. Taken together, the family, twin, and adoption studies provide solid evidence that genes influence the susceptibility to schizophrenia. At the same time, they underscore the fact that the concordance rate in monozygotic twins is much less than 100%.

Researchers also often attempt to infer the means in which a genetic disorder is passed on from one generation to the next using segregation analysis. This approach tests whether the observed incidence of disease matches up with the expected incidence, under different assumptions (using models that test a single major locus, or many loci, specifying various inheritance modes, penetrances, and disease allele frequencies). Although there are reports of rare single locus effects in isolated families with schizophrenia, in general segregation models that assume a single major locus do not explain either the worldwide prevalence or the 50% discordance rate in monozygotic twins. On the contrary, it does appear that several genes with minor effects, possibly acting in combination with each other (through epistasis), or in combination with environmental factors, may be important in determining the degree of vulnerability for developing the illness ***(3)***.

1.2. Approaches to Genetic Disease

In any given search for disease-causing mutations, there are a number of steps that need to be undertaken to ensure that the sample can adequately address the experimental question. Here, we review some of the major criteria and variables that researchers should consider for inclusion.

1.2.1. Choosing the Study Population

In genetic studies that seek to define the chromosomal location of a disease-causing mutation, one must first ascertain an informative study population. One common practice is to focus on geographically isolated regions with a high rate of familiality of the disease. By doing so, the study will benefit from the power provided by a more genetically homogeneous population and possibly a more homogeneous disease phenotype or in other words the same causative genetic defect. However, it is also important to consider the major drawback to using only a population isolate; that is, it may be difficult to generalize and externally validate your results in other worldwide populations. In our studies of schizophrenia and bipolar disorder, we have made extensive use of the population isolate approach, with a sample that is highly representative of European and American Caucasians populations.

1.2.2. Single Versus Multi-Family

If the aim of a study is to define a set of candidate genes that could be generalized across populations, or at least within samples from the same ethnic background, then it is often more appropriate to employ a multi-family study design. To study genetic diseases and find evidence of transmission of risk genes in a family structure, there are two main methods of analysis—family-based association and linkage. In these studies, the goal is to ascertain as many sets of parents and their affected and unaffected children and grandchildren as possible. The family-based association testing merely screens for evidence of overtransmission of specific markers to affected versus unaffected subjects. Family-based linkage analysis, on the contrary, attempts to define the chromosomal regions that have been inherited across one or more generations in all the affected subjects within one or more families.

1.2.3. Case–Control

If one does not have access to family structure, but can ascertain a large number of affected and unaffected subjects, then it is also possible to implement a population-based case–control study. In this study design, the goal is to define the set of markers that are most over-represented in the group of affected subjects compared with the unaffected subjects, but one need not search for evidence of transmission of these alleles. This study is also referred to as a linkage disequilibrium (LD) study.

1.2.4. Choosing the Variables

In any genetic linkage or association study, there are two types of data—those assessing the trait of interest and those assessing the DNA sequence. Trait assessments are highly variable, because they are tailored to the disease of interest. However, a distinction is often made between a qualitative trait (having a disease or not having a disease) and a quantitative trait (degrees of impairment or functioning on a continuous scale). The two most commonly used marker types in genetic studies are microsatellites and single-nucleotide polymorphisms (SNPs). Microsatellites are defined as highly polymorphic DNA sequence repeats that are usually present in noncoding regions surrounding genes. SNPs, on the contrary, are a change in a particular nucleotide in the DNA that may be fairly rare or more common, and these can occur in intergenic, intronic, and exonic regions of the genome. More recently, it has become a common practice to convert SNP data into haplotype data, which refers to the specific sequence present in a set of two or more SNPs located in close physical proximity on the same DNA strand. This, in essence, makes SNP data as informative as microsatellite data. In addition to SNPs and microsatellites, which are discrete variables, it is also possible to measure and consider gene expression data (the quantity of mRNA generated from a specific gene locus) and DNA copy number data (if deletions of amplifications have occurred) as continuous variables.

1.2.5. Boosting Statistical Power

To accurately identify chromosomal regions or alleles that distinguish affected from unaffected members of a population requires a study design with adequate power for detecting such events. The power of a genetic screen is strengthened by using homogeneous and informative populations, where the variation in DNA sequence or RNA expression values due to factors unrelated to the disease can be minimized, and the signal-to-noise ratio related to the disease can be greatly enhanced. Customarily, power analyses are performed before initiating sample collections. For gene expression screens, there are two types of power to consider—nominal and biological. The nominal power to detect meaningful changes in expression in a whole genome expression screen depends on three factors—the sensitivity of the assay (usually a DNA microarray) to detect the presence of an mRNA in a sample, the variance of that noise in a given set of samples, and the size of the dynamic range of change that can be measured in that expression level. On the contrary, the biological power of an expression assay for identifying gene effects is limited by three additional related factors—whether the potential candidate genes are regulated at the transcriptional level (versus translational or post-translational level), the amount of transcriptional change that can occur in a cell, and whether the candidate genes are expressed in the biological tissue that is available to measure. The latter of these factors may be the most limiting of all factors, in view of the fact that, for most cells in the body, only approximately 40–60% of the mRNA transcripts appear to be expressed at any given time. Thus, investigators should carefully consider the biological tissue being assayed in any given study. In genetic linkage studies, the statistical power is largely dependent on two factors—the ability of the marker set to capture the recombination events that have taken place (called the informativeness) and the actual amount of recombination that has taken place in chromosomal regions flanking the disease alleles in affected subjects. To boost the former, investigators can use highly polymorphic markers (microsatellites or SNP haplotypes), and to boost the latter, it is common practice to include several generations of subjects in a pedigree. In genetic association studies, the power to detect meaningful associations is largely determined by the allele frequency of the mutation (or markers that co-segregate with the mutation) in the case population compared with the control population, as well as the degree of difference in allele frequencies of markers unrelated to the disease in the two study populations which can result in false positives. In order to maximize the difference between these two, investigators try to match the cases and controls as much as possible in terms of genetic background and often employ "supernormal" controls that have been thoroughly screened to minimize the frequency of disease-related alleles. One way around the genetic matching problem for association studies is to utilize discordant family members as controls, which forms the basis for family-based association studies. Here, the power is then mostly limited to the degree that the families being studied share the same risk alleles. Efforts are thus made to carefully match families from the same geographic regions or with the same phenotypic features.

1.3. Genetic Marker Analysis

There are two primary methods of discovering causal mutations in a disease. One method involves direct re-sequencing of the DNA to discover known or novel mutations in affected individuals. The second method involves performing genotyping of DNA to test whether subjects with a disease possess specific known DNA sequence alterations. If the specific DNA sequence alteration being tested is an amino acid-altering polymorphism, then causality can be inferred and tested. The other method of genotyping does not seek to identify a specific DNA sequence change but instead genotypes common polymorphisms, which are used as markers, to determine the smallest segment of a chromosome that is likely to contain disease-causing mutations. The first method of genetic mapping we discuss is genetic linkage analysis.

1.3.1. To Define Recombination Events—Linkage Studies

In a typical whole genome linkage analysis, at least one generation of affected and unaffected individuals from single or multiple pedigrees are genotyped with a common set of markers. The markers are used to define the regions of chromosomes that are shared by individuals with the disease and not shared by the subjects without the disease. The regions that are shared among affected individuals are said to be linked. The standard measure of significance in a linkage study is the limits of detection (LOD) score, which is the logarithm of the odds favoring linkage compared with no linkage. Parametric linkage analysis tests whether sets of markers are passed on to affected subjects, but not unaffected subjects, and whether you need one copy or two copies to best explain the occurrence of the disease. If a single copy of the marker(s) is sufficient, then that allele is said to be inherited in a dominant fashion. If two copies are required, then it is more likely to fit a recessive model. If one cannot predict the mode of inheritance, or sees obvious signs of mixed modes of inheritance, then one method that can overcome this is what is termed a nonparametric linkage analysis. In a nonparametric linkage analysis, the allele sharing is calculated among affected members of the different pedigrees, and the metric most often reported is the nonparametric linkage Z score (NPL Z score). One important consideration in interpreting LOD scores and NPL Z scores is the amount of confidence that should be assigned to different score levels. The best practice recommendation for conferring genome-wide significance is to use a cutoff that is equal to the score one could expect to obtain with randomized genotyping data in the same pedigree sets at a frequency of once in every 20 genome scans. On the contrary, genome-wide suggestive significance can be inferred when one attains a score that is equal to the maximal score expected with randomized genotyping data at a frequency of once per whole genome scan. Another important feature to consider with linkage studies that combine multiple markers (termed multipoint linkage) is the degree of co-segregation that exists between adjacent markers. If the SNPs are considered independent markers but are actually inherited together (are in LD with each other), then the statistical score estimating the amount of linkage occurring in affected subjects can be artificially enhanced. There are two main methods around this potential issue. The first is to reduce the number of SNPs that are being used in the linkage calculations, by trimming the list to include only makers that are not in LD with each other. This generally requires an intermarker D' score of less than 0.7 or an R^2 value of less than 0.1. Once these stretches of SNPs in high LD are identified, one can then either choose a single SNP from among them or deduce haplotypes and choose a unique haplotype-tagging SNP for the linkage analysis. Another method to reduce linkage score inflation is simply to construct the haplotype map to begin with, which will use all of the SNP and pedigree information to infer the recombination breakpoints and estimate haplotypes and their frequencies based on these data.

1.3.2. Genome-Wide Linkage Studies of Schizophrenia

Many investigators have concluded that schizophrenia has a complex, multifactorial etiology and that epistatic interactions among some of the common susceptibility genes is likely ***(4,5)***. Thus, to obtain the most accurate picture of the susceptibility genes, it is necessary to define the most important candidate genes. A large number of linkage studies have been performed in an attempt to define these candidate gene regions in different populations, and numerous genome-wide significant linkage peaks have been reported. However, almost none of these peaks occur in the same genomic regions. Meta-analyses have begun to clarify these conflicting data. For example, Lewis and colleagues ***(6)*** applied Genome Scan Meta-Analysis to 20 schizophrenia genome scans; the study group comprised a total of 1208 pedigrees and 2945 affected subjects from more than 12 different population sources. Evidence for linkage to chromosome 2q met their genome-wide level of statistical significance, whereas a number of other regions also appeared to show evidence as strong candidates, including 5q, 3p, 11q, 6p, 1q, 22q, 8p, 20q, and 14p.

1.3.3. Candidate Gene Studies of Schizophrenia

There have been an impressive number of candidate gene studies of schizophrenia performed to date. These studies have typically used small samples with relatively sparse coverage of the target gene. Looking across these studies, small but significant associations have been reported between schizophrenia and various neurotransmitter receptor genes, including the dopamine D2 and D3 receptor genes, the catechol *o*-methyltransferase gene, and the serotonin 2A receptors. Other candidate genes have been implicated through fine mapping studies of linkage peaks. No genome-wide association studies have been performed to date, however, using dense SNP marker maps in a comprehensive fashion.

1.4. To Define Cytogenetic/Chromosomal Abnormalities

Changes in the copy number of genes or alleles have now been well established as a mechanism for causing neuropsychiatric disorders. Examples of this include microdeletions, that result in only a single copy of one allele being present, and uniparental isodisomy, where a region of a chromosome contains only DNA representative of a single parent. In either of these situations, any disease-causing recessive alleles would be manifest, because of the absence of the normal dominant allele that would offset it. Gene amplification, on the contrary, has also been reported to cause neuropsychiatric disease, including Parkinson's disease among others *(7)*. It is now possible with microarray technology to screen for such changes in copy number in a comprehensive and unbiased manner.

1.4.1. Chromosomal Abnormalities in Schizophrenia

By far, the most direct evidence for chromosomal abnormalities in schizophrenia comes from the study of velocardiofacial syndrome (VCFS). VCFS is a condition caused by a loss of one copy of a small piece (2–3 Mb) of chromosome 22 (*see* **Fig. 1**). The symptoms of VCFS are complex and include cognitive deficits, mood instability, hearing defects, and craniofacial abnormalities, among others. Moreover, subjects with VCFS are approximately 25 times more likely to develop schizophrenia or bipolar disorder than the general population, and conversely, a much higher percentage of subjects with schizophrenia appear to have VCFS, compared with the general population *(8)*. Thus, a strong relationship exists between these two clinical and genetic entities. In nearly all cases of VCFS, a child is the first person in a pedigree who has it, meaning the causal microdeletion appears spontaneously in the absence of any family history. The microdeletion becomes heritable, however, as any hemizygous allele would be, with a 50% chance of the proband passing the deletion on to his or her children. The same is true of partial or complete isodisomy, where rather than losing a copy of a piece of a chromosome, there are two copies of either the maternal or paternal chromosomal segment. Inheriting two copies of a long piece of a chromosome from a single parent means that you would then display traits associated with any disease-causing recessive alleles. We have detected such changes occurring in at least one proband with major depressive disorder *(9)*. In addition, it is also possible with two copies of a chromosomal segment from a single parent to incur gene-silencing effects, because methylation patterns are inherited, and the piece that is duplicated may have been the silenced copy. This is what causes Prader–Willi and Angleman syndromes, among others. Finally, gene amplification is

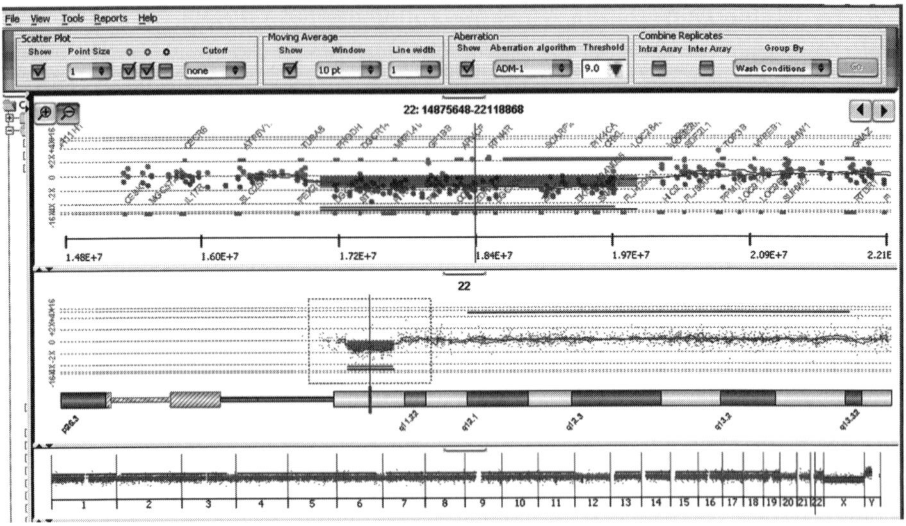

Fig. 1. Region of the velocardiofacial syndrome (VCFS) deletion on chromosome 22 detected by comparative genome hybridization (CGH). The figure shows a screenshot of CGH Analytics software. The three windows show all chromosomes (bottom window), chromosome 22 (middle window), and the hemizygous microdeleted region with its associated genes. Red and green points are probes on the array with measured intensities. The deletion is clearly identified on the q-arm of chromosome 22, in the 17–20 Mb region. A gene list can be saved from all the genes in the deleted region and imported into GeneSpring GT or GeneSpring GX. (*See* Color Plate 1.)

also a known mechanism for disease pathogenesis, particularly in the cancer field. To date, however, other than the study of VCFS, there have not been any definitive studies establishing changes in copy number or the parental origin of the copies in schizophrenia.

1.5. To Define Abnormal Expression Level—Functional Genomic Markers

To help accelerate the discovery of disease-causing genes, it can often be helpful to directly screen for changes in the transcript level of different genes. The rationale for choosing such an approach is obvious from work done on transgenic and knockout mice—the gene that has been mutated is almost always identified by a well-controlled functional genomic assay. The limitations on the ability to detect such differences in expression are often the simple question of whether the gene of interest is expressed in the tissue that is available for study. In brain disorders, the two most commonly used tissues are postmortem brain tissue and peripheral blood leukocytes (or transformed lymphoblasts).

1.5.1. Postmortem Tissue

The analysis of brain tissue obtained from subjects who have passed away with a disease has been used extensively to establish abnormal patterns of gene expression. In these studies, researchers often try to enrich their RNA samples with the brain regions that are deemed most critical to the disease process. The benefits of using such a tissue are that it can be assumed that the findings will have greater validity because of the use of the tissue where the disease process is known to occur. However, there are often severe drawbacks on studies performed with postmortem tissue, including brain pH, agonal state, medication effects, postmortem interval (autolysis time), and availability, to name only a few. Over the past 6 years, this approach has become widely adopted in the study of schizophrenia ***(10–16)***.

1.5.2. Biomarker Tissues

The definition of a biomarker tissue has evolved to generally mean any readily accessible peripheral tissue source that can provide a window into a disease or pathological process taking place in that tissue or another tissue in the body. The use of transcriptional profiling of peripheral blood leukocytes has now begun to be more commonplace in studies of brain disorders. The greatest limitation on these studies is the simple fact that the gene of interest which may be harboring a mutation may not be expressed in the blood. In our studies, for example, we have observed that at least 60% of the genes that are expressed in the cerebral cortex are also robustly expressed in the peripheral blood leukocyte population. On the contrary, it is much easier to match samples from living subjects, in terms of their cellular composition, medication effects, diet, and other variables that are not easily controlled for in postmortem studies. Only recently have researchers begun to utilize biomarker tissues such as peripheral leukocytes or transformed lymphoblasts in the study of schizophrenia, but already these studies have yielded very promising results ***(17,18)***.

1.6. Bioinformatics Data Sources

The ability to identify bona fide candidate genes in any genetic study is limited by one's knowledge of the function of the genes that are determined to harbor the most risk of that disease. With more than 30,000 full-length genes and tens of thousands of transcript variants, micro-RNAs, and noncanonical open reading frames (ORFs), the human genome annotation is undergoing constant development and refinement. Much of this information is now publicly accessible in major database interfaces, including the Gene Ontology, InterPro, pFam, KEGG, UCSC, and NCBI sites. Additional information, of course, can be found in actual published accounts. One major challenge is to develop tools that can cross-reference the results of genetic and functional genomic studies with these massive data sources, to develop knowledge networks for inference testing. Through free software (such as Cytoscape) and commercial software (such as Ingenuity and Pathway Assist), this goal is now within reach. In fact, these software applications can even use natural language-based text mining algorithms to directly integrate author statements with experimental data.

2. Example of Systems Biology Approach to Schizophrenia

2.1. Portuguese Island Collection

In the search for complex disease genes, it is often highly advantageous to perform studies of geographically isolated populations. Indeed, our combined genetic and functional genomic studies have made extensive use of a geographic population isolate, consisting of individuals living in the Madeiran and Azorean islands or their direct descendants. The Madeiran and Azorean islands were settled over 500 years ago almost exclusively by the Portuguese. The islands had no native population when they were first settled in the early 1400s. The settlement of the islands was methodically programmed with groups of families being awarded land and the right to settle different areas. The current population of Madeira is approximately 300,000 and that of the Azores approximately 250,000. Each set of islands is served by a modern centralized health system. Church records on the islands are excellent and provide a mechanism for tracing families for many generations.

Since 1995, we have been extensively ascertaining and collecting blood and DNA samples from the vast majority of familial cases of schizophrenia and bipolar disorder. We have achieved close to full ascertainment of subjects with schizophrenia in Sao Miguel, the most populated of the Azorean islands. On this island, 240 patients with schizophrenia (129 males, 111 females) were identified in a base population of 149,000, with an onset time-span of 62 years (1938–2000). A prevalence of 0.228% was calculated in a population of 105,152 adults over the age of 15 years. Familial cases were 68.9% and had an earlier age at onset than sporadic cases ($24.4 + 8.3$ SD versus $27.7 + 9.8$ SD) ($t = -2.315$, $P = 0.02$), with no influence of gender. These findings suggest that in this isolated population, the prevalence of schizophrenia is lower than reported in other populations. On the contrary, the familiality of schizophrenia, approximately 70%, appears to be far higher than the 10–15% often reported.

These observations strongly suggest that the Portuguese island population offers a stronger potential for discovering genetic causes of schizophrenia. Moreover, the greater genetic homogeneity of the population and potentially greater disease homogeneity should accelerate that process. In fact, we have hypothesized that the geographic isolation of the Portuguese islands has resulted in a higher percentage of families similarly sharing the same genetic risks of schizophrenia, possibly the same forms of illness, than in less isolated populations ***(19)***.

Importantly, despite the fact that the Portuguese islands are geographically isolated, the population inhabiting them appears to be highly representative of Caucasian populations. We have explored the question in two fundamental ways: genome-wide genetic diversity and local structure of haplotypes and LD. Each of the two angles was pursued with multiple analytic approaches and the end conclusion of all these analyses is that there is no significant difference detected between Azorean samples and those from mainland Portugal or of European descent in general. We conclude from this that the population of the Azores, despite their relative isolation and unique history, is not genetically distinct from mainland Europe in ways that would alter strategies for positional cloning in that population. On the contrary, we also conclude that, for the purposes of common disease, discoveries in this population are likely to carry over into most/all populations of European descent.

2.2. Description of Study

In the sections that follow, we briefly review our multi-faceted approach to gene discovery for schizophrenia in the Portuguese population. The approach we use includes new high-throughput tools for SNP genotyping and haplotyping, with subsequent linkage and family-based association analysis of these data. Next, we describe our functional genomic screens, using peripheral blood leukocytes from the same subjects used in the genetic screens, and our copy number analysis screens, using comparative genome hybridization (CGH) arrays. Finally, our method of bioinformatics integration of these data that have helped implicate novel pathways in schizophrenia is outlined.

2.2.1. Genetic Markers for Linkage and Family-Based Association

In the past 3 years, technological developments have made it possible to perform highly accurate and rapid genotyping of tens of thousands to hundreds of thousands of SNPs in individual DNA samples. These SNP genotypes have proven to be as useful for genetic analyses as more traditional microsatellites, which take much longer to genotype. We have previously shown ***(20)*** that there is greater power to detect linkage using high-density SNP genotyping panels [such as the Affymetrix 10K Human Mapping Assay (HMA)] compared with traditional 10-cM microsatellite-based scans. Specifically, in a linkage study of bipolar disorder, we obtained genome-wide significance using scans of exactly the same families and individuals who failed to attain this level of significance using microsatellites. We have suggested that the most likely explanations for the reduced power of microsatellite panels are the presence of prominent gaps in coverage and the reduced information content ***(20)***. Other researchers reported similar findings. In a study of 157 families segregating for rheumatoid arthritis, John and colleagues ***(21)*** used the same Affymetrix 10K assay we used in our earlier studies and compared the linkage results obtained for this platform with those found using a 10-cM microsatellite assay. Like our study, they obtained a genome-wide significant linkage peak with the SNP assay that failed to achieve this level of significance with the microsatellite panel. Moreover, four regions attained nominal significance in the SNP scan that had not been detected by the microsatellite scan. The SNP map also decreased the width of the 1-LOD-support intervals under linkage peaks and thus greatly reduced the number of potential candidate genes for follow-up study. Finally, Schaid and colleagues ***(22)*** compared the linkage results obtained for whole genome analysis of 467 men with prostate cancer from 167 families, using a panel of 400 microsatellites and the 10K HMA. They reported a small number of linkage peaks in that study with LOD scores exceeding 2.0 (up to 4.2) that were obtained with the 10K HMA but not detected in the microsatellite-based analyses, which they attributed to increased information content of the assay. The conclusions from these studies and ours, based on empirical studies of large pedigree sets in complex diseases, have been shown to have theoretical bases. In another study, using a series of simulations with different intermarker intervals, for affected sib-pair studies, Evans and Cardon ***(23)*** showed that linkage analysis with dense SNP maps extracts much more information than a 10 cM microsatellite map. They concluded "... the very low values of information content associated with sparse panels of microsatellite markers suggest that previous linkage studies that have employed these panels would benefit substantially from reanalysis with a dense map of SNPs. This is particularly true for sib-pair studies in which parents have not been genotyped (p. 691)."

2.2.2. Analysis of Dense SNP Data Sets

In the studies we have performed, we focus on the data obtained using genotypes from either the Affymetrix 10K or 50K HMAs for both family-based linkage and association. To date, there are very few software programs available that can incorporate dense SNP data and pedigree structures for both of these approaches. One free software platform that can accomplish these tasks is Multipoint Engine for Rapid Likelihood Inference (MERLIN) designed by Abecasis and colleagues ***(24)***. Another commercial software program that was specifically designed for these tasks is GeneSpring GT (Agilent). The basic workflow for genetic analysis in GeneSpring GT is outlined in **Box 1**. Briefly, after loading the genotyping data and pedigree structures into this program, we first clean the data, by checking for SNPs that violate Hardy–Weinberg Equilibrium, or contribute to a disproportionate number of inheritance errors (greater than 0.1%). Then, we use an expectation–maximization (EM) algorithm to deduce haplotypes and construct a map of these haplotypes, using the complete set of pedigrees available. Once this is accomplished, it is possible to proceed with haplotype-based single-point (single haplotype block) and multi-point (multiple blocks) downstream linkage analysis. Because of uncertainties regarding the mode of inheritance, disease allele frequency, and penetrance, we always begin our analyses using nonparametric linkage analysis. If a specific chromosomal region appears to be strongly implicated in a nonparametric linkage, then it should show even greater linkage using parametric testing, if a suitable inheritance model can be specified (dominant or recessive, with varying degrees of penetrance), and the specific families that share that linkage can be identified. Finally, family-based association testing is performed on these same families (or subsets of the families) using the same haplotype map used for linkage analysis, with the aim of identifying specific risk haplotypes that are over-transmitted to affected subjects from their affected parents.

The types of linkage results that can be obtained from the analysis of complex disorders are illustrated in **Fig. 2**, which present data based on a study of 25 Portuguese pedigrees segregating for bipolar disorder ***(20)***. Our original whole genome scan performed using a total of approximately 11,200 SNPs indicated that chromosome 6q contained the genome-wide maximally linked region in these pedigrees. We choose to reexamine the linkage signals in

Box 1. Analysis Workflow Within GeneSpring GT

Data cleaning (error removal)
Haplotype map construction
Nonparametric haplotype-based linkage
Parametric haplotype-based linkage
Haplotype-based family-based association
Defining genes with major and minor effects

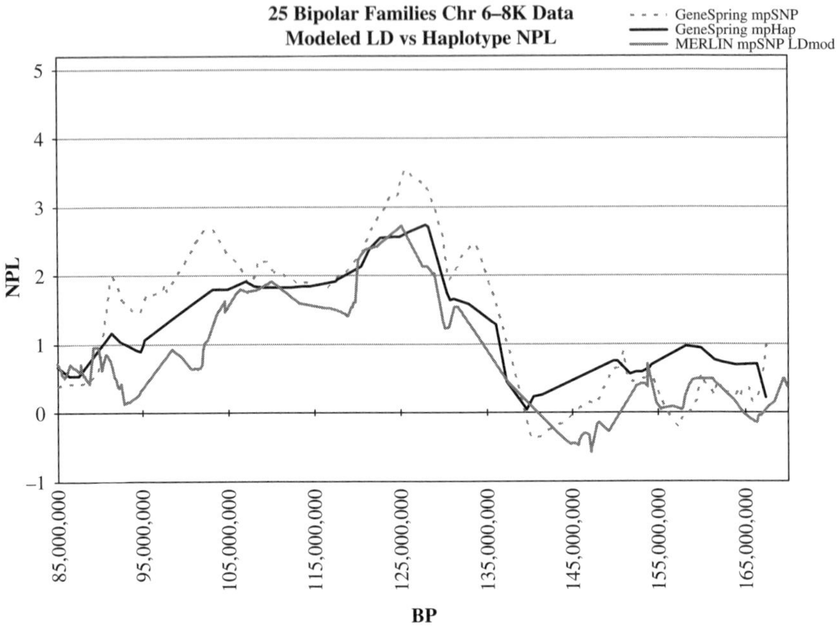

25 Bipolar Families Chr 6–8K Data
Modeled LD vs Haplotype NPL
GeneSpring mpSNP
GeneSpring mpHap
MERLIN mpSNP LDmod
NPL
5
4
3
2
1
0
–1
85,000,000
95,000,000
105,000,000
115,000,000
125,000,000
135,000,000
145,000,000
155,000,000
165,000,000
BP

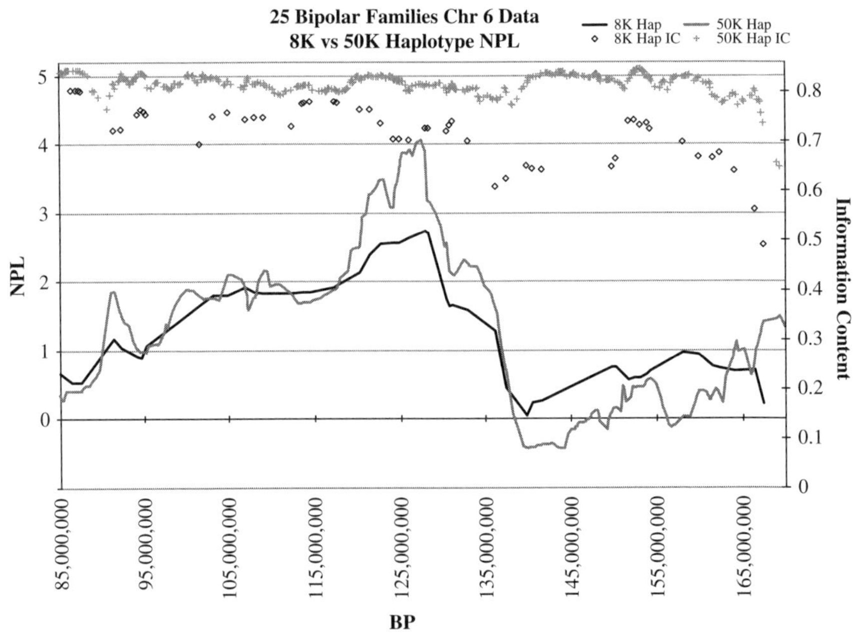

25 Bipolar Families Chr 6 Data
8K vs 50K Haplotype NPL
8K Hap
50K Hap
8K Hap IC
50K Hap IC
NPL
5
4
3
2
1
0
Information Content
0.8
0.7
0.6
0.5
0.4
0.3
0.2
0.1
0
85,000,000
95,000,000
105,000,000
115,000,000
125,000,000
135,000,000
145,000,000
155,000,000
165,000,000
BP

these same pedigrees at slightly lower (approximately 7800) and much higher (approximately 58,000) density of SNP coverage, using two different methods to control for intermarker LD (LD modeling in MERLIN and haplotype-based analysis in GeneSpring GT). The results from multipoint SNP linkage with and without LD modeling and multipoint haplotype-based linkage are shown for the region with the maximal score (chromosome 6q) in a study of 25 pedigrees segregating for bipolar disorder (*see* **Fig. 2**, top). Interestingly, we found that the peaks obtained by LD modeling in MERLIN and haplotype-based analysis in GeneSpring GT are remarkably similar, while multipoint SNP results alone clearly appeared to inflate the linkage score. On the basis of this direct comparison, we conclude that the use of EM-derived haplotypes is at least a first step toward controlling for potential inflation of linkage scores. In addition to this, we also noticed a clear effect of moving from an 8K density haplotype map to a 50K density haplotype map (*see* **Fig. 2**, bottom). As might be expected, by having more SNPs available, it was possible to construct more informative haplotypes that better captured the recombination events in the different pedigrees. The multipoint linkage results obtained with these more informative haplotype blocks greatly enhanced the signal on chromosome 6q, to an NPL Z score above 4.0. This represented more than a full Z score increase relative to the results obtained with the haplotype map constructed from 8K data. Thus, we conclude that in this population, the 50K data are far superior for the construction of informative haplotypes that can be used for linkage analysis.

In addition to boosting the linkage score and possibly improving the localization of the peaks, another advantage of having informative haplotypes is also the ability to test these for association. We employed the haplotype-based

◄

Fig. 2. Nonparametric linkage results for bipolar disorder. In the top panel, the results from multipoint single-nucleotide polymorphism (SNP) linkage with and without linkage disequilibrium (LD) modeling and multipoint haplotype-based linkage are shown for the region with the maximal score (chromosome 6q) in a study of 25 pedigrees segregating for bipolar disorder. These data are from a set of 7736 (8K) SNPs. Note that the peaks obtained by LD modeling in MERLIN and haplotype-based analysis in GeneSpring are quite similar, while multipoint SNP results alone appear to inflate the linkage score. In the lower panel, the results of going from 8K to 50K SNP coverage in construction of the haplotype map clearly provides more information and boosts the linkage signal beyond 4.0 in the peak region, even with the intermarker LD correction provided by the use of haplotypes. BP, base pair; Hap, haplotype; IC, information content; mp, multipoint; NPL, nonparametric linkage Z score.

haplotype relative risk (HHRR) algorithm in our set of 25 bipolar families and an additional 15 bipolar families to test for overtransmission of specific risk haplotypes using family-based association in GeneSpring GT. The results from this analysis are shown only for the peak chromosome linkage region on chromosome 6 (*see* **Table 1**) but provide direct support for the presence of several putative risk haplotypes in these bipolar families.

2.2.3. Markers for Gene Expression in Same Subjects and Families

While genetic studies often lead to the identification of putative candidate genes and risk haplotypes, this is only the first step, and greater support for specific candidates can be obtained by consideration of the functional consequences of genetic variability. One method of assessing function is to quantify the levels of expression of individual genes. We have been performing systematic studies on the utility and feasibility of high-throughput gene expression analysis of leukocyte samples from the Portuguese Island Collection (PIC) collection as assays of gene function in families with known patterns of linkage ***(20,25,26)***. As a result, we have developed an extensive database of gene expression data on more than 55 discordant sib-pairs from families with schizophrenia and bipolar disorder. These studies use primary leukocyte samples subjected to differential white blood cell

Table 1
Significant Family-Based Association Blocks in the Peak Linkage Region

Chromosome	Base pair	Marker	D	D′	Rho	Chi square	*P* value
6	123901852	rs4307191	0.07	0.32	0.28	6.67	0.0098
6	123903977	rs10499125	0.07	0.32	0.28	6.67	0.0098
6	124984287	rs9375346	0.03	1.00	0.23	5.28	0.0216
6	124984549	rs9321003	0.03	1.00	0.23	5.28	0.0216
6	125278562	rs781748	0.04	0.38	0.29	7.04	0.0296
6	125292115	rs6910745	0.04	0.38	0.29	7.04	0.0296
6	125310084	rs4896606	0.04	0.38	0.29	7.04	0.0296
6	125364046	rs10485345	0.04	0.38	0.29	7.04	0.0296
6	129744486	rs6899448	0.04	0.31	0.26	6.34	0.0419
6	129754042	rs4599678	0.04	0.31	0.26	6.34	0.0419
6	130143572	rs10484280	0.08	0.81	0.88	25	0.0016
6	130144652	rs9321180	0.08	0.81	0.88	25	0.0016
6	130161191	rs2326869	0.08	0.81	0.88	25	0.0016
6	130236569	rs10499165	0.08	0.81	0.88	25	0.0016
6	130284551	rs7753901	0.08	0.81	0.88	25	0.0016

counts. Only gender-matched, age-matched, and blood cell composition-matched discordant sib-pairs are included. One of our first general observations about this body of work was that both the brain and leukocytes express approximately half of the >22,000 genes present on this array and that there is considerable overlap (mean 59%; up to 92%) in the genes that are expressed in these different sources depending on the loci or the functional gene group being studied. The commonly expressed genes are present at all cytogenetic loci. At the functional level, every neurotransmitter signaling group we examined was also found to express many genes in common in the blood and the brain, and most groups even expressed a few genes in the blood that were not detectable in the brain (*see* **Table 2**). Other cellular function groups, such as the ribosomal group, and certain metabolic pathways expressed almost all of the same genes (>85%), as might be expected, given their essential roles in general cellular functions. We conclude that leukocyte expression analysis can provide information on the expression patterns of numerous genes that may play a critical role in brain function. This

Table 2
Functional Group Overlap in Expressed Genes from Brain and Blood

Functional gene group	Present in blood	Present in brain	Overlap in expression
Acetylcholine signaling	4	5	3/6 (50%)
Enzyme-Linked signaling	131	153	101/183 (55%)
Serotonin signaling	5	9	5/9(55%)
Integrin signaling	57	46	35/68 (51%)
GABA signaling	17	31	16/32 (50%)
Glutamate signaling	25	46	24/47 (51%)
Dopamine receptors	3	4	3/4 (75%)
Ligand binding/carriers	126	168	105/189 (56%)
Beta amyloid	27	39	23/43 (53%)
Synucleins	4	6	4/6 (75%)
Ribosomal	85	89	83/97 (86%)
Caspases	19	11	8/20 (40%)
Vacuolar	35	36	32/38 (84%)
Proteasome	53	56	52/57 (91%)
Oxidative phosphorylation	11	12	11/12 (92%)
Unfolded protein response	38	40	36/42 (86%)
Transcription factors	413	434	326/521 (63%)
Ubiquitin	111	113	97/127 (76%)
All mitochondria-related	298	368	273/393 (69%)

overlap implies that if any genetic variation affects the level of expression of a gene, then we would have approximately 60% power to detect changes in brain-enriched genes, merely through the screening of blood expression patterns.

The utility of the primary leukocyte expression data to actually classify or differentiate affected subjects from control subjects is evident in our studies of 33 age- and gender-matched discordant sibs with schizophrenia ***(18)***. In this study, of the 22,283 probe sets on the U133A GeneChip, a list of 302 significantly changed genes was obtained after Robust Multiarray Averaging (RMA) and pairwise Mann–Whitney testing with Benjamini–Hochberg correction for multiple comparisons. This list was further refined to a set of 20 genes that could correctly classify each of the 33 control subjects and 30 of 33 subjects with schizophrenia, thus surpassing 95% diagnostic accuracy (63/66 samples correctly diagnosed). Among the genes with the most consistent alterations in schizophrenic subjects were several with interesting pathophysiological roles in schizophrenia, including specific neuregulin 1 (NRG1) variants and a number of neurotransmitter-receptor transcripts, including several related to γ-aminobutyric acid (GABA)ergic, glutamatergic, and serotoninergic transmission. Furthermore, the data on the NRG1 expression changes were extended to a larger haplotype analysis in a larger set of families from the PIC collection and identified a pattern of polymorphisms associated with altered expression ***(25)***. Moreover, in another parallel study, we have recently shown even more significant evidence of promoter haplotypes regulating the levels of GABA receptor subunit expression in the peripheral leukocytes of schizophrenic subjects ***(26)***.

3. Bringing it All Together—Extended Comparisons of Expression Profiling, Comparative Genome Hybridization, and Genotyping

In an extended analysis, a nonparametric whole genome linkage scan of schizophrenia pedigrees using GeneSpring GT identified 322 genes located within 2 Mb of SNPs that showed evidence of linkage in a set of schizophrenia pedigrees analyzed at 10K density. Expression analysis using GeneSpring GX identified 2545 differentially expressed genes between the unaffected individuals and the affected siblings (*see* **Box 2**). Interestingly, the intersection of these two gene lists contained only 15 common genes.

Of the genes implicated by both linkage and association approaches, four genes had been previously associated with neuropsychiatric diseases. This gene list included *GAD2*, the gene that codes for the 65-kDa isoform of glutamic acid decarboxylase, an enzyme that synthesizes GABA from glutamate. Interestingly, *GAD65* has been strongly implicated in recent postmortem brain

Box 2. Analysis Workflow Within GeneSpring GX
Data cleaning (removing hypervariable and unexpressed genes)
Pairwise or unpaired parametric testing
Correction for multiple comparisons
Class prediction analysis
Gene group analysis (biological and mathematical clusters)
Defining genes and gene groups with major and minor effects

studies of subjects with schizophrenia *(27)*, and mice lacking *GAD2* have been shown to display prominent behavioral changes consistent with those seen in schizophrenia *(28)*.

Further analysis looked at potential biological pathways involved in schizophrenia. The union of the gene lists obtained through expression analysis and linkage analysis was constructed. Pathway enrichment analysis of this list showed that a significant number of these genes segregated to four pathways (*see* **Fig. 3A**). The pathway with the most significant Z score (a value of 4.77) was the HD pathway. Ten out of 25 genes from the pathway were differentially expressed. Importantly, one gene had a significant SNP in its vicinity that could potentially serve as neuropsychiatric disease biomarker.

3.1. Network Building

In addition to the HD pathway, the three other identified pathways were those for galactose metabolism, regulation of p27 phosphorylation during cell cycle progression, and purine metabolism. To explore relationships and illustrate genes in common between the pathways, we used the Agilent Literature Search tool (available as a free plugin from Cytoscape—http://www.cytoscape.org) (*see* **Fig. 3B**). A large literature-based association network was generated from approximately 450,000 PubMed abstracts retrieved for a set of disease-related queries (neuropsychiatric diseases such as schizophrenia, bipolar disorder, etc., and immune- and inflammation-related diseases such as atherosclerosis, type I diabetes, Alzheimer's disease, rheumatoid arthritis, etc.). These abstracts were processed by the Agilent Literature Search tool to identify approximately 39,000 biomolecular associations among 5400 genes. The shortest path among every pair of genes in the four canonical pathways identified above was extracted from the comprehensive association network to represent a literature-based extension of the four pathways (*see* **Fig. 3C**). The extended network was further filtered as follows to yield an "interesting" sub-network. This sub-network was constructed in three steps. First, SNP LOD scores and gene expression *P*-values were combined

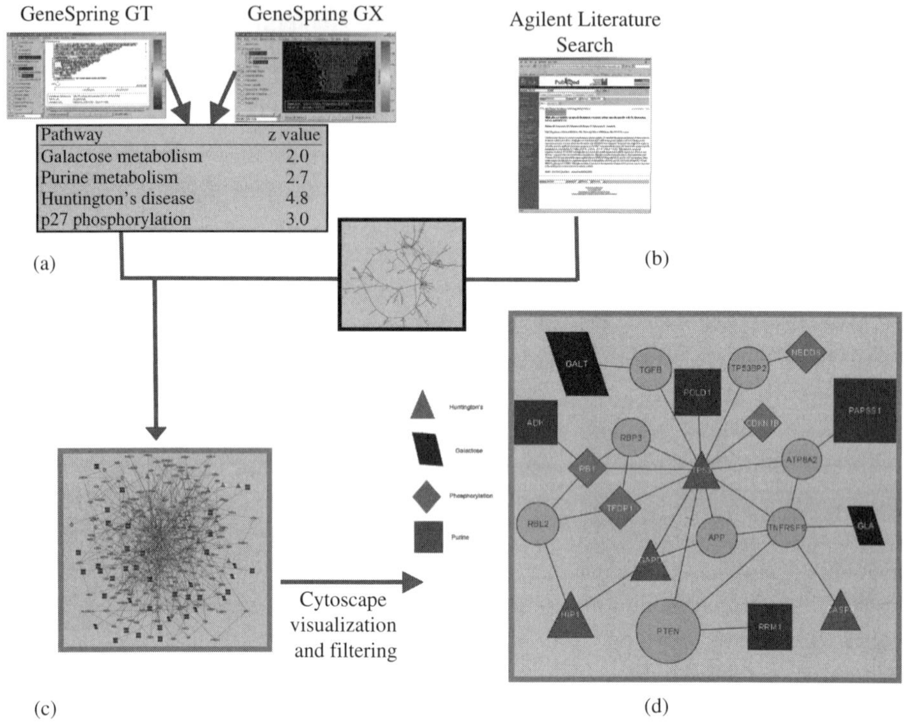

Fig. 3. Workflow for integrating gene expression profiling and genotype analysis. **(A)** Combined gene expression and genotype gene list-based pathway enrichment analysis identifies four canonical pathways. **(B)** Agilent Literature Search tool extracts associations from literature for disease-related genes and proteins. **(C)** Literature-based extended association network for the four canonical pathways identified by combined genotype and gene expression datasets. **(D)** "Interesting" sub-network identified from the extended network in **(C)** using combined single-nucleotide polymorphism (SNP) LOD score and gene expression *P*-values. (*See* Color Plate 2.)

for each gene to yield a single significance score (Z score). Then, an average Z score for a gene was computed based on the average of the Z scores of all first neighbors (directly connected by an edge). Finally, the network was filtered to extract all edges whose end nodes (genes) passed an average Z score threshold *T*. Cytoscape was then used to visualize the "interesting" sub-network (*see* **Fig. 3D**). Node shape and color attributes were set based on the canonical pathway membership of a gene. Node size was set to the average Z score for the gene based on the combined SNP LOD scores and gene expression *P*-values.

Eight genes were identified that connect the HD, galactose metabolism, purine metabolism, and p27 phosphorylation regulation pathway. These nodes (in pink) were identified as potentially significant genes/proteins for further study in the context of neuropsychiatric disease based on two properties—if their immediate neighbors on different canonical pathways were potential points of interaction between disease pathways, and if their immediate neighbors on the same canonical pathways are potentially additional nodes on other canonical pathways that have not yet been identified. In both cases, these kinds of nodes may be potential therapeutic targets. This could also be an indication that metabolic and signal transduction pathways are involved in neuropsychiatric disease progression.

3.2. Comparative Genome Hybridization

The evidence that chromosomal abnormalities play an important role in schizophrenia prompted us to analyze a subset of the individuals used in the genetic and functional genomic screens for DNA copy number changes. For this analysis, we used the Agilent CGH platform (44K and 95K arrays). As a positive control, and to establish the sensitivity of the instrument, we used a DNA sample from a subject with VCFS (kindly provided by W. Kates and R. Shprintzen) that showed a deletion on chromosome 22 (*see* **Fig. 1**). Indeed, as expected, this subject's hemizygous microdeletion (loss of one copy) was accurately mapped by the CGH Analytics software. To screen for DNA aberrations in the schizophrenia cases, we analyzed a total of 10 sib-pairs (20 individuals) for whom genetic linkage and gene expression data had also been acquired previously. In this analysis, a nonstringent aberration calling setting (moving 3 point average with a Z score threshold of 2.0) was applied to identify all regions that were amplified or deleted in each sib-pair. Next we collected all the probes that showed aberrations (amplifications and deletions) and identified their corresponding genes (3397 genes). This gene list was used in a pathway enrichment analysis similar to the Gene Expression/Genotyping gene list. We identified a number of neuropsychiatric pathways with high Z scores. The pathway with the highest Z score (7.0) was the "Role of erk5 in neuronal survival (ERK5)" pathway. Comparisons of the ERK5 pathway with the HD pathway revealed an intersection of these pathways through the *GRB2* gene (Growth Factor Receptor-Bound Protein 2). Two of the functions of *GRB2* are growth factor regulation of the cytoskeleton and DNA synthesis. However, owing to its ability to cause complex formation between proteins involved in growth signaling pathways, *GRB2* has been shown to play a central role in neutrophin signaling in neurons and consequently neuronal differentiation and survival.

We further used a GeneSpring Network building plug-in (*see* **Fig. 4**) to better understand the associations between genes in the Huntington and ERK5 pathways. The network building plug-in constructs a Cytoscape network, starting with the combined gene lists of the Huntington and ERK5 pathways, and gathering information on known interactions from multiple protein–protein interaction databases (cPath, Reactome, BIND), pathway information (GenMAPP and KEGG), and the scientific literature (Agilent literature search). A connection between two genes is built if any of the databases contains information about known interactions between them.

Figure 5 shows the resulting, new network based on database information for known associations between genes from the Huntington and ERK5 pathways. Lines between genes are color coded to indicate the database source of their associations. The thickness of the lines indicates the confidence in this interaction based on the number of sentences in the scientific literature that supports these associations.

Interestingly, the identification of the *GAD2* gene from the previous expression and linkage pathway analysis integrates nicely with the identification of pathways related to neurotrophin signaling in terms of potential involvement in schizophrenia. Indeed, recent studies of postmortem brain tissue have implicated both GABAergic and neurotrophic factor changes in the same populations of cells in schizophrenia *(29)*. Specifically, Lewis and colleagues concluded

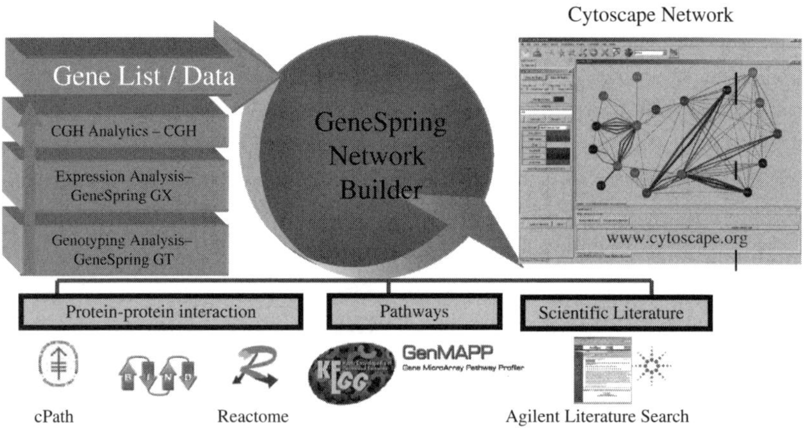

Fig. 4. GeneSpring Network builder. Schematic representation of the workflow for the plug-in. A gene list is selected and the selected databases are searched for gene–gene interactions. A network is built using the Cytoscape viewer. (*See* Color Plate 3.)

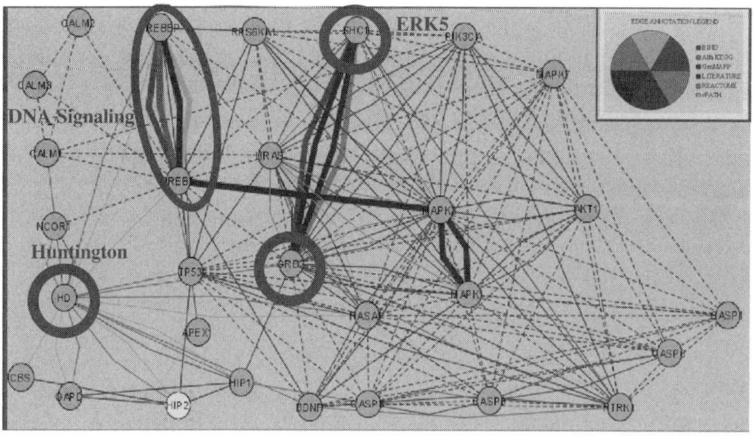

Fig. 5. The Huntington and ERK5 gene lists from the gene expression, genotyping, and comparative genome hybridization (CGH) experiments are used to build an interaction network using the GeneSpring Network builder. New and known interactions can be identified. *GRB2* is one of the connector genes between the genes of the ERK5 pathway and the Huntington pathway. (*See* Color Plate 4.)

that "... a deficiency in signaling through the TrkB neurotrophin receptor leads to reduced GABA synthesis in the parvalbumin-containing subpopulation of inhibitory GABA neurons in the dorsolateral prefrontal cortex of individuals with schizophrenia." Given our evidence for alterations in GABAergic signaling in peripheral blood of subjects with schizophrenia, particularly those who display certain risk haplotypes ***(26)***, we find compelling reasons to continue to focus on this central biological pathway in our studies of schizophrenia.

4. Conclusions

We have developed a straightforward and streamlined algorithm for combining the results of at least three different types of assays—ploidy (copy number), function genomic (gene expression), and genetic linkage and association studies. When the results of these approaches are combined with available bioinformatics tools, they can rapidly lead to the discovery of new candidate genes and biological pathways. We strongly believe that this systems biological approach offers far greater promise for discovering genes of major and minor effects that collectively interact with environmental variables to produce complex disease states.

References

1. Faraone, S.V., Glatt, S.J., and Taylor, L. (2004) The genetic basis of schizophrenia, in *Early Clinical Intervention and Prevention in Schizophrenia*, M.T. Tsuang, Editor. Humana Press: Totowa, New Jersey, pp. 3–25.
2. Kendler, K.S., Gruenberg, A.M., and Kinney, D.K. (1994) Independent diagnoses of adoptees and relatives as defined by DSM-III in the provincial and national samples of the Danish Adoption Study of Schizophrenia. *Arch Gen Psychiatry* **51**(6):456–68.
3. Faraone, S.V., and Tsuang, M.T. (1985) Quantitative models of the genetic transmission of schizophrenia. *Psychol Bull* **98**:41–66.
4. Gottesman, I.I., and Moldin, S.O. (1997) Schizophrenia genetics at the millennium: cautious optimism. *Clin Genet* **52**(5):404–7.
5. Risch, N. (1990) Linkage strategies for genetically complex traits. I. Multilocus models. *Am J Hum Genet* **46**(2):222–8.
6. Lewis, C.M., Levinson, D.F., Wise, L.H., et al. (2003) Genome scan meta-analysis of schizophrenia and bipolar disorder, part II: Schizophrenia. *Am J Hum Genet* **73**(1):34–48.
7. Singleton, A.B., Farrer, M., Johnson, J., et al. (2003) Alpha-synuclein locus triplication causes Parkinson's disease. *Science* **302**(5646):841.
8. Pulver, A.E., Nestadt, G., Goldberg, R., et al. (1994) Psychotic illness in patients diagnosed with velo-cardio-facial syndrome and their relatives. *J Nerv Ment Dis* **182**(8):476–8.
9. Middleton, F.A., Trauzzi, M.G., Shrimpton, A.E., et al. (2006) Complete maternal uniparental isodisomy of chromosome 4 in a subject with major depressive disorder detected by high density SNP genotyping arrays. *Am J Med Genet B Neuropsychiatr Genet* **141**(1):28–32.
10. Mirnics, K., Middleton, F.A., Marquez, A., Lewis, D.A., and Levitt, P. (2000) Molecular characterization of schizophrenia viewed by microarray analysis of gene expression in prefrontal cortex. *Neuron* **28**(1):53–67.
11. Mirnics, K., Middleton, F.A., Lewis, D.A., and Levitt, P. (2001) Analysis of complex brain disorders with gene expression microarrays: schizophrenia as a disease of the synapse. *Trends Neurosci* **24**(8):479–86.
12. Hakak, Y., Walker, J.R., Li, C., et al. (2001) Genome-wide expression analysis reveals dysregulation of myelination-related genes in chronic schizophrenia. *Proc Natl Acad Sci USA* **98**(8):4746–51.
13. Bahn, S., Augood, S.J., Ryan, M., Standaert, D.G., Starkey, M., and Emson, P.C. (2001) Gene expression profiling in the post-mortem human brain – no cause for dismay. *J Chem Neuroanat* **22**(1–2):79–94.
14. Vawter, M.P., Crook, J.M., Hyde, T.M., et al. (2002) Microarray analysis of gene expression in the prefrontal cortex in schizophrenia: a preliminary study. *Schizophr Res* **58**(1):11–20.

15. Middleton, F.A., Mirnics, K., Pierri, J.N., Lewis, D.A., and Levitt, P. (2002) Gene expression profiling reveals alterations of specific metabolic pathways in schizophrenia. *J Neurosci* **22**(7):2718–29.
16. Middleton, F.A., Peng, L., Lewis, D.A., Levitt, P., and Mirnics, K. (2005) Altered expression of 14-3-3 genes in the prefrontal cortex of subjects with schizophrenia. *Neuropsychopharmacology* **30**(5):974–83.
17. Tsuang, M.T., Nossova, N., Yager, T., et al. (2005) Assessing the validity of blood-based gene expression profiles for the classification of schizophrenia and bipolar disorder: a preliminary report. *Am J Med Genet B Neuropsychiatr Genet* **133**(1):1–5.
18. Middleton, F.A., Pato, C.N., Gentile, K.L., et al. (2005) Gene expression analysis of peripheral blood leukocytes from discordant sib-pairs with schizophrenia and bipolar disorder reveals points of convergence between genetic and functional genomic approaches. *Am J Med Genet B Neuropsychiatr Genet* **136**(1):12–25.
19. Pato, C.N., Azevedo, M.H., Pato, M.T., et al. (1997) Selection of homogeneous populations for genetic study: the Portugal genetics of psychosis project. *Am J Med Genet* **74**(3):286–8.
20. Middleton, F.A., Pato, M.T., Gentile, K.L., et al. (2004) Genomewide linkage analysis of bipolar disorder by use of a high-density single-nucleotide-polymorphism (SNP) genotyping assay: a comparison with microsatellite marker assays and finding of significant linkage to chromosome 6q22. *Am J Hum Genet* **74**(5):886–97.
21. John, S., Shephard, N., Liu, G., et al. (2004) Whole-genome scan, in a complex disease, using 11,245 single-nucleotide polymorphisms: comparison with microsatellites. *Am J Hum Genet* **75**(1):54–64.
22. Schaid, D.J., Guenther, J.C., Christensen, G.B., et al. (2004) Comparison of microsatellites versus single-nucleotide polymorphisms in a genome linkage screen for prostate cancer-susceptibility Loci. *Am J Hum Genet* **75**(6):948–65.
23. Evans, D.M., and Cardon, L.R. (2004) Guidelines for genotyping in genomewide linkage studies: single-nucleotide-polymorphism maps versus microsatellite maps. *Am J Hum Genet* **75**(4):687–92.
24. Abecasis, G.R., Cherny, S.S., Cookson, W.O., and Cardon, L.R. (2002) Merlin – rapid analysis of dense genetic maps using sparse gene flow trees. *Nat Genet* **30**(1):97–101.
25. Petryshen, T.L., Middleton, F.A., Kirby, A., et al. (2005) Support for involvement of neuregulin 1 in schizophrenia pathophysiology. *Mol Psychiatry* **10**(4):328, 366–74.
26. Petryshen, T.L., Middleton, F.A., Tahl, A.R., et al. (2005) Genetic investigation of chromosome 5q GABAA receptor subunit genes in schizophrenia. *Mol Psychiatry* **10**(12):1057, 1074–88.
27. Dracheva, S., Elhakem, S.L., McGurk, S.R., Davis, K.L., and Haroutunian, V. (2004) GAD67 and GAD65 mRNA and protein expression in cerebrocortical regions of elderly patients with schizophrenia. *J Neurosci Res* **76**(4):581–92.

28. Heldt, S.A., Green, A., and Ressler, K.J. (2004) Prepulse inhibition deficits in GAD65 knockout mice and the effect of antipsychotic treatment. *Neuropsychopharmacology* **29**(9):1610–9.
29. Lewis, D.A., Hashimoto, T., and Volk, D.W. (2005) Cortical inhibitory neurons and schizophrenia. *Nat Rev Neurosci* **6**(4):312–24.

19

Alzforum

E-Science for Alzheimer Disease

June Kinoshita and Timothy Clark

Summary

The Alzheimer Research Forum Web site (http://www.alzforum.org) is an independent research project to develop an online community resource to manage scientific knowledge, information, and data about Alzheimer disease (AD). Its goals are to promote rapid communication, research efficiency, and collaborative, multidisciplinary interactions. Introducing new knowledge management approaches to AD research has a potentially large societal value. AD is among the leading causes of disability and death in older people. According to the Alzheimer's Association, four million Americans currently suffer from AD. That number is expected to escalate to over 10 million in coming decades. Patients progress from memory loss to a bedridden state over many years and require near-constant care. In addition to imposing a heavy burden on family caregivers and society at large, AD and related neurodegenerative disorders are among the most complex and challenging in biomedicine. Researchers have produced an abundance of data implicating diverse biological mechanisms. Important factors include genes, environmental risks, changes in cell functions, DNA damage, accumulation of misfolded proteins, cell death, immune responses, changes related to aging, and reduced regenerative capacity. Yet there is no agreement on the fundamental causes of AD. The situations regarding Parkinson, Huntington, and amyotrophic lateral sclerosis (ALS) are similar. The challenge of integrating so much data into testable hypotheses and unified concepts is formidable. What is more, basic understanding of these diseases needs to intersect with an equally complex universe of pharmacology, medicinal chemistry, animal studies, and clinical trials. In this chapter, we will describe the approaches developed by Alzforum to achieve knowledge integration through information technology and virtual community-building. We will also propose some future directions in the application of Web-based knowledge management systems in neuromedicine.

Key Words: Alzheimer disease; knowledge management.

From: *Methods in Molecular Biology, vol. 401: Neuroinformatics*
Edited by: Chiquito Joaquim Crasto © Humana Press Inc., Totowa, NJ

1. A Brief History of Alzforum

The Alzheimer Research Forum (ARF) Web site was founded in 1996 to explore the application of Web technology to accelerate Alzheimer disease (AD) research. Its aims are to deliver high-quality, timely scientific information, support cross-disciplinary communication, and provide a Web-based open forum for debate and discussion of scientific controversies. June Kinoshita created the Web site concept and managed the Web development effort with assistance from a group of consultants and a commercial development company. She currently serves as Executive Editor and oversees a team of specialists with expertise in scientific writing and editing, data curation, information architecture, project management, and technology (*see* **Fig. 1**).

The Web site, popularly called Alzforum by the scientific community, was set up as a not-for-profit, independent organization. During the planning phase, the team considered whether the Web site should be hosted at an academic institution, but initial research indicated that individual and institutional rivalries

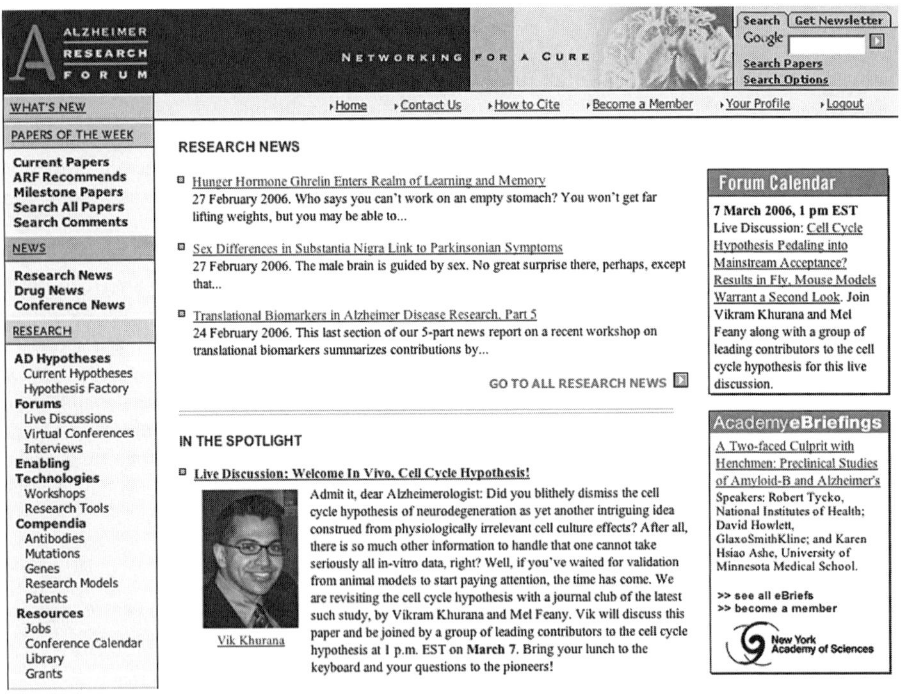

Fig. 1. Alzforum, a "tabloid" Web site for Alzheimer research.

could undermine the open community ideals of the resource. The Alzforum was therefore established as an independent Web site.

Without an institutional affiliation, the Alzforum faced the challenge of establishing its credibility within the scientific community. The Alzforum recruited a scientific advisory board[1] comprised of prominent research leaders representing diverse scientific interests and institutions. The composition of the board conveyed multiple signals to the AD community: that the site was dedicated to high scientific standards; that it was neutral with regard to any specific hypothesis or dogma; and that perspectives from multiple disciplines were welcome. The Alzforum made its debut in 1996 at the International Conference on Alzheimer's Disease and Related Disorders in Osaka, Japan.

At its launch, the Alzforum's content featured a "papers of the week" list of peer-reviewed publications in the AD field, slide and audio of scientific presentations, and a Milestone Papers list of seminal publications dating back to Alois Alzheimer's landmark 1907 publication. Over the first several years, the site registered around 100 new members per month. In surveys and informal communications, scientists were highly enthusiastic about the concept of a Web forum but reserved in their assessment of how such a new medium for scientific communication could enhance their progress.

A central aim of the Alzforum has been to identify and exploit opportunities to add value to the information resources that are available to AD researchers. To state the obvious, this has been a rapidly moving target over the past decade. For example, PubMed is today an essential and ubiquitous resource used by scientists everywhere to search the medical and scientific literature. In 1996, PubMed did not exist. (PubMed was launched in June of 1997.) Medical researchers had access through their institutions to Medline, but in order for Alzforum to produce its Papers of the Week listings, the editor made arrangements through the Countway Medical Library at Harvard Medical School to receive weekly text documents listing newly indexed AD papers and abstracts. The Alzforum hired a curator to paraphrase each abstract so that this information could be posted without violating journal copyrights. These documents were manually edited, sent out in a weekly email to the advisors for comments, and posted as static HTML documents on the site. Searching the Papers of the Week was done by using the "Find" function.

[1] Founding scientific advisory board members were Eva Brach, Joseph Coyle, Peter Davies, Bradley Hyman, Gerald Fischbach, Zaven Khachaturian, Kenneth Kosik, Virginia Lee, Elliott Mufson, Don Price, John Olney, and Robert Terry.

Looking back, the entire process seems as anachronistic as the hand-copying of manuscripts in the Middle Ages.

In 1996, the prospect of developing a dynamic, data-driven Web site seemed too costly and risky. It also seemed more important to invest first in some key content areas that would demonstrate the value of the Alzforum as a repository for shared community information. Thus, among the next resources to be developed were those that provided comprehensive information about the mutations that cause hereditary AD and a related hereditary neurodegenerative disorder linked to mutations in the tau gene, frontotemporal dementia with Parkinsonism (FTDP-17). Key to the success of these gene mutation resources was the fact that they were initiated and produced by scientists who had strong personal incentives for organizing this information. Geneticists John Hardy and Richard Crook, both then at the Mayo Clinic Jacksonville, approached Alzforum about posting a comprehensive list of familial AD mutations in the amyloid precursor protein (APP), presenilin-1 and presenilin-2. Jennifer Kwon at Washington University in St. Louis volunteered to curate a knowledge resource on FTDP-17 tau mutations. Michael Hutton of the Mayo Clinic Jacksonville also helped to edit and update the FTDP-17 tau mutation list. These resources were well received. Scientists began to contact Alzforum for permission to reproduce the presenilin-1 diagram in their lectures and papers. "These tables and reviews are incredibly useful for people like me, who don't closely follow the genetics field," observed one well-known AD biochemist.

In 2000, the Alzforum team decided the time was ripe to convert to a data-driven server. The average number of Alzheimer papers indexed by Medline/PubMed had grown from around 50 per week to 200. What is more, PubMed had come into existence, so that Alzforum could design a system to run automated searches and link out to abstracts posted on the PubMed site. Another extremely important change was the development of an administrative back-end that placed the updating of content directly under the control of Alzforum editors. This change had an immediate impact, making it possible to upload news stories, comments, job postings, conferences, and other content continually and without the considerable cost and delays that had previously been incurred, when server updates had been in the hands of an outside Web development company.

With the server under its control, Alzforum team is able to continually improve the server and add new functions and databases. This flexibility is essential to the Alzforum's ability to identify and address new ways to provide value-added services and to keep up with the rapidly evolving landscape of Web resources for scientific research (*see* **Fig. 2**).

Mutation & Exon	Effect on Exon 10 splicing & Biochemistry	Clinical	Neuropathology	References
3'Ex10+11 T to C *Posted 10/1/05*	Modifies alternative splicing of exon 10, no effect on MT assembly or filament formation.	FTDP-17. Age of onset 48, duration 6 years. Initial symptoms mental retardation, myoclonus, parkinsonism (age 10). Other symptoms dementia of frontal type, Mask-like face, PSP	Neuronal loss and gliosis. Globus pallidus, thalamus, substantia nigra, dentate nuc affected. And frontal and temporal lobes. No ballooned neurons	▫ Kowalska et al., 2002 ▫ Miyamoto et al., 2001
Exon 10 +12 "Kumamoto pedigree"	Marked reduction in melting temperature of the tau exon 10-splicing regulatory element RNA	FTD	Behaves like the 5' splice site mutations, with increments of 4R tau. Tau aggregates in degenerating neurons and glia. Twisted ribbon morphology, with hyperphosphorylated 4R tau.	▫ Yasuda M., et al. 2000.

Fig. 2. Tau mutations table (first two rows).

2. The Alzforum Today

Since the relatively simple server that was launched in 1996, the Alzforum has grown into a rich knowledge environment for neurodegenerative disease research. The Web site is free and accessible to the public. Registration is required for individuals who wish to post comments or vote in the various online polls produced by the Alzforum. We estimate that a substantial fraction of the worldwide community of AD researchers consists of registered members, and a majority of all AD researchers visit the site.

The site statistics for 2005 are as follows:

- 50,000–70,000 sessions served per month
- Nearly 3,900 registered members
- More than 4000 subscribers to the weekly e-newsletter
- More than 43,000 articles in the citation database
- 60–90 commentaries from scientists per month
- Referenced by more than 131,000 Web sites
- Cited regularly in scientific lectures and publications
- Number one Google search result under "Alzheimer research."

A guiding principle in designing the new homepage was to think of the Alzforum as "the daily tabloid for AD research." Our aim was to induce

AD researchers to choose Alzforum as their homepage and to visit it at least weekly, if not more. To accomplish this goal, the homepage has to be dynamic, useful, and entertaining. The content is updated almost daily with the latest science news, live discussions, conference reports, community activities, opinion polls, and other offerings. In addition, a weekly e-newsletter highlights all new content posted during the past week and prompts readers to visit the Web site.

The Alzforum team strives to make the Web site an essential resource for scientists by focusing on adding value to information that is already available in the public domain. How we pursue this goal is illustrated by specific examples of our major content areas.

2.1. Papers of the Week

This citation database is the core content area. Many AD researchers keep up with the literature by browsing the Papers of the Week rather than PubMed because it provides a high-quality list tailored to their interests and saves the individual researchers from having to conduct multiple searches on PubMed. In addition, the "POW" citations are enriched with news stories and expert commentaries, as well as links back to earlier related articles. High-impact articles are designated as "ARF Recommended" papers and "Milestones." For many scientists, the citation database supplies the context that is missing from the pure journal publication, namely real-time reactions by peers.

Papers of the Week uploads some 100–200 new citations per week. An automated search function finds and downloads new citations every night from PubMed. The search strategy is designed to capture all Alzheimer papers in basic and clinical research fields along with papers published in high-impact journals on key AD molecules (e.g., APP, presenilin, apolipoprotein E, tau, alpha-synuclein, etc.), Parkinson disease, dementia with Lewy Bodies, amyotrophic lateral sclerosis, polyglutamine disorders, stroke and trauma, neurodevelopmental disorders, amyloidoses, AIDS dementia, etc.

The Papers of the Week is also essential for the knowledge management role of the Alzforum and drives much of the content development on the site. The citations are screened by editors for news and to send data to curators for the AlzGene genetics database, Telemakus AD biomarkers database, mutations directory, research models database, antibody database, and so forth. The scientific advisory board reviews and annotates new citations on a weekly cycle. Thus, the "firehose" of PubMed citations is channeled systematically into multiple content streams and helps ensure that all of the Alzforum's knowledge management resources are kept up to date.

2.2. Research News

One of the most important ways in which the Alzforum adds valuc to scientific publications is by providing original reporting and analysis of news on AD, related neurodegenerative disorders, basic science, and biotechnology. The writing is geared to scientists and is at a technical level comparable with the news sections in journals such as *Science* and *Nature*. The news writers aim to place new findings in context by describing relevant previous findings or discussing the relevance of new developments in other fields to AD research. News stories are linked to previous related news and publications so that readers can easily trace the history of a specific line of investigation.

2.3. Commentaries and Live Discussions

Another essential function of the Web site is to provide scientists with a forum to respond quickly and publicly to new findings. Readers can post commentaries on any Papers of the Week citation or news story through a "Vote/Submit Comment" text-entry box. Alzforum editors frequently invite individual scientists to comment on news or journal articles. The response rate is very high, probably close to 75% or better. We attribute this high response rate to the skill with which queries are tailored to scientists' interests, the trust that Alzforum has gained within the scientific community, and the perception that comments posted on Alzforum are being read and discussed by the contributors' colleagues. With the growing visibility and legitimacy of the Alzforum as a medium for publication, scientists have begun to post commentaries spontaneously.

Over the years, many scientists have remarked on how effective the Alzforum has been in driving rapid and productive discussion of their ideas and findings. For example, in 2005 Vincent Marchesi, a cell biologist at Yale University, published an alternative hypothesis of the amyloid hypothesis that under normal circumstances might have been read by a small number of largely skeptical readers. The paper was featured by Alzforum and elicited lengthy, detailed, and productive commentaries by 17 colleagues *(1)*. "The postings on the Alzforum site regarding my PNAS paper *(2)* have been incredibly rewarding for me, and I suspect, for many of the others that participated," wrote Marchesi. "I don't see how so many candid exchanges could have taken place any other way. Four stars for the Forum." (*see* **Fig. 3**).

2.4. Databases

Data about key research findings and reagents are curated and loaded into databases designed by Alzforum. These data are published in disparate articles

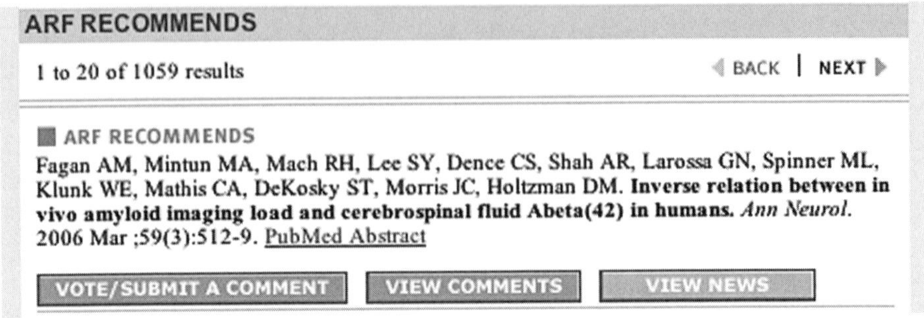

Fig. 3. Adding value through comments and news.

and formats, and scientists expend much time and effort to keep up to date. Because there is little incentive for individual scientists to carry out and share this task on behalf of the scientific community at large, the Alzforum identified the development and maintenance of open databases as a high priority. Databases on Alzforum include the following:

- Familial AD mutations: All published mutations in the APP, presenilin-1 and presenilin-2, as well as tau mutations that cause FTDP-17. Individual mutations are displayed in a table along with clinical, pathology, and biochemical data and primary publications. (http://www.alzforum.org/res/com/mut/default.asp)
- AlzGene: All published genetic association studies for late-onset AD. The database can be browsed by chromosome or searched by gene, polymorphism, protein, keyword, or author. Each gene is summarized in a table listing all published studies along with study details, including the ethnicity, size, female–male ratio and average age of AD and control populations, and the size of the observed effect (*see* **Fig. 4**).(http://www.alzforum.org/res/com/gen/alzgene/default.asp)
- Antibodies: 6000+ antibodies to proteins that are widely studied by AD researchers. The database includes noncommercial and commercial antibodies. Antibodies are

1. CASE-CONTROL STUDIES (BY ETHNIC GROUP)

				AD Cases				Normal Controls		
Study	Population	Source	# Polys	# Subjects (% women)	DX	Mean Onset Age (range)	Mean Age (range)	# Subjects (% women)	Mean Age (range)	Result
CAUCASIAN										
Cook, 2005	UK (I)	CL	2 (detail)	215 (-)	N	-	-	320 (-)	-	Positive
Cook, 2005	UK (II)	CL	1 (detail)	179 (-)	C	-	-	175 (-)	-	Negative
Harold, 2003	UK (I)	CL	14 (detail)	131 (-)	C	72.5 + 6.5 (-)	-	118 (-)	78.2 + 7.2 (-)	Positive
Harold, 2003	UK (II)	CL	8 (detail)	135 (-)	C	74.3 + 6.2 (-)	-	135 (-)	75.3 + 6.4 (-)	Negative

Fig. 4. AlzGene data table.

displayed in a table summarizing data that researchers can use to identify products of interest. (http://www.alzforum.org/res/com/ant/default.asp)
- Research models: All published mouse models of interest to AD researchers. These include mice in which genes of interest have been manipulated to enable the study of putative molecular mechanisms related to AD, other neurodegenerative conditions, and aging. (http://www.alzforum.org/res/com/tra/default.asp)
- Drugs in clinical trials: All published drugs that have reached Phase II clinical trials and beyond, including drugs that were discontinued following clinical trials. This database is searchable by drug name and keyword and includes data on development status, mechanism, role in AD, and company. Our current plan is to re-design this database to include preclinical compounds and additional data of value to researchers other than clinicians. (http://www.alzforum.org/drg/drc/default.asp)

3. User Demographics and Feedback

The Alzforum is freely accessible to the public, so we do not have statistics on our entire user population. However, nearly 4000 individuals have registered as members, of whom more than 2500 have also filled out the "researcher profile" form. Thus, we assume a lower limit of 2500 on scientists and clinicians who use the site and estimate that around the same number are using the site without registering. This implies that 30–50% of the global community of AD researchers is made up of regular visitors. We find that at major international meetings of AD researchers, most attendees are familiar with Alzforum. In addition, growing numbers of researchers from closely related areas such as Parkinson disease, amyotrophic lateral sclerosis, and Huntington disease are using the Web site.

Feedback has been strongly positive. A recent survey of our scientific advisory board members revealed that the average frequency of visits to the Web site is around one to three times per week. "[Alzforum] is the local newspaper for Alzheimer research," writes John Hardy, Director of the Laboratory of Neurogenetics at the National Institutes of Aging. "I visit it one to two times a week just to see what's going on . . . to check up on recent papers, . . . to see who's hiring people and so on. I read people's comments on papers, and I go from there to PubMed for anything I've missed. I think pretty much everyone in the field uses it in the same way, and I have often seen my informal reviews on the site cited."

Scientists mention a variety of reasons why they find the Alzforum valuable. One is that the Alzforum enriches published papers with news analysis and rapid peer commentary. "This is the major e-forum for AD ideas," observes Jeffrey Cummings, a neurologist at the University of California in Los Angeles. "The discussion forums have shaped and sharpened my ideas. It's a great way to get a grasp of the literature and to follow emerging events in real time."

Many researchers value the breadth of the Alzforum's coverage, which is intended to communicate diverse developments that no specialist could possibly keep up with. "Instead of relying only on published papers and meetings, you provide rapid insights into new developments and introduce us to areas that are related to our work but yet we fail to notice were it not for you," writes Gunnar Gouras of Weill Medical College of Cornell University. The databases also are frequently mentioned as resources that help scientists stay abreast of advances in fields outside their own.

Another important aspect of Alzforum is its community-building function. Through commentaries and live discussion forums, the Web site provides a neutral ground for scientists to get comfortable with one another. Scientists have agreed to share reagents and explore collaborations following one of the Alzforum's live chat room events. The following excerpt from a live discussion on "Protofibrillar A-beta in Alzheimer Disease" includes an agreement to share a new protocol and discussion of unpublished data *(3)*:

Dennis Selkoe: "...How precisely do you make the ADDLs [amyloid-derived diffusible ligands] from A-beta42?"
William Klein: "ADDLs used to be made always with clusterin, but now we use the 'cold' ADDLs prep ...F12 medium at 4 degrees. A big deal is to use HFIP to monomerize."
Dennis Selkoe: "Can you send us a detailed protocol?"
William Klein: "Delighted! And I'd be happy to send some ADDLs. They now travel well."
Dean Hartley: "That would be great!"
William Klein: "Have you had a chance to look at LTP?"
Dennis Selkoe: "No. However, we are currently dissecting the increase in electrical activity with PFs. More specifically looking at NMDA and AMPA."
June Kinoshita: "Bill, what have you worked out as far as how ADDLs affect LTP? What appears to be the pathway?"
William Klein: "Pathways are speculative. The response is hugely reliable and occurs in under 45 minutes ...We are intrigued by the possible involvement of Fyn, which is anchored to PSD95 and modifies NMDA receptors. But it's work in progress."

The Alzforum appears to be having an impact on speeding up the dissemination of ideas that could lead to novel treatment strategies. William Klein, the Northwestern University cell biologist and co-discoverer of ADDLs, writes:

"Our 1998 PNAS paper has now been cited 385 times, and our 2003 PNAS paper has already been cited 25 times (in less than 12 months). The Alzforum, by raising the collective consciousness regarding our work's importance, has played a truly major role in advancing the field. Meeting highlights as well as online dialogs have played significant roles. Thanks to ... the Alzforum, there are multiple research groups in academia and big-pharma hotly pursuing A-beta oligomers as prime targets for therapeutics and diagnostics. I would say the field is 3–4 years more advanced than it would have been without the Alzforum's contributions."

4. Lessons Learned

The Alzforum Web site is a thriving online community that resulted from the convergence of information technologies with traditional editorial functions performed by science media and journals. This convergence offered new opportunities to make scientific work more efficient and to create new forms of scientific publication and discourse.

Various skill sets—editorial, design, programming, and data curation—are needed to create and operate the Alzforum. A professional editorial team screens pre-embargo journal articles and new PubMed citations, evaluates the importance of a broad range of developments, and identifies the expert commentators who are likely to contribute an interesting perspective on a story. The editorial philosophy of Alzforum is to be "nonpartisan" and cover any development that is of scientific interest, whether it conforms to prevailing hypotheses or not. Editors frequently cover advances in basic biology or methodology that could have applications in AD research. As at any publication, the Alzforum editors are not passively waiting for "stuff to happen" but actively generating community participation. Commentators are invited to contribute critiques, context, and insights and to speculate on future directions. Their comments in turn often fuel further discussions. A crucial part of the editorial function is to screen and edit all postings to ensure that the language is respectful and clear and that the contents are relevant and useful to readers.

The Alzforum employs information specialists who develop knowledge structures to handle the diverse types of information and maintain standards for accreditation, provenance, and stability of texts and data on the site. This is essential if the contents are to be cited as part of the scientific literature. In addition, this team designed the Alzforum member registration system, which is required for any person who wishes to post a comment or vote on papers or site polls. The registration process ensures accountability and transparency. Our information specialists also develop search strategies, including those used to automatically retrieve PubMed citations and to search the Alzforum site itself, and also design data structures and ontologies for the databases.

Alzforum's software programmers work closely with the editors and information architects to develop all aspects of the server: a front end that is transparently easy for biologists to use and an administrative back end that enables the Alzforum team to manage and post content to the site. Finally, there are curators and outside collaborators who work with Alzforum to develop resources such as the AlzGene database, research models directory, drug database, and so on.

It is important to note that Alzforum is not simply a technology-meets-editors story. The success of the Alzforum rests squarely on the relationship of trust developed over time between the editors and the target scientific community. To earn that trust, the Alzforum had to demonstrate that it was editorially and technologically dependable. Both the content and the method of delivery had to be of the highest quality and stable over time so that scientists would be willing to invest in the resource. We were extremely fortunate to be sponsored by a philanthropic foundation that was willing to make a long-term commitment to supporting the Alzforum.

5. Future Directions

The digital revolution has been a boon to scientists, but databases and pdf documents by themselves barely begin to tap the potential of information technology to transform research. Scientific findings need to be put into contexts that are relevant to specific individuals and communities. Studies that go unread might as well not have been done at all. One way to think about the cost-effectiveness of Alzforum is that the annual budget of Alzforum, an open knowledge base for the entire research community, is roughly equivalent to the funding needed to generate one average research paper.

Various neuroscience research communities are beginning to take notice. In 2005, the Schizophrenia Research Forum launched a Web site (http://www.schizophreniaforum.org) with funding from the National Institutes of Mental Health and sponsored by the National Alliance for Research on Schizophrenia and Depression (NARSAD). This Web site uses Alzforum software and architecture, and several Alzforum consultants helped to set up the server. The scientific community responded with enthusiasm, and more than 1000 people registered as members within a few weeks of the launch.

Around the same time, the Michael J. Fox Foundation awarded a grant to Alzforum and Lars Bertram of the Massachusetts General Hospital (MGH) to build a Parkinson Disease gene resource modeled on the AlzGene database. "PDGene" will take the shell of the AlzGene database and populate it with data from published genetic association studies of Parkinson disease. Soon after, NARSAD approached Alzforum and MGH about creating a similar resource for schizophrenia genetic studies.

One challenge we now face is how to leverage this growing family of databases to take advantage of the scientific connections among them. Do we simply crosslink among genes that have been studied for association with more than one disease? Or do we develop a single, unified database? Also, how can we build up a richer body of knowledge around each gene to help elucidate its possible role in various diseases?

There is nothing disease specific about the ARF approach. We look forward to collaborating with those seeking to extend the model to diseases with overlapping knowledge bases. These could include other neurologic diseases such as Parkinson's, Huntington's, and multiple sclerosis, as well as diabetes and cardiovascular disease, which are known risk factors for AD.

To prepare for the possibility of additional sister Web sites in the future, we have developed the concept of "Neurobase." The Neurobase would be a portal for multiple communities that tap into a central knowledge base containing data and information about pathology, disease phenotypes, genes and proteins, molecular pathways, animal models, neuroimaging, drug chemistry, clinical trials, and so forth. This approach makes sense because so much basic data is applicable to multiple disease contexts. Each community Web site would be funded by an organization with an interest in a specific disease, and maintained by its own editorial team, which would develop news, comments, databases, and other content tailored to its readers but easily sharable with other Neurobase Web sites. Managing this complex federation of knowledge bases will be a fascinating challenge.

6. Semantic Web Applications in Neuromedicine

A perennial challenge for the Alzforum is to find new ways to apply information technology to significant problems in AD research. The most ambitious effort to date is the Semantic Web Applications in Neuromedicine (SWAN) project, a collaboration between Alzforum and the Center for Interdisciplinary Informatics at the Massachusetts General Hospital (MGH). SWAN is developing a semantically-structures, Web-compatible framework for scientific discourse using Semantic Web technology ***(4–6)***, applied to significant problems in AD research. The initial concept for SWAN was proposed in a talk at the W3C Semantic Web in Life Sciences workshop, October 2004 *(7)*.

One of our key observations in developing SWAN is that existing knowledge bases, even those as rich and diverse as Alzforum, still remain little more than collections of documents and data. For all their hyperlinks, they are not embedded in a knowledge model. The human user carries the knowledge model inside his or her head. When the human reads a paper or follows a link, he or she fills in the contextual blanks, such as "this paper supports assertion X made by scientist Y in her paper Z," or that "this paper describes a novel method that I would like to use in my next experiment to test hypothesis Q." With SWAN, we intend to provide scientists with a tool to embed their documents, data, and other digital materials in a knowledge model, and then to share the entire model with other scientists and communities. Semantic interoperability in SWAN is based on a common set of software and a common ontology of scientific discourse. This ontology (or vocabulary) is specified in an Resource Description Framework (RDF) schema available on the Web *(8)*.

Another key concept informing the design of SWAN is that scientific ideas, documents, data, and other materials evolve within a "scientific ecosystem." Scientists ingest publicly available information, digest and use it to inform hypotheses and experiments, carry out experiments, modify hypotheses, and eventually publish that information back into the public space. The SWAN models this scientific ecosystem in ways that should accelerate the ecological turnover. Thus, SWAN will incorporate the full biomedical research lifecycle in its ontological model, including support for personal data organization, hypothesis generation, experimentation, laboratory data organization, and digital pre-publication collaboration. Community, laboratory, and personal digital resources may all be organized and interconnected using SWAN's common semantic framework.

In addition, SWAN will allow users to import and use various ontologies, ranging from standard ontologies such as Gene Ontology and Unified Medical Language System (UMLS) to personal ontologies ("my working hypothesis," "my collaboration with Brad"). We have proposed an "Alzforum ontology" that describes the etiological narrative inherent in pathway models and hypotheses. Using this ontology, scientists can tag a piece of information as being an "initial condition," "perturbation," "pathogenic event," or "pathologic change" in the disease process. In addition, scientists can tag information with evidence types (e.g., genetic, epidemiologic, pathophysiologic, biomarker, clinical trial, etc.). The goal is to populate this knowledge matrix with research findings in order to reveal interesting convergences and critical gaps in scientific knowledge.

We plan to deploy a prototype of SWAN that includes the following components:

- A MySWAN client that resides on individual scientists' personal computers (or possibly on a password-protected Web server);
- An Alzforum SWAN that will enable Alzforum editors to edit and publish knowledge models on the Alzforum Web site;
- A LabSWAN that will be used by scientists within a laboratory (or collaboration).

Individuals will use SWAN software as a personal tool to find, filter, and organize information, to extend their knowledge, to motivate discoveries, and to form and test hypotheses. At the community level, the same software and the same ontological framework can be used to organize and curate the research of a laboratory or of an entire research community. Therefore, elements of the personal knowledge base (KB) can be shared with the community at a low incremental effort in curation. What's more, community KB contents may be shared back with individuals and re-used in new contexts.

We intend to seed the curation effort by enlisting an Alzforum expert curator to convert the "AD Hypotheses" section of the Alzforum Web site into SWAN schema. Specifically, the curator will break down each hypothesis into its component assertions and supporting evidence. Assertions themselves can be associated with supporting or refuting assertions, each with its own author and supporting evidence. This process will provide a level of granularity that will enable the reader to examine the individual assertions that comprise the hypothesis narrative and to drill down to supporting evidence and their associated assertions, alternate hypotheses, news, and other content. On the front end, each hypothesis will be presented in an attractive, table-of-contents-like format that displays an abstract, links to full text articles and news, lists of assertions and comments, a knowledge matrix chart, and a link to the "SWAN view" where new comments, findings, and assertions can be added by readers.

Another goal of SWAN will be to semantically enrich documents by automatically linking genes, proteins, antibodies, drugs, animal models, etc. to entries in public databases. For example, individual gene entries in AlzGene could be linked to assertions about gene function (available as GeneRIFs through the PubMed portal and as assertions in the SWAN data store), and these gene functions would also be tagged with knowledge matrix terms and hypotheses, thereby providing a richer, disease-relevant context. Antibodies listed in the Methods sections of papers could be linked to the matching product in the Alzforum Antibody Database.

The editorial team's effort is made tractable because of the work of the community members. The community members, unlike those in a process such as Wikipedia ***(9)***, are principally concerned with advancing their own research program. The incremental effort required to share knowledge from the team to the community is intended to be relatively small and in many way can be seen as an enhancement of the standard publication process for scientific literature.

At the same time, we believe the kind of knowledge sharing facilitated by SWAN is a qualitative step beyond the current infrastructure and can result in the creation of the kind of highly facilitative knowledge-sharing networks argued for by the leadership of neuroscience research institutes at NIH ***(10)***.

We believe our approach will enable the construction of a distributed AD knowledge base as an operation piggybacked on the existing scientific knowledge ecology and driven principally by the self-interest of participants. Under this assumption, the requirement for specialized curators is minimal and corresponds more to the notion of editors, who would operate at the research community level.

The initial SWAN software prototypes were developed by a collaborative team from the MassGeneral Institute for Neurodegenerative Disease (MIND) Center for Interdisciplinary Informatics (http://www.mindinformatics.org), the ARF (http://www.alzforum.org), and MIT's Computer Science and Artificial Intelligence Laboratory (CSAIL) Decentralized Information Group (http://groups.csail.mit.edu/dig/).

Acknowledgments

We thank Yong Gao and Gabriele Fariello of the SWAN team at the MGH and members of the ARF team: Elaine Alibrandi, Tom Fagan, Donald Hatfield, Hakon Heimer, Sandra Kirley, Colin Knepp, Paula Noyes, Nico Stanculescu, Gabrielle Strobel, and Elizabeth Wu. We are grateful to the Ellison Medical Foundation for its generous support of the SWAN project and are indebted to an anonymous foundation for its unstinting support of the ARF.

References

1. http://www.alzforum.org/res/for/journal/detail.asp?liveID=35
2. Marchesi, V. T. (2005) An alternative interpretation of the amyloid Abeta hypothesis with regard to the pathogenesis of Alzheimer's disease. *Proc Natl Acad Sci USA* **102**(26):9093–8.
3. Protofibrillar A-beta in Alzheimer Disease, an Alzforum Live Discussion with Brian Cummings, Dean Hartley, William Klein, Mary Lambert, Dennis Selkoe, et al. (2000) Accessed at http://www.alzforum.org/res/for/journal/protofibrillar/default.asp
4. Berners-Lee, T. (1998) Semantic Web Road Map. Available at http://www.w3.org/DesignIssues/Semantic.
5. Berners-Lee, T., Hendler, J., and Lassila, O. (2001) The Semantic Web. *Scientific American* **284**(55):28–37.
6. Hendler, J. (2003) Science and the Semantic Web. *Science* **299**:520–521.
7. Clark, T., and Kinoshita, J. (2004) A pilot KB of biological pathways important in Alzheimer's Disease. W3C Workshop on Semantic Web for Life Sciences, Cambridege, MA, USA.

8. Available at http://purl.org/swan/0.1/. Some browsers may require a "view source" operation to see the RDF.
9. Wikipedia: The Free Encyclopedia. Available at http://www.wikipedia.org.
10. Insel, T. R., Volkow, M. D., Li, T. K., Battey, J. F., Landis, S. C. (2003) Neuroscience Networks. PLoS Biol **1**(1): e17 doi: 10.1371/journal.pbio.0000017.

Index

1
1×N
  mapping, 28
129X1/SVJ
  mouse strain, 222

2
2D, 214
  anisotropic series, 221
  grid, 313
  image, 202
  series, 221
2DG, 201
  autoradiography, 197

3
3D, 214
  atlas, 220, 224
  atlas server, 212, 228
  coordinate system, 226
  data sets, 220
  atlas server, 228
  fMRI, 198
  motion, 221
  MRI, 229
  reconstruction, 211
  storage format, 228
  time series, *See* 4D
  viewer, 211
  volume, 221, 225
3D–3D alignment, 221

4
4D, 214

5
5-HT, 145, 147, *See* serotonin

A
abducent nerve, 74
A-beta, 374
Abstract Schema Language, 330
ABXD5F2
  mouse strain, 222
Access, 172, *See* MS Access
accuracy, 178
ACeDB, 58
acetylcholine, *See* ACh
  signaling, 355
acetylsalicylic acid, 49, *See* aspirin
ACh, 325, *See* acetylcholine
action potentials, 127
activation
  time-dependent, 137, 145
  time-independent, 137
  voltage-dependent, 145
activation function
  time-dependent, 143
  voltage-dependent, 137, 143
activation variable, 178
Active Data Objects, 31
activity profile, 177
activity-dependent facilitation, 148
activity-dependent Hebbian model, 324
AD, 315, 322, 323, 329, 366, *See* Alzheimer's Disease
adaptive parameter, 309
ADO, *See* Active Data Objects
afferent
  damage to pathways, 329
  synaptic transmission, 325
afferent/efferent
  neurons, 41
affirmers lexical, 17
affirming/negating, 5, 10
Affymetrix, 290, 294, 350
AFNI, 240, 241, 242, 253, *See* Analysis of Functional NeuroImages
AFP, 214
AFS, *See* Andrews File System
age, 54
Agilent, 351, 359
agonal state, 347

agonists, 49
AIR, 192, 240, 241, 252, *See* Automated Image Registration
AIR5, 187
aldehyde, 207
Align Warp module, 191
allele frequency disease, 339
alleles, 296, 341
  disease-related, 342
Allen Brain Atlas, 79, 281
alpha relaxation state, 328
alpha-amino-3-hydroxy-5-methyl-4-isoxazolepropionic acid, *See* AMPA
ALS, *See* amyotrophic lateral sclerosis
AltaVista, 68
alternate average creation, 185
Alzforum, 366, 371, *See* Alzheimer's Research Forum
  Antibody Database, 379
AlzGene, 370, 372
Alzheimer Research Forum, *See* ARF
Alzheimer's Disease, 7, 307, 357, *See* AD
ambiguity discrimination, 11
amino acid-altering polymorphism, 342
ammon horn, 69, 74
AMPA, 12, 319
amygdala, 74
amygdaloid nuclear complex, 74
amyloid precursor protein, 368
amyotrophic lateral sclerosis, 365
Analysis of Functional NeuroImages, 237
ANALYZE75, 185
anatomic variance, 184
Andrews File System, 241
Angleman Syndrome, 345
ANN, 159, *See* artificial neural network
antagonist, 49
anterior–posterior
  olfactory bulb, 197
antibody, 372
anticholinergic, 49
anti-inflammatory drugs, 49
antonym, 49
API, 70, *See* Application Programming Interface
Aplysia, 129, 139, 140, 145–147
apolipoprotein, 370
APP, *See* amyloid precursor protein
Arabidopsis, 289
archipallium, 308
archival tables, 14
area V1, 282
ARF, 377
Aricept, 322
artificial intelligence, 7
artificial neural network, 157
artificial neuron, 309
artificial scotoma, 319
ASCII, 128, 130
aspirin, 49
aspx, 70
associative memory models, 305
atherosclerosis, 357
atlas
  brain, 183
  digital, 184
  digital-canonical, 70
  minimum deformation, *See* MDA
  mouse brain, 184
  construction, 184
attributes, 45–46
attribute–value, 50
auto-launch, 92
automated, 25
  nonlinear registration, 223
Automated Image Registration, 237
automatic deposition, 8
autoradiographic, 276
autoradiographic mapping, 267
autoradiography, 197, 268, 277–278
autosomal dominant, 339
average
  intensity-based, 183
  activation frequency, 112
axial resistance, 117
axon, 111, 207
axonal
  branches, 98
  arborizations, 58
Azores, 348

B

backpropagation, 306
BAMM, *See* Brain Activation and Morphological Mapping
BAMMfx, 237
BAMS, 81
basal
  forebrain, 322
  ganglia, 317

basic statistics, 294
BatchFtp, 242
BaxGrid, 237
BBTsyntax, 58
behavioral traits, 301
Benjamini-Hochberg correction, 356
benzaldehyde, 43
beta amyloid, 355
BIC toolkit, 192
bimodal, 139, 144
BIMS, *See* Brain Information Management Systems
binary, 51
  branch tree, 58
BIND, 360
biochemical, 104
bioinformatics, 287, 337
Bioinformatics Research Network, 23
biological mechanisms, 306
biological pathway, 361
  schizophrenia, 357
biomarker, 378
  for neuropsychiatric disease, 357
bipolar
  disorder, 308, 340, 345, 357
  neurons, 80
BIRN, 28, 31, 214–215, *See* Bioinformatics Research Network
BIRN-CC, 31
BLAST, 6
Blue Brain project, 308
Boltzman, 130, 132
Bonfire, 33–34
Boolean, 30, 47–48, 50, 310
BPEL, *See* Business Process Execution Language
BPEL4WS, *See* BPEL
brain, 7
  atlas, 60, 183
    comprehensive, 73
    macaque, 84
    mouse, 84
    probabilistic, 183
  hierarchy, 71
  cortex, 196
  dynamics, 305
  functions, hierarchical approach to modeling, 315
  hemispheres, 268
  images, 235
  mammalian, 195
  map, 271
    primate, 73, 80
    mouse, 212
  mapping, 195
  parcellation, 281
  pH, 347
  region, 48, 54, 276
  regions, rat, 59
  slices, 268
  spatial normalization, 211
  structure, 69
  structures in, 39
  tissue, 197
Brain Activation and Morphological Mapping, 237
Brain Information Management Systems, 69
Brain Surface Extractor, 192
Brain Voyager, 237
BrainInfo, 67–70, 73–74, 79–80, 84
brain-mapping, 184
BrainMaps, 79
BrainML, 25, 56, 91, 101, 251
BrainML/XML, 28
BrainPharm, 7
BrainPy, 248
brainstem, 217
BrainSuite, 2, 185, 188, 237
BrainVISA/Anatomis, 237
BrainVoyager, 241, 250
  neuroimaging package, 241
branch point, 136, 147
BRENDA, 6
bridge
  table, in databases, 50
Bruker
  spectrometer, 197
BSE, 188
burst, 174
  duration, 168, 176
  period, 168, 176
  spiking, 320
burster, 110
  conditional, 146
bursting, 174
  activity, 147
  discharge frequency, 172
  neuron, 139
Business Process Execution Language, 243
BXA1
  mouse strain, 222

BXD, 289–291, 293, 298, 300
BXD1, 291
BXD100, 291

C
C++, 64, 170, 238
C57BL
  mouse strain, 290
C57BL/6J
  mouse strain, 212–215, 291
Ca1
  pyramidal cell, 92
CA1, 69, 80
CA2, 69, 80
Ca^{2+}, 139, 144, 147
  intracellular concentration of, *See*
  ion channel, 12
Ca2+-dependent K+ current, 146
CA3, 69, 80, 121
  cell model simulation, 108
  hippocampal, 121, 326
  pyramidal, 10
  pyramidal cell, 122
  cells, 10
Caenorhabditis elegans, 58
calcium
  intracellular concentration, 110
  imaging, 196
cAMP
  accumulation of, 147
  regulation of, 139
  bimodal regulation of, 144
cancer, 288
candidate gene, 337, 342
caption, 45
Caret
  brain mapping program, 198
cartesian coordinates, 116
CAS, *See* Chemical Abstracts Services
Caspase, 355
Castor, 64
Catacomb, 63, 101
Catalyzer, 60
catfish, 43
CCDB, 49, *See* Cell Centered Database
CDM, *See* Common Data Model
Celera, 291
cell
  morphology, 59
  processing, 24
  signalling, 24
  population, 276
Cell Centred Database, 31
CellML, 91
CellPropDB, 3–5, 7, 10, 15, 25
cellular composition, 347
central pattern generator, *See* CPG
cerebellar
  cortex, 31
  Purkinje neuron, *See* Purkinje
cerebellum, 10–11, 13, 17, 217, 290, 317
cerebral cortex, 198
cerebral palsy, 315
c-fos, 208
CGH, 359, *See* comparative genome hybridization
CGI-PERL, 11
channels, 40, 48
Chemical Abstract Services, 202
Chiron, 248, 254
chlorpromazine
  dopamine antagonist, 49
Choice_Set_Values, 44
Choice_Sets, 44
cholinergic, 322
  innervation, 325
chromosomal
  aberrations, 337
  regions, 341
circuit diagram, 137
C-language, 160
  vector-oriented, 160
class
  child, 47
  parent, 47
Class_Hierarchy, 47
classes, 46
Classes/Attributes, 51
classes/tables, 38
  bridge, 41
  many-to-many, 41
clinical trial, 378
cluster map, 293, *See* QTL cluster map
cluster tree, 299
clustering, 157
CNFT, *See* continuum neural field theory
CNS, 230
CoCoDat, 268, 269
CoCoMac, 81, 267–269, 276, 283
Cognex, 322

cognition, 24
cognitive computational neuroscience, 305
cognitive deficits
VCFS, 345
cognitive impairment, *See* Alzheimer's Disease
Collation of Receptor Data, *See* CoReDat
column/row, 164
combinatorial methods, 338
Common Data Model, 26
community-building, 374
comparative genome hybridization, 361
compare correlates, 293, 301
complex trait, 296
analysis, *See* trait analysis
modulation, 287
computational
linguistics, 7
model, 155, 330
in psychiatry, 305
psychiatry, 315
simulation, 168
speed, 178
Computer Science and Artificial Intelligence Laboratory, 380
concordance table, 51
Condor, 216, 253
conductance, 132, 174
cyclic nucleotide gated, 139
ion, 134
voltage-dependent, 132–133, 136
confocal imaging, 64
connectionism, 306
connectionist
mode, 309
model, 310
networks, 310
connexins, 316
context, 5, 7–8
statistical, 68
contextual, 16
continuum neural field theory, 306
control current, 132
control population, 342
controlled vocabulary, 37
copy number, 341
copy number analysis, 349
CORBA, 245, 253
CoReDat, 267–269, 276, 283
coronally sliced, 217
correlation matrix, 293–294, 301
cortex
global model, 306
cortical
areas, 319
lesions, 317
map reorganization, 318
neurons, 319
reorganization, 305
sublayer, 275
volume, 324
cortical columns, 328
cortico-basal-ganglia-thalamic-cortical loop, 317
cPath, 360
CPG, 129, 140, 147–148
CPU, 245
Linux, 252
cranial nerve, 74
craniofacial abnormalities
VCFS, 345
cryosection, 268
CSAIL, *See* Computer Science and Artificial Intelligence Laboratory
Csm, 145
curator, 70, 74, 367
current, *See* ionic, reverse potential
external, 132
hyperpolarization, *See*
intrinsic, *See* ion channel
voltage-dependent, 137, 146
cvapp, 58
cyclic AMP, *See* cAMP
cyclic attractor, 312, 317
cyclic nucleotide-gated conductance, 146
Cygwin, 104
Cygwin-X, 123
cytoarchitectonics, 268, 274
cytogenetic locui, 355
Cytoscape, 348, 358, 360
web link, 357

D

DAG, *See* Directed Acyclic Graph
data
analysis, 267
binding, 57
complexity, 38
conversion, 243
copy number, 341
definition language triggers, 44
derivation, 246
diversity, 38

elements, 38
extraction, 176
format conversion, 235
genotyping, 351
repository, 27, 212
haplotype, 341
heterogeneity, 23
integration, 27–28, 155
interoperability, *See* interoperability
lineage, 246
mediation, 28
mediator, 27
mining, 155–156
model, 38, 271
organization, 43
pedigree, 246
presentation, 44
provenance, 251
repositories, 27
temporal dependencies, 244
Data Source Server, 28–29
database, 24, 37, 159, 172
architecture, 4, 37, 271
construction, 177
construction and analysis, 176
data-mining, 156
design, 38
electrophysiology, 54
external, 27
federation, 27
genetics, 370
integration,
centralized, 27
mediated, 27
interface, 201
mediators, 23
neuronal, 40
programmer/manager, 70
relational, 37
schema, 249
simulation, 174–175
single neuron, 177
spatial, 67, 70
symbolic, 70
text-based, 267
transaction, 235
transcriptome, 290
databasing, 155
receptor distributions, 267
simulation-centric, 158
DBA/2J
mouse strain, 290–291
dbCollator, 277
DBMS, 26, 28, 44–46
dbSNP, 291
Decentralized Information Group, 380
decision tree, 13
decrease-conductance, 143–144
general equation, 142
delusion, 339
dendrite, 111, 319–320
compartment, 112
dendritic
arborizations, 58
compartments, 4
location, 161
morphology, 103
density mechanisms, 96
densocellular, 72
dentate gyrus, 11, 72, 325
depolarization, 142
slow, 146
deviation, 178
DICOM, 251
differential equation
second-order, 139
Diffusion Constant, 192
digital atlas, 184
dimensional stacking, 177
model neuron database, 173
Directed Acyclic Graph, 249
disease, 337, *See* infectious
disease susceptibility, 287
disease-induced, 184
dissected tissue, 290
DNA
alterations, 337
damage, 365
Document Object Model, 57
DOM, *See* Document Object Model
domain, 70
domain-dependent, 16
dopamine, 49, 292, 317
receptors, 355
dorsal, 200
dorsal root ganglion, 321
dorsolateral prefontal cortex, 361
Douglas syntax, 58
down-regulate, 145
Drosophila, 289

drug, 50
  anti-inflammatory, 49
  non-steroidal, 49
dsArch
  web link, 46
DSS, 29–31, *See* Data Source Servers
DTSearch, 8
duty cycle, 171
dyslexia, 314

E
EAV, 37, 41, 43, 51
  triple, 41
EAV triplets, 45
EAV/CR, 25–26, 28–29, 41–42, 45–46
EAV_Flag, 46
Edge Constant, 192
EEG, 316, 328
e-forum, 373
electrical
  coupling, 132
  non-synaptic communication, 316
electroencephalogram, *See* EEG
electrophysiology, 157
  button, dialog box, 108
  database, 54
Em, 117
enhancing
  score, 11
Ensembl, 294
  Genome Browser, 296
enthorinal, 325
Entities/Objects, 42
entity, 271
  neuroanatomic, 34, 71, 73, 81, 271, 345
entity–attribute–value, 41, *See* EAV, EAV/CR
entorhinal cortex, 322
Entrez, 41
  Gene, 293
  Nucleotide, 293
environmental risk factors, 288
epidemiologic, 378
epilepsy, 305, 315
epileptic-like discharge, 316
episodic memory, 314
epistasis, 339
EPSP, 142
equilibrium potential, 137, 143
equivalent electrical circuit, 137
erosion size, 192
ESRI GIS, 70
EUCOURSE
  web link, 105
Eutectic, 58
excitatory, 311
  neuron, 320
  synaptically activated channel, 111
  stellate, 320
executive functions
  disturbances in, 322
expert-mediated, 5
expression, 288
Extended Markup Language Document Object
    Model, 31
Extensible Markup Language, *See* XML

F
false-negatives, 9
false-positive, 68
familial AD mutations, 372
family-based association, 340
FAQ, 302
fast indexing, 16
FE, *See* forward Euler
feedback
  loop, 145
  negative, 146
  positive, 146
feedback connections, 316
feedforward, 306
Fiswidgets, 236–238, 240–242, 245
  workflow editor, 239
flow diagram, 8
fMRI, 195, 197, 201–202, 206–208, 235
FMRIB, 237
forebrain, 80
formalin-fixed, 72
FormatConvert, 242
forward Euler, 135
frequency
  average activation, 112
frontal cortex, 322
frontal lobe, 80
frontotemporal dementia, 368
FScell, 112, 117
FSecell, 112
FSL, 241
FTDP-17, 368, 372
functional genomics, 337, 348
functional magnetic resonance imaging, 200

functions
  time-dependent, 142
  voltage-dependent, 142

G

GABA, 319, 356
  neurons, 361
  receptor, 320, 356
  signalling, 355
  synthesis, 361
GABAergic, 360
  signalling, 361
galactose metabolism, 357, 359
gamma
  bursts, 326
  rhythm, 326
GenBank, 25
gene, 50, 337
  amplification, 345
  effects, 341
  expression, 291, 338, 357
    P-values, 357
    screens, 341
    traits, 293
  ontology, 287
  pleiotropy, 288
  variants, 50, 287
Gene Expression/Genotyping, 359
Gene Ontology, 39, 348, 378
GeneChip, 356
GeneNetwork, *See* GN
GEneral NEural SImulation System, *See* GENESIS
GENESIS, 101, 103, 111, 115, 121, 123, 148, 170, 319, 326, 330
  Reference Manual, 105
  scripting, 115
  simulator, 63
  software, 58
  syntax, 105
  web link, 104
GeneSpring GT, 351, 353–354, 356
GeneSpring Network, 361
genetic, 337
  analysis, 338
  association, 342
  complexity, 288
  correlations, 290, 297
  covariance, 288
  disease, 338
  makeup, 291
  manipulation, *See*
  markers, 349
  reference population, 287
  screen, 349
  variation, 291
GenMAPP, 360
genome
  linkage analysis, 337, 343
  projects, 156
Genome Scan Meta-Analysis, 344
  schzophrenia, 344
genomics, 24
genotype, 291
  phenotype relations, 289
genotyping, 342
Gensat, 281
gif, 70
Globus, 242
  Toolkit, 245
glomerular layer, 199, 207
glomerulus, 200
glutamate, 356
  receptors, 12
glutamine, 338
GN, 287–289, 293, 298, 302
GNU, 104
GO, *See* Gene Ontology
goodness-of-fit, 176
gradient-echo imaging, 197
grammatical analysis, 310
graphemes, 309
graphical interface, 106
gray scale intensities, 251
Grid
  architectures, 244
  environment, 252
GRID, 215, 216
Grid Services Flow Language, 249
GRID-based, 214
GridFTP, 242
Growth Factor Receptor Bound Protein, 359
GRP, 290, 291 *See* genetic reference population
GSFL, *See* Grid Services Flow Language
GUI, 112, 128, 133, 241, 248
gyri tracts, 72
Gyrus F1, 74

H

hallucination, 339
haplotype, 341, 351, 353, 361

Hardy–Weinberg Equilibrium, 351
HBP, *See* Human Brain Project
HD, 338, 357, *See* Huntington's Disease
HDF5, 242
hearing defects
  VCFS, 345
Hebbian
  model, 324–325
hematopoietic stem cells, 290
hemorrhagic stroke, 315
heptanal, 203
heterosynaptic plasticity, 139
HH, 128, 135, 137, 148
HHRR, 354
hierarchies, 17
  of modules, 330
hierarchy, 5, 7, 9, 80
  partitive, 80
hindbrain, 80
hippocampal, 121
  networks, 326
hippocampus, 74, 83, 290, 306, 316, 325
histological
  delineation, 282
  markers, 283
  mask, 268
histostained tissue sections, 204
HMA, *See* Human Mapping Assay
Hodgkin and Huxley, 130, *See* Hodgkin-Huxley
Hodgkin–Huxley, 112, 122, 131, 134, 136, *See* HH
hodological, 81
homogeneity
  population, 349
homology, 282
homonyms, 38
homosynaptic plasticity, 137, 144
Hopfield
  autoassociative memory, 324
  networks, 312
Hopfield network, 312
HTML, 54, 57, 70, 105, 247, 249, 302
HTTP, 31, 58, 214–215
HTTPS, 227
Human Brain
  Atlas of, 79
Human Brain Project, 23, 42
human genome
  annotation, 347
Human Mapping Assay, 350
human-readable, 73
human-recognizable, 73
  label, 74
Huntingtin, 338
Huntington, 365
  disease, 338
HyperCard
  Apple/MacIntosh, 81
Hyperdoc, 106
hyperpolarization current, 92
hypertext, 81
Hypertext Markup Language, *See* HTML
hypothalamus, 79
  paraventricular nucleus of, 79
hypothesis testing, 287
hypothetical neuron
  equivalent electrical circuit, 139

I
I/O, 148
IATR, *See* Internet Analysis Tools Registry
IBM, 308
Iclamp, 110
image
  alignment, 220
  analysis, 306
  -based query, 211
  brain, 24
  histological, 212
  human-recognizable, 73
  magnetic resonance microsopy, 183
  processing, 235
  radiological, 212
  registration, 211
Image Query
  web service, 226
imaging
  functional magnetic resonance, *See* fMRI
  modality, 211
  neuroanatomical, 183
impaled recording electrodes, 196
inbred strains, 287
increase-conductance, 143, 144
index
  hierarchical, 80
  string-based, 68
indexing, 3
  citation, 68
  neuroanatomical database, 211
  word, *See*
inductive database, 155
infectious disease, 289

inferior temporal cortex, 321
informatics, 287
information
  extraction, 3, 5, 7, 16
  processing, 309
  source map, 30
informativeness, 342
Ingenuity, 348
inheritance modes, 339
inhibitory, 311, 320
  neuron, 320
  postsynaptic potentials, *See* IPSP
  synaptically activated channels, 111
input/output, 131, *See* I/O
integrate-and-fire
  cell, 129, 132
integrator Servers, *See* IS
Integrin signalling, 355
Intel Xeon, 70
intensity-rescaling, 190
interface, 8
  constraints, 37
  graphical, *See* graphical, interface
intermarker D score, 344
Internet Analysis Tools Registry, 250
interneurons, 10
interoperability, 4, 25, 91, 148, 235
InterPro, 348
intersection
  XML schema approach, 55
inter-spike interval, 176
intracellular ion pool, 132
intracellular pool, 146
  multiple, 146
ion
  channel, 4, 167, 173, 316
  current, 174
  pool, 144, 146
    intracellular, 136
ionic
  channel, 320
  conductances, *See* channels
  current, 14, 134
    reversal potential, 132
IPSP, 142
IS, 28, 30
ischemic stroke, 315
ISM, *See* information source map
isodisomy, 345
isolate, 340

J
J2EE, 213, 216, 224, 227
Java, 57, 62, 64, 212–223, 240–241
  applet, 70
  virtual machine, *See* JVM
Java 2 Enterprise Edition, *See* J2EE
Java Beans, 216
Java Message Service, 245, *See* JMS
Java Remote Method Invocation (RMI), 245
JAXB, 64
Jftp, 242
JMS, 245
jpeg, *See* jpg
jpg, 70
jsp, 70
JVM, 129, 133

K
K
  channels, 122
KB, 379, *See* knowledgebase
KDD, 156, *See* knowledge discovery and data mining
KEGG, 348, 360
keywords, 5, 10
  score, 13
Kinetikit, 110–111
knockout, 287
knowledge
  base, *See* KB
  discovery, 155
  domain, 71
  management, 365
  repository, 73
knowledge base, *See* knowledgebase
knowledge discovery and data mining, 155
knowledgebase, 3, 5, 8, 11, 14, 16
knowledge-base, *See* knowledgebase
KnowMe, 34
Kohonen network, 313

L
laboratory data management system, 60
LabSWAN, 378
language skills
  degradation of, 322
lateral inhibition, 321
lateralization, 308
late-response, 196

layer, 72
  III, 320
  IV, 320
  V, 320
LD, 344, *See* linkage disequilibrium
leak conductance, 134
learning
  algorithm, 309
  supervised, 3, 8, 14
  unsupervised, 3, 11
left–right hemisphere, 308
lesions, 317, 318
leukocyte, 347, 349, 354
lexical, 5
  affirmers, 18
  constraints, 16
  negators, 18
ligand, 267, 269
ligand-binding, 268
likelihood ratio statistic, *See* LRS
limbic, 308, 322
limits of detection, 343
linguistics, 16
linkage, 338, 340
linkage calculations, 344
linkage disequilibrium, 340
Linux, 104, 123, 192, 217
LISP, 41
list association, 41
Load Simulation, 135
local voltage, 176
locus ceruleus/coeruleus, 74
LOD, 350, *See* limitsof detection
  score, 358
logical
  neuron model, 310
  typing, 242
long-term potentiation, 320
LONI, 79, 185, 190, 192, 224, 237, 251–253, 282
  Debabler, 187
  Pipeline, 189
lower bound, 48
LRS, *See* likelihood ratio stastistics
LTP, *See* long-term potentiation
Lycos, 68
lymphoblast, 347

M
Mac, 192
Mac OS, 223
Mac OS/X, *See* OS X
macaque, 79, 282
  atlas, 69
machine learning, 159, 306
MacOS, 217
MacroVoxel, 228
Madeira, 348
MAGE XML, 64
magnetic
  field strength, 197
  resonance imaging, *See* MRI
  resonance microscopy image, 183
Magneto and Electroencephalogram, 237
major depressive disorder, 345
Mann–Whitney
  testing, 356
map
  locus, 287
  odor, *See* odor maps
  server, 70
mapping, 267
markers, 340
Marquardt–Levenberg, 223
MAS5, 290
mask values, 251
MathML, 56, 62
MATLAB, 172, 240–241
Mauthner, 230
maximal conductance, 168
MBL, *See* Mouse Brain Library
MDA, 184, 185, 186, 187, 191, *See* minimim deformation atlas
MDT, 190, 191, 192, *See* minimum deformation target
MDT2, 191
mean, 157
mean-square-root
  deviation, 176
mechanoreceptors, 319
medial dorsal nucleus, 72
mediator, 67
Medical Subject Headings, *See* MeSH
medium spiny neuron, 12
MEG/EEG, *See* magneto and electroencephalogram
melanophores, 43
membrane, 7
  capacitance, 134, 174
  conductances, 132, 144–146
  current, 168

membrane potential, 177
  of neurons, 196
  simulation, 110
  soma, 116
memory, 306
  decline, 324
  loss, 365
Mendelian genetics, 288–289
Mercator, 208
MERLIN, 351, 353
MeSH, 17
Message Passing Interface, 240
metabolic-vascular state, 197
metadata, 12, 16, 23, 25–26, 30–31, 37, 42–44, 45, 47, 60, 156, *See* meta-data
meta-data, 250, 253
  fields, 251
meta-database, 30
meta-language, 164
metaphysical, 72
metathesaurus
  UMLS, 84
MGH, 376
MIAME, 64
microarray, 290, 294
  data sets, 288
Microbrightfield, 58
microdletions, 345
microsatellite, 291, 341, 349, 350
Microsoft Access, 45, 81
Microsoft IIS, 201
Microsoft Visual C++, *See* C++
microstructure of cognition, 306
midbrain, 80
middleware, 246
MINC, 187, 189, 251, 253
mind objects, 315
minimum deformation atlas, 183
minimum deformation target, 183, 187
misfolded proteins, 365
mitochondria, 355
MLP, 312
model, *See* Poirazi
  compartmentalized cell, 121
  computational, 129
  connectionist, 310
  hierarchical, 81
  multicompartment, 147
  networks, 177
  neural
    spiking, 314
  neural systems, 176
  neuron,
    high-dimensional, 178
  neurons, 177
  psychiatric, 72
  single neurons, 103
  systemic, 81
ModelDB, 91, 100, 129, 148
modeling
  neuronal, *See* neuronal, modeling
ModelView, 91, 93, 96, 100
MODm, 144–145
modulation, 144
modulatory
  function, 145
  relationships, 132
  synapse, 138, 144–145
  transmitter, 144–145
module
  Align Warp, 191
  Crop, 190
  Soft Mean
molecular layer, 72
mollusc, 129
Mollweide, 208
monozygotic twins, 339
mood instability
  VCFS, *See* velocardiofacial syndrome
mood swings, 308
MorphML, 53, 56, 58, 100–101
  schema, 59
  viewer, 60
morphology, 53, 184
  dendritic, 103
mossy fibers, 10
motor
  cortex, 317
  function
    disturbances in, 322
mouse brain, 212
mouse brain atlas, 281
Mouse Brain Library, 79, 217
  Project, 214
Mouse Pheromone Database, 291
movement disorders, 195
MPI, *See* Message Passing Interface
MRI, 183–184, 188, 197, 201, 235
MRM, 186–187, 191

mRNA, 341
  expression, 290
  transcripts, 342
MS Access, 25, 269
MS Access XP® ,, 271
MS Excel, 25
MS Windows, 70, 92, 104
MS-SQL, 70
Mulit-layer perceptron feedforward network, 311
multi-compartmental models, 310
multifactorial analysis, 269
Multilayer Perceptron, 311
Multiple Interval Mapping, 299
multiple mapping, 293
Multiple QTL Mapping, 300
multi-point downstream linkage analysis, 351
Multipoint Engine for Rapid Likelihood Inference, *See* MERLIN
multivariate analysis, 269
mutant
  mice, 183
mutation, 368
  disease-causing, 339
myeloarchitectonics, 268
MySql, 25
MySQL, 269
MySQL® 4,1,10, 271
MySWAN, 378

N
Na
  channel, 118, 122
nanoAmps, 110
narrow distribution, 176
N-ary, 43
NAS, *See* Network Attached Storage
National Alliance for Research on Schizophrenia and Depression, *See* NARSAD
National Center for Biotechnology Information, *See* NCBI
National Library of, 39
National Library of Medicine, 15, 84, 201
natural language processing, 35
Nature
  journal, 371
NCBI, 41, 348
NDG, *See* Neuroscience Database Gateway
negating
  contextually, 12
  score, 11
negators
  lexical, 17
neocortex, 306, 308, 316
neostriatum, 12
nerve bundle, 207
nerve cell, *See* neuron
nervous system, 23
NetCDF, 189
NetOStat, 223–226, 229
  Brain Atlas Navigator, 226
Netscape, 244
network
  graph, 293
  Hopfield, 312
  nodes, 309
  non-rhythmic, 174
  rhythmic, 174
  simulations, 161
Network Attached Storage, 242
network databases, 170
  model, 170
Network File System, *See* NFS
Neuoinformatics Information Framework, 23
neural
  activities, 198
  cell assemblies, 308
  circuits, 128, 196
  function, 127
  information processing, 310
  model, 309, 326
  network, 305, 311, 128, 140, 147, 155, 167
  networks, 24, 127, 130, 316
  plasticity, 306
  processes, 305
  query system, *See* NQS
  simulation, 103, 157–158, 314
  system, 127
    complex, 177
    function, 167
  systems, 24
Neural Simulation Language, 330
neuregulin1, *See* NRG1
neuroanatomical, 288
  informatics, 211
neuroanatomist, 79
neuroanatomy, 59, 67, 71–72
  hierarchy, 80
  database
    indexing, 211

microscopic, 72
texts and atlases, 69
Neurobase, 377
neuroConstruct, 63
neurodegenerative, 7
neurodegenerative disorders, 365, 371
neurodynamic, 315
neurodynamics, 315
neurogenetics, 287
neuroimaging, 185, 235, 246, 249
Informatics, *See* NIfTI
size and complexity of, data, 243
Neuroimaging Informatics Technology Initiative, *See* NIfTI
neuroinformatics, 3, 53–54, 67, 83, 129, 148, 211, 235, 305
Neurokit, 107, 111
neurological, 7
diseases, 195
terminology, 72
Neurolucida, 58
NeuroML, 53, 55–56, 101
Development Kit, 56, 60, 62
neuromodulation
acetylcholine, 326
neuromorphic devices, 306
neuron, 7, 14, 17, 30, 40, 48, 130
activation, 309
activity
bursting, 168
afferent, *See*
afferent/efferent, 41
biophysical and biochemical properties, 128
bipolar, 80
computational element, 136
electrical coupling between, 147
excitability, 162
HH-type, 132, 136
large-scale networks, 55
medium spiny, 34, *See*
pyramidal, 80
reticular, 4
thalamic reticular
tonically spiking, 176
tracing system (software), 58
two-compartment, 111
two-state, 312
NEURON, 91–92, 101, 148, 155, 157–160, 164, 170, 330
software, 58
neuron databases
model, 170
neuronal
activity
odor-stimulated, 196
compartment, 8, 10
compartments, 14
death, 322
membrane properties, 4
model, 55
modeling
morphology, *See* morphology
network, 155, 167
pathways, 49, 81
signals, 195
neuronal excitability, 134
NeuroNames, 33, 67, 69–72, 74, 79–80, 82, 84, 212, 224, 251
NeuronDB, 3–5, 7, 10, 15, 25, 48, 49
neuropharmacological, 288
neurophysiology, 155, 329
neuropil, 33
neuropsychiatric
diseases, 356
disorders, 345
neuropsychological syndromes, 305
neuropsychopharmacology, 314
neuroscience, 267, 305, 315, 330
ontologies, 67
Society For, 71
XML in, 55
Neuroscience
Journal of, 3, 8, 14
Neuroscience Database Gateway, 267
Neuroscience Information Framework, 71
neurosimulator, 130, 148
neurosimulators, 128
NeuroSys, 25, 27, 60
NeuroTerrain, 212, 224–225
Atlas Server, 228
mouse atlas, 229
NeuroText, 3, 5, 7– 8, 10–11, 13, 15–16, 19
web link, 3
neurotransmitter-receptor transcripts, 356
neurotransmitters, 4, 10, 14
neurotransmitters/neuromodulators, 48
Nevin, 58
nextToken, 57
NFS, 214, 241
NIF, *See* Neuroscience Information Framework

NIH, 379
NIH Blueprint for Neuroscience Research, 71
NIMH, 307
Nissl-stained, 72, 212, 230, 282
atlas, 223
NLP, 7, 18, *See* Natural Language Processing
NMDA, 319
channel, 320
receptor, 320–321, 374
NMOD, 158
nodes, 306
nodulation, 138
nomenclature, 70
Nomina Anatomica, 72
non-declarative memory, 327
nonparametric linkage, 343
analysis, 351
nonspecific hybridization, 294
nonspiking neurons, 307
non-steroidal drugs, 49
noradrenaline, 38
norepinephrine, 38
normal
genetic pathways
disruption, 183
maps, 320
probability plot, 294
normalized
correlation, 205
fluorescence, 294
NPAIRS, 237
NPL Z score, 353, *See* nonparametric linkage Z score
NQS, 155, 157–161, 163–164
NTDB, 6
NT-SDK, 229
numerical
classification, 157
simulation, 178

O

OB, 198, 207–208, *See* Olfactory Bulb
obesity, 289
object, 271
class, 55
objects/entities, 50
octanal, 203
odds ratio test, 14
ODE, 148, *See* ordinary differential equation
odor, 43
map, 195, 198, 200–202, 204–205
mapping, 197–198
stimuli, 196, 202, 207
OdorDB, 201
OdorMapBuilder, 198, 200, 206–207
OdorMapComparer, 195, 202–203, 206
OdorMapDB, 195, 201–202, 206
olfaction, 24
olfactory
bulb, 195, *See* OB
receptor, 207
interactions with odor, 202
signals, 208
olivary nucleus, 39
OME, *See* open microscopy environment
OMIM, 292, 293
ontological, 23
ontology, 27, 29, 33, 50, 67–68, 157, 287
neuroanatomical, 72
neuroscientific, 68
XTM databases, 82
Ontology Server, *See* OS
Open Microscopy Environment, 242
open reading frames, 347
OpenGL, 185
Openmicroscopy, 64
OpenPBS
workflow, 216
optimization algorithm
stacking order, 177
Oracle, 25, 27, 44
ORDB, 201
ordinary differential equation, 136
ORF, *See* open reading frames
orthographical representations, 314
OS, 28, 30–31
OS X, 92
overtransmission, 340
oxidative phosphorylation, 355

P

pair-scan, 297
pair-scan mapping, 296
paleomammalian, 308
PANTHER, 292
Parallel Distributed Processing, *See* PDP

parameter
  change
    minimization, 178
  combination, 174–175
  space, 174
parametric linkage, 343
paranoid delusions, 322
paraventricular nucleus, 75–76, 79
Parkinson, 365
Parkinson's, 7
Parkinson's disease, 315, 329, 345, 373
Parkinsonism, 315, 368
parsing
  contextual, 3
  fine-grained, 11
  lexical, 3
  semantic, 3
  syntactic, 6
partitive, *See* partitive, hierarchy
parvalbumin, 361
pathogenesis
  Alzheimer's Disease, 323
pathogenic event, 378
pathologic change, 378
pathophysiologic, 378
Pathway Assist
  commercial software, 348
pattern recognition, 306
PDGene, 376
PDNN, 299, *See* position-dependent nearest neighbor
PDP, 306
PDP++, 330
per burst
  interval, 176
peripheral
  leukocytes, 347, 356
  nervous system, 196
  receptor, 207
PERL, 8, 64
Perlegen/NIEHS
  resequencing project, 291
perturbation, 378
pFam, 348
pH, 347
phantom limb, 320, 321
phase relations, 310
phenothiazine
  chlorpromazine, 49
phenotype, 288, 291, 299–300, 337, 342
phonemes, 309–310
phonological, 314
phonology, 314
phosphorylation, 357
  regularion pathway, 359
photomicrographs, 184
PHP, 57, 64
physiological traits, 301
physiopathology, 24
PIC, *See* Portuguese Island Collection
pipeline, 211, 214, 228, 235, *See* LONI
  image_processing, 214
  NeuroTerrain, 213
piriform, 110
pivoting, 43
plasticity
  function, 140
  heterosynaptic, 145
  homosynaptic, 142, 145
plateau potentials, 148
pleiotropic, 299
plexiform layer, 199
PNAS
  journal, 371
point attractor, 312
point processes, 96
Poirazi model, 93
polygenic
  disease, 288
  traits, 289
polymorphism, 50, 356
  amino acid-altering, 342
population isolate, *See* isolate
Portuguese Island Collection, 354
position-dependent nearest neighbor, *See* PDNN
posterior parietal cortex, 321
PostgreSQL, 25, 213, 215
post-processing, 10, 11
postscript, 249
postsynaptic
  cell, 139, 145
  potentials, 143
    multicomponent, 142
  receptor, 161
potassium channel, 110, 112, 118
  gene, 292
POW, 370
PowerBuilder, 45
PowerPoint, 302
Prader-WIlli Syndrome, 345

precision, 8
prefrontal cortex, 317
presenilin-1, 368, 372
presenilin-2, 368, 372
presynaptic
  activity, 138, 142
  cell, 161
    spikes, 142
primates, 281
probabilistic brain atlas, 183
probe, 294
process
  non-linear, 167
process management, 248
processing
  natural language, 3
projection pathways, 196
Protégé, 27
protein interactions, 55
proteomics, 24
protocols
  standardized, 24
prototype
  library, 118
provenance, 253
pruning, 323
PSD95, 374
PSM, 140
PSP, 142, 143, *See* postsynaptic potential
  time-dependent, 143
  voltage-dependnet, 143
psychological, 315
psychology, 315
psychotic features, 322
PubMed, 3, 15, 17, 25, 74, 290, 357, 367, 375, 379
pull-down, 45
purine metabolism, 357, 359
Purkinje, 10, 13, 17, 110
  dendrites, 33
  neurons, 33
putamen, 317
P-value, 298
pyramidal neurons, 80, 316
Python, 64, 237

Q
QC, 219, 220
QIS, 28, 30–31
QTL, 287–288
  cluster map, 293, 299
quasi-stable states
  attractor networks, 315
query
  language, 158
  translation, 28
  web-service, 214
Query Integration System, 29
Query Integrator System,
    See QIS

R
radial separation, 204
radioactively labeled, 268
radioactivity, 268
random number variable, 137, 143
rate-constant equation
  HH-type, 132
RDBMS, 28, 212, 215–216, 227
RDF, 16, *See*
Reactome, 360
readcell, 116
reading, 306
recall, 8
receptor, 4, 10, 14, 40, 43, 48, 278
  binding, 276
  ligands, 268
  table, 279
recessive alleles
  disease-causing, 345
recombinant inbred, *See* RI
recombination events, 342
recording-site, 54
red/green/blue, 201
reference atlas, 211
regulators
  intracellular, 139
regulatory agents
  intracellular, 137, 142
relational database, 267
relational database management system, *See*
    RDBMS
relationship
  binary, 41
  many-to-many, 40
resequencing project, 291
Resource Description Framework, *See* RDF,
    See RDF
resting conductance, 142
resting potential, 171

retrieval, 68
  information, 74
  precision, 68
reversal potential, 132
RGB, 203, *See* red/green/blue
rheumatoid arthritis, 357
rhythmic network, 176
RI, 297–298, *See* recombinant inbred
ribosomal, 355
rigidity, 316
RMA, 356, *See* robust multichip analysis or robust multiarray averaging
RMI, *See* remote method invocation
Robust Multiarray Averaging, *See* RMA
robust multichip analysis, 290
rodents, 281
Roget's Thesaurus, 10, 18
RPPL, 248
Ruby scripts, 216
rule-based systems, 310
Run Pipeline in Parallel, *See* RPPL
run simulation, 135

S
SAMBA, 214
Sao Miguel Azores, 348
SAX, 31, *See* Simple Application Programming for XML
  XML parser, 63
SBML, 56
scalability, 246
Scalable Vector Graphics, 31
schema, 40, 44
  complex-database, 39
  continually evolving, 38
  global, 28
  mixed, 45
  RDF, 378
  simplification, 40
  SWAN, 379
schematic, 139
schizophrenia, 289, 316, 337–339, 345–346, 349, 356–357, 361
Schizophrenia Research Forum, 376
Science
  journal, 371
SCIL, 216
scotoma, *See* artificial scotoma
screen-scraping, 25
scrolling lists, 14
SDP, *See* strain distribution pattern
second messenger, 132, 138, 145
  pool, 144
second messengers, 144
secondary database, 176
secondary dendrites, 230
sectioning, 184
segmentation, 185
  errors, 219
  layer, 223
Selbst-Erhaltungs-Therapie, *See* Self-Maintenance Therapy
self organized mapping networks, 318
Self-Maintenance therapy, 327
self-organizing, 309
  networks, 321
Self-Organizing Mapping, *See* SOM
semantic, 5
  -based indexes, 68
  data models, 23
  data-types, 251
  inverse/reciprocal, 49
  metadata annotations, 23
  phrases, 11
  reciprocal, 49
  relationship, 11
  Web, 35
  web services, 26
  relationships, 8
Semantic Web Applications in Neuromedicine, *See* SWAN
Semantic Web in Life Sciences, 377
senile plaque, 325
SenseLab, 4, 7, 14–15, 42, 46, 48–49, 92, 158
sensitivity, 8, 9, 10
sensory perceptions
  abnormal, 195
sentence processing, 306
sequence, 24, 50
serotonin, 139, *See* 5-HT
  signalling, 355
shared cluster, 175
short oligomer microarrays, 290
sigmoidal, 310
signal
  analysis, 306
  encoding, 195
  neuronal, *See* neuronal signals

silent
discharge frequency, 172
neuron model, 175
Simple Application Programming Interface, 57
Simple Application Programming Interface for XML, 31
simple interval mapping, 295
Simple Object Access Protocol, *See* SOAP
SimToolDB, 158
simulation, 118–119, 155, 159, 168
full network, 178
large networks, 104
program, 174
time, 174
simulator for neural networks and action potentials, *See* SNNAP
single
hidden layer, 311
neuron excitability, 161
spikes, 309
single-chromosome Map Viewer, 296
single-nucleotide polymorphism, *See* SNP
single-point
downstream linkage analysis, 351
slower synaptic learning, 318
SMART ATLAS, 224
smell
perception of, 196
SNNAP, 127–128, 130, 133, 135–136, 143, 147–148
Tutorial Manual, 129, 133
web link, 129
SNP, 291, 338, 341, 351, 353, *See* single nucleotide polymorphism
database, 291
genotype, 349
genotyping, 349
haplotype, 342
haplotyping, 349
LOD scores, 357–358
SOAP, 26
sodium
channel, 112
Soft Mean module, 190
SOM, 313, 318
soma, 111–112, 117, 319
somatosensory cortex, 318–319, 320
spatial
correlation, 205
coefficient, 203
discretization, 94
distributed markers, 268
frequency, 220
indexer, 211
normalization, 183–184, 212, 220
species, 54
specificity, 10
speech perception, 306
spike, 112, 174
activity, 145–146, 161, *See*
count, 176
duration, 142
frequency, 171, 310
histogram, 173
generator, 112, 118
per burst, 171
propagation, 136
spiketrain, 115
spiking
discharge frequency, 172
membrane conductances, 174
neural models, 314
neurons, 314
SPM, 241, 248
SQL, 39, 43–44, 81, 82, 160
server, 25
squashing functions, 310
squid, 110
Squid_Na, 130
squid-like channels, 112
SRB, 214, 227, *See* Storage Resource Broker
S-shape, 294, *See* sigmoidal
stacking order, 177
staining, 184
staining-method, 54
standard deviation, 157
startElement, 63
statistical evaluation, 290
statistics, 157, 306
steady state, 175
stereological, 288
stimulus
modulatory, 132
stimulus current
extrinsic, 134
storage
information, 74
Storage Request Broker, *See* SRB
Storage Resource Broker, 242
strain, 291

strain distribution pattern, *See* SDP
striatum, 290
structure
  anatomical, 184, 195
  chemical, 50
  dendritic, 111
  genotyping, 351
  long name, 79
  pedigree, 351
  primary, 71
  superficial, 72
Structured Query Language, *See* SQL
study population, 340
sub-cellular, 103
sub-nuclei, 72
sub-ontologies, 84
sub-schema, 39
sub-structures, 72
Sun Grid Engine, 253
superior frontal gyrus, 74
susceptibility gene, 344
SVG, *See* Scalable Vector Graphics
SWAN, 377–378, 380
  schema, 379
Sybase, 44
SymAtlas, 292
symbolic
  description, 315
  information processing, 310
synapse, 161
  chemical, 132–134, 142
  database, 292
  electrical, 134
  modulatory, 133
  strength, 174
  time-dependent, 132
  voltage-dependent, 132
synapses, 130
  network, 128
synaptic, 311
  activation, 143
  compensation, 323
  conductances, 143
    time-dependent activation, 139
  connection, 142
    exponential growth of, 325
  connections, 323
    per unit cortical volume, 324
  current, 143
  deletion, 323
  input, 112
  integration, 92
  isolation, 177
  loss, 326
  plasticity, 128
  potentials and integration, 137
  receptors
    activation of, 167
  strength, 139
  transmission, 128, 142
  transmissions, 196
synaptic runaway model, 325
synaptophysin, 215
synchrony, 316
SynDB, 292
synonym, 5, 10, 42
synonyms, 38, 49, 72
syntax, 68
synuclein, 355
systematic reasoning, 310
systemic, 81
systems
  biology, 337, 338
  genetics, 288
  theory
    non-linear, 168
Systems Biology Markup Language, *See* SBML

T

T1, 187
  MRI component, 197
T1-weighted, *See* T1
T2
  MRI component, 197
table
  bridge, 40
  column, 161
  create/alter, 44
  lookup, 40
  objects, 41
  reciprocal attributes, 49
tables/classes, 44
tactile, 319
tags, 54
tal
  Fiswidgets application interface, 241
tau, 372
  gene, 368
  mutation, 368
taxonomic, 157

TCP/IP, 26
Telemakus, 370
template workflow, 249
temporal gradients
  memory decline, 324
terminological, 67
Terms/Synonyms, 50
Tesla, 197
text
  human-readable, 73
  mining, 3, 348
  -based databases, 267
TEXTAREA, 48
thalamic, 4, 318
  interneuron, 319
  nuclei, 319
  relay cell model, 319
thalamo-cortical, 318
thalamus, 75, 319, 320
theta rhythm, 326
tiff, 216, 225
time constant, 139
tonic spikers, 173
tonically spiking, 174
  neuron, 176
Tools/FXParse, 238
Topic Maps, 82
topographical, 309
  map, 313, 318
trait
  analysis, 287
  regulation, 299
transcript expression, 287
transcriptional
  change, 342
  level, 342
  profiling, 347
transcriptome database, 290
transcripts, 290
transgenic, 184, 287
transmitter, 132, 136, 144
  pool, 146
transQTL, *See* QTL
tremor, 316
TrialDB
  web link, 43
trinucleotide repeat, 338
triplets, *See* EAV triplets
triune, 308
true-negatives, 14
true-positives, 14
t-test, 157
tutorial, 103, 106
two-state neurons, 312

U
Ubiquitin, 355
UCSC, 294, 348
  Genome Browser, 296
UMLS, 17, 29, 31, 33–34, 39, 41, 50, 84
UniGene, 292, 293
uniparental isodisomy, 345
unirelational
  concepts, hierarchical models, 81
Unix, 216, 223, 228, 240
UNIX, 104, 107, 123, 192, 228
UNIX/Linux, 92
upper bound, 48
up-regulate, 145
URL, 25, 30
user, 45
  interface, 38, 70

V
validation, 8, 13
  constraints, 37
values, 42
VCFS, 359, *See* velocicardiofacial syndrome
velocardiofacial syndrome, 345
ventral posterior lateral, *See* VPL
ventroposterolateral, *See* VPL
vibration, 319
View/Query Designer, 34
virtual datasets, 249
Virtual Knife, 225, 229
Virtual RatBrain Project, 59
VisANT, 34
Visome, 329
visual cortex, 282, 319
visualization
  dimensional stacking, 173
Vm, 117
vocabulary
  default, 74
  controlled, 39, 73
VOI, 225
voltage
  maxima, 174
  minima, 174
voltage trace, 171, 173

per burst, 171
voltage-activated
channels, 112
voltage-dependent gating, 167
volumes of interest, *See* VOI
VoxBo, 237
voxel, 251
VPL, 79, 319

W
W3C, 26, 377
WAMBAMM
web link, 105
warping algorithm, 204
wavelet-based registration, 222
Web Service Description Language, *See* WSDL
Web Services Interoperability Organization, *See* WS-I
WebGestalt, 299
WebQTL, 295, 300
Map Viewer, 296
WebStart, 224
Wellcome-CTC, 291
WFMS, 235–236, 238, 241–242, 245, 248, 252
white blood cell, 354
Whole Brain Atlas, 79
Wikipedia, 379
Windows, 123, 217
Windows 2000, 185
Windows XP, 185
WIRM, 25
word morphology, 306
word-indexing, *See* indexing
workbench, 236
workflow, 235, 244
for genetic analysis, 351
management system, *See* WFMS
neuroimaging, 235
verification, 252
working memory, 314
wrapper function, 240
WS-I, 227

X
X Windows, 107
XDTM, 243, 252
XML, 16, 26, 29–31, 53, 54, 62, 82, 100–101, 238, 251
libraries, 57
schema, 55
XML Dataset Type and Mapping, *See* XDTM
XML DOM, 31
XML Pull Parser, 57
XODUS, 106, 121, 123
web link, 104
XPP, *See* XML Pull Parser
XTM, 76, 82

Z
Z score, 343, 357, 358

Printed in the United States of America